General Topology and Applications

LECTURE NOTES

IN PURE AND APPLIED MATHEMATICS

1. *N. Jacobson,* Exceptional Lie Algebras
2. *L. -Å. Lindahl and F. Poulsen,* Thin Sets in Harmonic Analysis
3. *I. Satake,* Classification Theory of Semi-Simple Algebraic Groups
4. *F. Hirzebruch, W. D. Newmann, and S. S. Koh,* Differentiable Manifolds and Quadratic Forms (out of print)
5. *I. Chavel,* Riemannian Symmetric Spaces of Rank One (out of print)
6. *R. B. Burckel,* Characterization of C(X) Among Its Subalgebras
7. *B. R. McDonald, A. R. Magid, and K. C. Smith,* Ring Theory: Proceedings of the Oklahoma Conference
8. *Y.-T. Siu,* Techniques of Extension on Analytic Objects
9. *S. R. Caradus, W. E. Pfaffenberger, and B. Yood,* Calkin Algebras and Algebras of Operators on Banach Spaces
10. *E. O. Roxin, P.-T. Liu, and R. L. Sternberg,* Differential Games and Control Theory
11. *M. Orzech and C. Small,* The Brauer Group of Commutative Rings
12. *S. Thomeier,* Topology and Its Applications
13. *J. M. Lopez and K. A. Ross,* Sidon Sets
14. *W. W. Comfort and S. Negrepontis,* Continuous Pseudometrics
15. *K. McKennon and J. M. Robertson,* Locally Convex Spaces
16. *M. Carmeli and S. Malin,* Representations of the Rotation and Lorentz Groups: An Introduction
17. *G. B. Seligman,* Rational Methods in Lie Algebras
18. *D. G. de Figueiredo,* Functional Analysis: Proceedings of the Brazilian Mathematical Society Symposium
19. *L. Cesari, R. Kannan, and J. D. Schuur,* Nonlinear Functional Analysis and Differential Equations: Proceedings of the Michigan State University Conference
20. *J. J. Schäffer,* Geometry of Spheres in Normed Spaces
21. *K. Yano and M. Kon,* Anti-Invariant Submanifolds
22. *W. V. Vasconcelos,* The Rings of Dimension Two
23. *R. E. Chandler,* Hausdorff Compactifications
24. *S. P. Franklin and B. V. S. Thomas,* Topology: Proceedings of the Memphis State University Conference
25. *S. K. Jain,* Ring Theory: Proceedings of the Ohio University Conference
26. *B. R. McDonald and R. A. Morris,* Ring Theory II: Proceedings of the Second Oklahoma Conference
27. *R. B. Mura and A. Rhemtulla,* Orderable Groups
28. *J. R. Graef,* Stability of Dynamical Systems: Theory and Applications
29. *H.-C. Wang,* Homogeneous Branch Algebras
30. *E. O. Roxin, P.-T. Liu, and R. L. Sternberg,* Differential Games and Control Theory II
31. *R. D. Porter,* Introduction to Fibre Bundles
32. *M. Altman,* Contractors and Contractor Directions Theory and Applications
33. *J. S. Golan,* Decomposition and Dimension in Module Categories
34. *G. Fairweather,* Finite Element Galerkin Methods for Differential Equations
35. *J. D. Sally,* Numbers of Generators of Ideals in Local Rings
36. *S. S. Miller,* Complex Analysis: Proceedings of the S.U.N.Y. Brockport Conference
37. *R. Gordon,* Representation Theory of Algebras: Proceedings of the Philadelphia Conference
38. *M. Goto and F. D. Grosshans,* Semisimple Lie Algebras
39. *A. I. Arruda, N. C. A. da Costa, and R. Chuaqui,* Mathematical Logic: Proceedings of the First Brazilian Conference

40. *F. Van Oystaeyen,* Ring Theory: Proceedings of the 1977 Antwerp Conference
41. *F. Van Oystaeyen and A. Verschoren,* Reflectors and Localization: Application to Sheaf Theory
42. *M. Satyanarayana,* Positively Ordered Semigroups
43. *D. L. Russell,* Mathematics of Finite-Dimensional Control Systems
44. *P.-T. Liu and E. Roxin,* Differential Games and Control Theory III: Proceedings of the Third Kingston Conference, Part A
45. *A. Geramita and J. Seberry,* Orthogonal Designs: Quadratic Forms and Hadamard Matrices
46. *J. Cigler, V. Losert, and P. Michor,* Banach Modules and Functors on Categories of Banach Spaces
47. *P.-T. Liu and J. G. Sutinen,* Control Theory in Mathematical Economics: Proceedings of the Third Kingston Conference, Part B
48. *C. Byrnes,* Partial Differential Equations and Geometry
49. *G. Klambauer,* Problems and Propositions in Analysis
50. *J. Knopfmacher,* Analytic Arithmetic of Algebraic Function Fields
51. *F. Van Oystaeyen,* Ring Theory: Proceedings of the 1978 Antwerp Conference
52. *B. Kedem,* Binary Time Series
53. *J. Barros-Neto and R. A. Artino,* Hypoelliptic Boundary-Value Problems
54. *R. L. Sternberg, A. J. Kalinowski, and J. S. Papadakis,* Nonlinear Partial Differential Equations in Engineering and Applied Science
55. *B. R. McDonald,* Ring Theory and Algebra III: Proceedings of the Third Oklahoma Conference
56. *J. S. Golan,* Structure Sheaves over a Noncommutative Ring
57. *T. V. Narayana, J. G. Williams, and R. M. Mathsen,* Combinatorics, Representation Theory and Statistical Methods in Groups: YOUNG DAY Proceedings
58. *T. A. Burton,* Modeling and Differential Equations in Biology
59. *K. H. Kim and F. W. Roush,* Introduction to Mathematical Consensus Theory
60. *J. Banas and K. Goebel,* Measures of Noncompactness in Banach Spaces
61. *O. A. Nielson,* Direct Integral Theory
62. *J. E. Smith, G. O. Kenny, and R. N. Ball,* Ordered Groups: Proceedings of the Boise State Conference
63. *J. Cronin,* Mathematics of Cell Electrophysiology
64. *J. W. Brewer,* Power Series Over Commutative Rings
65. *P. K. Kamthan and M. Gupta,* Sequence Spaces and Series
66. *T. G. McLaughlin,* Regressive Sets and the Theory of Isols
67. *T. L. Herdman, S. M. Rankin, III, and H. W. Stech,* Integral and Functional Differential Equations
68. *R. Draper,* Commutative Algebra: Analytic Methods
69. *W. G. McKay and J. Patera,* Tables of Dimensions, Indices, and Branching Rules for Representations of Simple Lie Algebras
70. *R. L. Devaney and Z. H. Nitecki,* Classical Mechanics and Dynamical Systems
71. *J. Van Geel,* Places and Valuations in Noncommutative Ring Theory
72. *C. Faith,* Injective Modules and Injective Quotient Rings
73. *A. Fiacco,* Mathematical Programming with Data Perturbations I
74. *P. Schultz, C. Praeger, and R. Sullivan,* Algebraic Structures and Applications Proceedings of the First Western Australian Conference on Algebra
75. *L. Bican, T. Kepka, and P. Nemec,* Rings, Modules, and Preradicals
76. *D. C. Kay and M. Breen,* Convexity and Related Combinatorial Geometry: Proceedings of the Second University of Oklahoma Conference
77. *P. Fletcher and W. F. Lindgren,* Quasi-Uniform Spaces
78. *C.-C. Yang,* Factorization Theory of Meromorphic Functions
79. *O. Taussky,* Ternary Quadratic Forms and Norms
80. *S. P. Singh and J. H. Burry,* Nonlinear Analysis and Applications
81. *K. B. Hannsgen, T. L. Herdman, H. W. Stech, and R. L. Wheeler,* Volterra and Functional Differential Equations

82. *N. L. Johnson, M. J. Kallaher, and C. T. Long,* Finite Geometries: Proceedings of a Conference in Honor of T. G. Ostrom
83. *G. I. Zapata,* Functional Analysis, Holomorphy, and Approximation Theory
84. *S. Greco and G. Valla,* Commutative Algebra: Proceedings of the Trento Conference
85. *A. V. Fiacco,* Mathematical Programming with Data Perturbations II
86. *J.-B. Hiriart-Urruty, W. Oettli, and J. Stoer,* Optimization: Theory and Algorithms
87. *A. Figa Talamanca and M. A. Picardello,* Harmonic Analysis on Free Groups
88. *M. Harada,* Factor Categories with Applications to Direct Decomposition of Modules
89. *V. I. Istrăţescu,* Strict Convexity and Complex Strict Convexity: Theory and Applications
90. *V. Lakshmikantham,* Trends in Theory and Practice of Nonlinear Differential Equations
91. *H. L. Manocha and J. B. Srivastava,* Algebra and Its Applications
92. *D. V. Chudnovsky and G. V. Chudnovsky,* Classical and Quantum Models and Arithmetic Problems
93. *J. W. Longley,* Least Squares Computations Using Orthogonalization Methods
94. *L. P. de Alcantara,* Mathematical Logic and Formal Systems
95. *C. E. Aull,* Rings of Continuous Functions
96. *R. Chuaqui,* Analysis, Geometry, and Probability
97. *L. Fuchs and L. Salce,* Modules Over Valuation Domains
98. *P. Fischer and W. R. Smith,* Chaos, Fractals, and Dynamics
99. *W. B. Powell and C. Tsinakis,* Ordered Algebraic Structures
100. *G. M. Rassias and T. M. Rassias,* Differential Geometry, Calculus of Variations, and Their Applications
101. *R.-E. Hoffmann and K. H. Hofmann,* Continuous Lattices and Their Applications
102. *J. H. Lightbourne, III, and S. M. Rankin, III,* Physical Mathematics and Nonlinear Partial Differential Equations
103. *C. A. Baker and L. M. Batten,* Finite Geometries
104. *J. W. Brewer, J. W. Bunce, and F. S. Van Vleck,* Linear Systems Over Commutative Rings
105. *C. McCrory and T. Shifrin,* Geometry and Topology: Manifolds, Varieties, and Knots
106. *D. W. Kueker, E. G. K. Lopez-Escobar, and C. H. Smith,* Mathematical Logic and Theoretical Computer Science
107. *B.-L. Lin and S. Simons,* Nonlinear and Convex Analysis: Proceedings in Honor of Ky Fan
108. *S. J. Lee,* Operator Methods for Optimal Control Problems
109. *V. Lakshmikantham,* Nonlinear Analysis and Applications
110. *S. F. McCormick,* Multigrid Methods: Theory, Applications, and Supercomputing
111. *M. C. Tangora,* Computers in Algebra
112. *D. V. Chudnovsky and G. V. Chudnovsky,* Search Theory: Some Recent Developments
113. *D. V. Chudnovsky and R. D. Jenks,* Computer Algebra
114. *M. C. Tangora,* Computers in Geometry and Topology
115. *P. Nelson, V. Faber, T. A. Manteuffel, D. L. Seth, and A. B. White, Jr.* Transport Theory, Invariant Imbedding, and Integral Equations: Proceedings in Honor of G. M. Wing's 65th Birthday
116. *P. Clément, S. Invernizzi, E. Mitidieri, and I. I. Vrabie,* Semigroup Theory and Applications
117. *J. Vinuesa,* Orthogonal Polynomials and Their Applications: Proceedings of the International Congress
118. *C. M. Dafermos, G. Ladas, and G. Papanicolaou,* Differential Equations: Proceedings of the EQUADIFF Conference
119. *E. O. Roxin,* Modern Optimal Control: A Conference in Honor of Solomon Lefschetz and Joseph P. Lasalle
120. *J. C. Díaz,* Mathematics for Large Scale Computing
121. *P. S. Milojević,* Nonlinear Functional Analysis
122. *C. Sadosky,* Analysis and Partial Differential Equations: A Collection of Papers Dedicated to Mischa Cotlar

123. *R. M. Shortt,* General Topology and Applications: Proceedings of the 1988 Northeast Conference
124. *R. Wong,* Asymptotic and Computational Analysis: Conference in Honor of Frank W. J. Olver's 65th Birthday

Other Volumes in Preparation

General Topology and Applications

Proceedings of the 1988 Northeast Conference

edited by

R. M. Shortt
Wesleyan University
Middletown, Connecticut

Marcel Dekker, Inc. New York and Basel

Gladly and respectfully dedicated
to Melvin Henriksen on the
occasion of his 60th birthday

Library of Congress Cataloging-in-Publication Data

General topology and applications : proceedings of the 1988 Northeast conference / edited by Rae Michael Shortt.
p. cm. -- (Lecture notes in pure and applied mathematics ; 123)
Proceedings of the Northeast Conference on General Topology and Applications, held at Wesleyan University in June 1988 in honor of Melvin Henriksen.
ISBN 0-8247-8349-2 (alk. paper)
1. Topology--Congresses. I. Shortt, Rae Michael. II. Henriksen, Melvin. III. Northeast Conference on General Topology and Applications (1988 : Wesleyan University) IV. Series.
QA611.A1G44 1990
514--dc20 90-2707
CIP

This book is printed on acid-free paper.

MARCEL DEKKER, INC.
270 Madison Avenue, New York, New York 10016

Current printing (last digit):
10 9 8 7 6 5 4 3 2 1

PRINTED IN THE UNITED STATES OF AMERICA

Editor's Note

These proceedings comprise papers by authors who met at Wesleyan University in June 1988 to celebrate the 60th birthday of Mel Henriksen. This event, the Northeast Conference on General Topology and Applications, was sponsored by Wesleyan, the City College of New York, the College of Staten Island, the New York Academy of Sciences, and a grant from the National Science Foundation. *

The editor wishes to express his thanks to the other conference organisers, W. Wistar Comfort, Anthony Hager, Ralph Kopperman, Prabudh Misra, Richard Resch and Aaron Todd without whom this conference could not have come to pass. Thanks are also due Maria Allegra, Brian Black and the staff of Marcel Dekker, Inc. for their kind assistance and expert completion of this project. Lastly, the editor is indebted to the contributors of these research papers for their mathematical efforts and their cooperation. And, of course...Happy Birthday, Mel!

R. M. Shortt

*Grant number DMS–8801172

Preface

My Friend Mel Henriksen

Remarks at the banquet in his honor

LEONARD GILLMAN Department of Mathematics, The University of Texas, Austin, Texas

H e l l o! My name is Len, and I'm going to be your master of ceremonies for this evening.

Welcome to Verm—. Well, back where I come from we can't keep track of all your little states up around here. Whichever one we're in—welcome! I come from another great state. We have great governors, too, like the one who started a speech at the state penitentiary with "My fellow citizens." The hollow laughter reminded him that his audience were not 100% citizens—they don't have the vote, and I don't think they're permitted to travel freely. So he started over: "My fellow convicts." When the roar had subsided, he tried once more: "Well I guess I don't know exactly what to call you—*but it certainly is good to see so many of you here.*"

Well, with no analogy intended—it certainly is good to see so many of you here.

> *Introduction of the head table.* On my right, saving the guy at the end till later: **Arthur Stone, Louise Henriksen.** On my left: **John Isbell, Dorothy Maharam Stone,** and, finally—what did you say your name was?—oh, **Mel Henriksen.** [Applause for all and an ovation for Mel.]

Three of the nice things about Henriksen are the standard traps in spelling his name—Hen*d*, ri*ck*, and s*o*n. Once after seeing it in the Notices as *Hendrikson*, I sent him a note: "Dear Mel: See latest Notices, p. 189—kindly explain why the *c* is missing from your name." He wrote back resignedly that the Lafayette telephone book had his name wrong for the second year in a row. (He subsequently brought them to their senses by returning all bills unopened, marked "not at this address".)

Introduction of the organizing committee for the conference. **Wis Comfort, Tony Hager, Chip Neville, Lewis Robertson,** and, the one we omitted before, the chairman of the committee, **Rae Shortt.** [Applause for all and an ovation for Rae.]

Poem by ***Wis Comfort,*** consisting of a dozen couplets such as:

Send a rabbi, or priest, bearing all extreme unctions
If Mel should run out of continuous functions.

In the fall of 1952, Mel and I arrived at Purdue as instructors and were assigned to the same office, a long narrow affair, which we shared with two others. (One was Gordon Walker, later Executive Director of the AMS, and the other was a retired chap who rarely came in.) So it was easy to start talking with each other about mathematics—especially since that was what Mel was always talking about. I had been working in operations research for the Navy; it was interesting work and paid well, but I was glad to have escaped to a university. Mel had just spent a year teaching at the University of Alabama and was even gladder to have escaped.

It may surprise you to learn that in those days, Mel was bustling with energy. He had recently completed a dissertation on the ring of entire functions, and he lost no time trying to interest me in the subject. I in turn had just finished a dissertation on ordered sets and tried to defend myself by talking about *that.* Then one day Mel happened to use the word "interval", and I perked up. About the same time, he got talking about Ed Hewitt's big paper on rings of continuous functions. This led to our first joint paper (the one on P-spaces), submitted in the spring of 1953. (Once when 90% of the calls to the Gillmans were coming from Mel, Reba answered one with a cheery, "Mrs. P-space.") Within two years, we had submitted four more. One of the five was a "triple" paper with Meyer Jerison and another one was with Paul Erdös.

The administration recognized from the beginning that we were doing mathematics every spare moment, and they promoted us to assistant professor for the fall of 1953. At the meeting that summer (in Kingston, Ontario), Mel introduced me to Ed Hewitt and Ernie Michael. Mel and I were pretty happy about our promotions and had managed to leak the news out. Ed asked how one gets promoted at Purdue, and I replied, "The dean has a blacklist[1], and either you're on it or you're off it." Ed asked, "How can you tell?" At that moment, Ernie looked up to ask, "Is it posted?"

For a while, Mel was maintaining a regimen of submitting a paper to every national and midwestern AMS meeting. That way he could get travel expenses to present a paper and meet lots of mathematicians. An important goal was to lessen the danger of being remanded to Alabama. I called it his Alabama complex. Back in those days I was 10 years older than Mel (actually, I still am), and one day I told him in a fatherly tone that he could quit that, as he was now an established mathematician. He conscientiously skipped the next meeting and admitted to a feeling of relief.

Speech by ***Neil Hindman*** in praise of the help and encouragement Mel gives to young mathematicians.

Mel organized a seminar in rings of continuous functions in the spring of 1953 and probably the next year as well. During 1954–1955, Jerry and I joined him in conducting a more formal seminar with four graduate students: Joe Kist, Carl Kohls, Maynard Mansfield, and Bob McDowell. It was a rewarding, no-holds-barred affair, as we were all learning together. The four students eventually ended up with Ph.D.'s with one or the other of us.

[1]The original terminology was somewhat earthier.

Today, Mel has about five dozen papers.They weigh 3.4 kilograms and occupy a volume of 8.3 liters, making them 41% as dense as water.

I am proud to have helped with five of those papers. Writing a joint paper is a rewarding experience. Two things are necessary for it to work: neither author could do it alone, and each one contributes 75% to the effort. You work anywhere you can. Our F-space paper owes thanks to a table outside a fast-food spot (where we worked at night with moths flying all about us), the Gillmans' dining room table, and the telephone.

Our paper with Jerison was originally titled something like *On a theorem of Gelfand and Kolmogoroff*, but the editor requested something more descriptive. As we wanted to cash in on the name-dropping, we simply tacked on the description: *On a theorem of Gelfand and Kolmogoroff concerning maximal ideals in rings of continuous functions*. Our punishment is renewed every time any of us has to type up a bibliography.

By now, Mel also has 11 "children" and some grandchildren.

> *Call for Mel's children.* There were two: Don Johnson and Frank Smith.
>
> *Call for all those who have written a joint paper with Mel.* [Huge response.]
>
> *Call for everyone's "Henriksen number"* (the length of the smallest chain linking one to Henriksen via joint papers. Henriksen's Henriksen number is 0). There were lots of 1's and 2's, some 3's, a couple of 4's, and a 5.
>
> *Call for all those who have published a paper whose bibliography includes a paper of Henriksen's.* (Bev Diamond suggested this one.) The vast majority responded, to loud applause.
>
> *Call for those who have published at least one paper but have* ***never*** *listed a paper of Mel's in their bibliography.* A handful of people rose, to a rousing chorus of boos.
>
> *Speech by* ***Charles W. Neville.*** Chip had two presents for Mel. The first was a brass ring engraved with the symbols sin(z), cos(z), and e^z. With the variable thus chosen as z, this **ring of continuous functions** also served as a **ring of entire functions**. The second present was an ultra oil filter that had been purchased on a "Buy One, Get One Free" deal, thus a **free ultrafilter**. Chip had also hoped to provide a **measurable cardinal** "but the silly bird wouldn't hold still for the tape."

Mel Henriksen has always been a person of total integrity. He can smell sham a mile away; and at Purdue there was plenty to smell. We would often find ourselves using valuable research time plotting against the administration's ethical and professional lapses. To help us keep to the mark, I suggested the slogan: *Teach the courses, flunk the students, write the papers.* Mel was delighted, and it stood us well as a spirit-lifting rallying cry.

Speaking of integrity, I have always criticized authors who use "For a related result, see ... " as a ploy for a self-plug in the bibliography. When I mentioned the ploy to Mel, he quickly filed it in his list of crimes. But it is hard to resist, and some of our own references may be suspect. I also proposed its counterpart: "For an *un*related result, see ... ." (Today I rarely refer to anyone *other* than myself, as my own papers are the ones I remember.)

In the fall of 1958, I left Purdue for the Institute and then Rochester, and we continued our discussions by correspondence. Because Mel is such a colorful writer, his letters were enjoyable despite the gloom:

> (Mel, October 1958) [The administrators'] function is to keep things running smoothly. Only academic idiots like us concern ourselves with why we are running anything at all. In view of this, fighting them should be compared with punching a mattress.

(Mel, May 1959) Evidently, [Purdue President] Hovde, like Eisenhower, seems to believe that if he ignores a problem long enough, it will go away. In the case of the mathematics department, the latter may be correct.

(Len, June 1959) "Every few years to show who is boss the administration ruins the department."—P. Erdös.

(Mel, December 1959) I think that [Dean] Ayres regards us in much the same way that a zoo keeper regards his animals. They should be reasonably well fed, and their cages should be kept clean, but you don't consult the animals about how to run the zoo.

(Mel, April 1960) The fact that the dean throws parties for the [military] officers and does not trouble to even offer congratulations to staff members for scholastic honors is particularly disgusting. You complain of the failure to congratulate you for receiving a Guggenheim and an NSF. As far as I could tell, the administration regarded my Sloan grant as a pain in the neck[2].

Speech by ***Don Johnson*** in praise of how Mel works with doctoral students.

Mel is a very sociable person, but mathematics comes first. I remember my brother's first visit to Lafayette. *Len*: "Mel, this is my brother Bob." *Mel*, shaking his hand: "Hello," then turning to me: "Oh, Len, suppose X is a P-space, … ."

Mel is warm, kind, and loyal. When Gordon Walker left Purdue, he sold his house to Mel, who had been putting up with substandard university quarters. Mel was ecstatic, but remarked to me, "You know, it's wonderful having this house. But I'd rather have Gordon still here."

The Gillmans enjoyed a circle of close friends that included the Henriksens, Jerisons, Golombs, and Shankses. If Reba left town for a few days, they would divvy me up for dinners. Rochester turned out to be different. Once while Reba was off on an extended trip, I bumped into a colleague with whom we were on fairly warm social terms. *He*: "Your wife is away, isn't she?" *I*, hopefully: "Yes … ." *He*: "When she gets back we'll have you both over to dinner."

One day Caspar Goffman and Mel—Santa-Claus-shaped gentlemen both—came up with the practical idea of founding FMU (*Fat Man's University*). The key was a simple evaluation scheme: don't bother weighing the research papers—just weigh the professor.

Speech by ***John Isbell.*** John presented Mel with a gift of a hippo (a favorite toy of Mel's), supported by references to several literary figures, including T. S. Eliot and Hillaire Belloc. The latter was represented by the couplet:

I hunt the hippopotamus with bullets made of platinum,
Because if I use leaden ones his hide is sure to flatten 'em.

By 1959 I was actively looking for a job elsewhere. But Mel was ambivalent:

(Mel, December 1959) I am still as mixed up as ever [about leaving Purdue]. Our difference of opinion is best illustrated by a conversation I had with Reba last summer. After a lengthy review of the well known strengths and weaknesses of Purdue, I said, "Well, all universities seem to have some garbage associated with them. The advantage of the Purdue garbage is that I know and understand most of it." Reba's reply was, "Lenny feels quite differently. He's looking for fresh garbage."

Mel and I had often wondered whether running the department ourselves would save us time. When Rochester offered me the chairmanship (for the fall of 1960), I grabbed it. After a while, Mel got interested in such possibilities too, and even went for interviews. But he had strict stan-

[2] The locale he actually suggested was lower.

dards and had coined an epigram to guide him: *Every silver lining has a cloud.*

> (Mel, April 1964) I'm definitely out of the chairman business.
> (Mel, December 1964) I decided to accept the chairmanship at Case.

Speech by ***Ralph Kopperman***, who had organized three earlier Northeast Topology Conferences at CUNY, practically single-handed. Ralph expressed special thanks to Mel for having encouraged him back into research after a period devoted to other interests.

Jerison was unable to attend the conference and sent a letter expressing his regret. As Mel had been telling me for years how much he has learned from Jerry, you can imagine how moved he was to read:

> You have inspired a lot of people to do mathematics. I am one of many who would have done far less if we had not known you and worked with you.

In the summer of 1955, Mel (and I) attended the Topology Institute in Madison, Wisconsin. On the return trip, with his wife and two small children, Mel wanted to stop over in Ann Arbor. Dean Ayres personally searched the schedules to find him the cheapest route: train to Detroit, then bus to Ann Arbor getting in at 3:00 a.m.

We wondered whether the dean would support our travel to the 1956 summer meeting in Seattle, and Mel calculated that we could save him some money by sending ourselves by mail. One day Mel was holding forth on this at a crowded party. As the appreciative laughter died down, some measured tones came wafting across from the other end of the room: "As I see it, the primary obstacle would be getting you through the slot."

Contents

Contributors

CHARLES E. AULL Virginia Polytechnic Institute and State University, Blacksburg, Virginia

RICHARD BALL Boise State University, Boise, Idaho

PAUL BANKSTON Marquette University, Milwaukee, Wisconsin

ANDREAS R. BLASS University of Michigan, Ann Arbor, Michigan

ALAN DOW York University, North York, Ontario, Canada

LEONARD GILLMAN University of Texas at Austin, Austin, Texas

ELISE M. GRABNER Slippery Rock University, Slippery Rock, Pennsylvania

ANANDA V. GUBBI Southwest Missouri State University, Springfield, Missouri

ANTHONY W. HAGER Wesleyan University, Middletown, Connecticut

T. R. HAMLETT East Central Oklahoma State University, Ada, Oklahoma

KLAAS PIETER HART Delft University of Technology, Delft, The Netherlands

MELVIN HENRIKSEN Harvey Mudd College, Claremont, California

NEIL HINDMAN Howard University, Washington, D. C.

H. H. HUNG Concordia University, Montreal, Quebec, Canada

DRAGAN JANKOVIĆ East Central Oklahoma State University, Ada, Oklahoma

EFIM KHALIMSKY College of Staten Island, City University of New York, Staten Island, New York

T. Y. KONG City College, City University of New York, New York, New York

RALPH KOPPERMAN City College, City University of New York, New York, New York

JOHN MACK University of Kentucky, Lexington, Kentucky

JAMES MADDEN Indiana University at South Bend, South Bend, Indiana

KENNETH D. MAGILL State University of New York at Buffalo, Buffalo, New York

ROBERT A. McCOY Virginia Polytechnic Institue and State University, Blacksburg, Virginia

P. R. MISRA College of Staten Island, City University of New York, New York, New York

A. MOLITOR Wesleyen University, Middletown, Connecticut

CHARLES NEVILLE Central Connecticut State University, New Britain, Connecticut

JACK R. PORTER University of Kansas, Lawrence , Kansas

MARLON RAYBURN University of Manitoba, Winnipeg, Manitoba, Canada

R. M. SHORTT Wesleyan University, Middletown, Connecticut

ARTHUR H. STONE Northeastern University, Boston, Massachusetts

ANDRZEJ SZYMANSKI Slippery Rock University, Slippery Rock, Pennsylvania

JAN VAN MILL Vrije University, Amsterdam, The Netherlands

R. GRANT WOODS University of Manitoba, Winnipeg, Manitoba, Canada

HUEYTZEN J. WU Texas A&I University, Kingsville, Texas

On Well-Embedding

C. E. Aull

Department of Mathematics
Virginia Polytechnic Institute & State University
Blacksburg, Virginia 24061

In 1960 Gillman and Jerison [GJ, 1960] introduced a necessary and sufficient condition for a C^*-embedded subset to be C-embedded. Later (1970) W. Moran [Mo, 1970] called this condition well-embedding.

In 1974, Blair and Hager [BH, 1974] showed that this condition was a necessary and sufficient condition for a z-embedded subset to be C-embedded. In this paper we attempt to make a systematic study of this important embedding property.

Definition 1. A set S is well-embedded in a space $X, S \subset X$, if S is completely separated from every zero set of X contained in $X \sim S$.

In this paper all spaces considered will be Tychonoff. The following results are easily proved and many are probably known.

THEOREM 1. *The following are satisfied.*

(a) *A set S is dense and well-embedded in a space X iff either (i) every nonempty zero set of X intersects S or (ii) if f is continuous on X and $\frac{1}{f}$ is continuous on S then $\frac{1}{f}$ is continuous on X.*

(b) *A space S is well-embedded in any space in which it is embedded (as a dense set), [as a closed set] iff S is pseudocompact (iff S is pseudocompact) [iff S is pseudocompact].*

(c) *A space X is a Z-space or a δ-normally separated space (see [Z, 1969], [M,*

1970] and [A, 1983]) iff every closed set is well-embedded in X.

(d) *A space X is an almost P-space [L, 1977] (space where non-empty zero sets have non-empty interiors) iff every dense set of X is well-embedded in X.*

(e) *A space X is a P-space [GJ, 1960] iff either (i) every cozero set of X is well-embedded, or (ii) every set of X is well-embedded in X.*

(f) *Finite unions of sets well-embedded in X are well-embedded in X.*

PROOF OF CLOSED SET CASE OF: (b)

A pseudocompact subset is always completely separated from any disjoint zero set (implicitly in [I, 1967] and explicitly in [S, 1975]). Let S be not pseudocompact and T be a space such that $\beta T \sim T$ is exactly one point "p" such as W. Let $X = (S \times (p)) \cup (\beta S \times T)$. The space X is pseudocompact since $\beta S \times T$ is a pseudocompact dense set. Thus X is C-embedded in $\beta X = \beta S \times \beta T = \beta(\beta S \times T)$. The last equalities follow from the fact T is pseudocompact [GJ, 1960] a result of Glicksberg. Assume S is well-embedded in X; since S is C^*-embedded in βS and βS is C-embedded in βX then S is C-embedded in βX and hence S is C-embedded in βS which implies S is pseudocompact. So any set that is well-embedded in every space in which it is embedded as a closed set is pseudocompact.

2. TRANSITIVITY.

If $S \subset T \subset X$ and if S is C-embedded in T and T is C-embedded in X then S is C-embedded in X; i.e. C-embedding is transitive: we may substitute C^*-embedding, z-embedding [BH, 1974] or v-embedding [B, 1976] in the above statement. With well-embedding, we have to qualify the statement.

Example 1. In the Tychonoff Plank T [GJ, 1960] N the right edge is well-embedded in $N \cup W$, where W is the top edge, and $N \cup W$ is well-embedded in T but N is not well-embedded in T.

Example 2. A countable well-embedded copy of N is C-embedded since it is z-embedded [BH, 1974]. So a countable subset M of D in ψ, [GJ, 1960] is not well-embedded in ψ but M is well-embedded in D and D is well-embedded in ψ, where D is the set of nonisolated points of ψ.

THEOREM 2. *Let $S \subset T \subset X$ with S well-embedded in T and T well-embedded in X. Then S is well-embedded in X if either (a) S is dense in T or (b) T is C-embedded in X.*

PROOF: (a) Let Z be a zero set of X such that $Z \subset X \sim S$. Since S is dense and well-embedded in T, $Z \subset X \sim T$. There exists a zero set H of X such that

$T \subset H$ and $H \cap Z = \phi$ by the well-embedding of T in X. Since $S \subset H$, well-embedding is transitive.

(b) Let Z be a zero set of X such that $Z \subset X \sim S$. There exists a zero set H of T such that $S \subset H$ and $H \cap (Z \cap T) = \phi$ by the well-embedding of S in T. There exists a zero set $E(H)$ of X with $E(H) \cap T = H$ by the z-embedding of T in X. Then $E(H) \cap T \cap Z = \phi$. There exists a zero set H' of X such that $H' \cap (E(H) \cap Z) = \phi$ and $T \subset H'$ by the well-embedding of T in X. Then $H' \cap E(H)$ is a zero set of X such that $(H' \cap E(H)) \cap Z = \phi$ and $S \subset H' \cap E(H)$. So S is well-embedded in X.

3. STRONG PARATRANSITIVITY OF WELL-EMBEDDING.

Definition 2. An A-embedding is strongly paratransitive (is paratransitive) if for $S \subset T \subset X$, S dense in T, S A-embedded in X (and T) implies T is A-embedded in X.

THEOREM 3. *Well-embedding is paratransitive.*

PROOF: Let Z be a zero set of X such that $Z \subset X \sim T$. There exists a zero set H of X such that $S \subset H$, $H \cap Z = \phi$ by the well-embedding of S in X. Since S is dense and well-embedded in T, $T \subset H$, so T is well-embedded in X.

From Theorem 4 in the next section we will be able to drop the stipulation in Theorem 3 that S is well-embedded in T and say that well-embedding is strongly paratransitive.

We note in Theorem 3, S dense in T is needed; for if $x \in T \subset X$, $\{x\}$ is well-embedded in X and T but T may not be well-embedded in X. The same argument may be applied to $z-, C^*-$, or C-embeddings. We note z-embedding does not satisfy a weaker condition than that of Theorem 3 metatransivity. To define metatransivity we replace S dense in T by S dense in X in the definition of paratransitivity.

4. HEREDITARY PROPERTIES OF WELL-EMBEDDING.

If S is C-embedded in X and $S \subset T \subset X$ then S is C-embedded in T. (i.e. C-embedding is hereditary). Similar results hold for C^*-embeddings and z-embeddings. We do not know if the above result holds in general for well-embeddings. However we have the following results.

THEOREM 4. *Let $S \subset T \subset X$, S well-embedded in X; then S is well-embedded in T if either (a) S is dense in T, or (b) T is z-embedded in X or (c) $T \sim S$ is σ-compact.*

We will need the following definition and lemma.

Definition 3. A set S is quasi-z-embedded in a space X if for Z a zero set of S, there exists a family of zero sets of X, $\{Z_a;\ a \in A\}$, such that $S \cap (\cup(Z_a : a \in A)) = Z$.

LEMMA 4. *If $S \subset X$, X completely regular, then S is quasi-z-embedded in the space X.*

PROOF OF LEMMA: The set $S \sim Z$ is an F_σ-set of S and can be extended to an F_σ-set M of X such that $M \cap S = S \sim Z$. If $x \notin M$, there exists a zero set Z_x such that $x \in Z_x \subset X \sim M$, by the complete regularity of X. Then $Z = S \cap (\cup Z_x : x \in X \sim M)$ and thus S is quasi-z-embedded in X.

PROOF OF THEOREM: (a) Let Z be a zero set of T such that $Z \subset T \sim S$. Then there exists a zero set H of X by Lemma 4 such that $H \subset X \sim S$ and $H \cap (T - S) \neq \phi$. Since S is well-embedded in X, there is a zero set of $X, H', S \subset H'$ and $H' \cap H = \phi$. Since S is dense in $T, T \subset H'$, contradicting $H' \cap H = \phi$, since $H \cap (T - S) \neq \phi$.

(b) Let Z be a zero set of T such that $Z \subset T \sim S$. By the z-embedding of T in X, there exists a zero set $E(Z)$ of X such that $E(Z) \cap T = Z$. Furthermore there exist a zero set H of X such that $S \subset H$ and $H \cap E(Z) = \phi$ by the well-embedding of S in X. Then $H \cap T$ is a zero set of T with $S \subset H \cap T$ and $(H \cap T) \cap Z = \phi$.

(c) Again let Z be a zero T such that $Z \subset T \sim S$. The set $T \sim S = \cup K_n$ where K_n is compact. The set $T \sim Z = \cup F_k$, F_k closed in T. Then $X \sim Z \subset \cup E(F_k)$, where each $E(F_k)$ is closed in X and $E(F_k) \cap T = F_k$. There exists a zero set Z_n of X satisfying $Z_n \supset Z \cap K_n$ and which is disjoint from $\cup E(F_k)$; then there exists a zero set H_n of X such that $S \subset H_n \subset X \sim Z_n$. Then $S \subset \cap H_n \cap T$ which is a zero set of T disjoint from Z. So S is well-embedded in T.

COROLLARY 4. *If S is dense and well-embedded in X, then P is well-embedded in R where $S \subset P \subset R \subset X$.*

This justifies the addition to the proof in [AS, 1988] concerning full well-embedding.

The question remains as to whether well-embedding is in general hereditary. i.e. (1) If $S \subset T \subset X$ and S is well-embedded in X, then is S well-embedded in T. Closely related is the following question (2) if X is a Z-space is every closed subspace of X a Z-space? If (2) has a negative answer then so has (1).

BIBLIOGRAPHY

[A, 1983] C.E. Aull, 'On Z- and Z^*-spaces', *Topology Proc.* **8** (1983), 1-19.

[A, 1988] C.E. Aull, Some embeddings related to C^*-embeddings, *J. Austral. Math. Soc.* (Series A), **44** (1988), 88-104.

[AS, 1988] C.E. Aull and J.O. Sawyer, The pseudocompactification αX, *Proc. Am. Math. Soc.*, **102** (1988) 1057-1064.

[B, 1976] R.L. Blair, 'Spaces in which special sets are z-embedded', *Canad. J. Math.* **28** (1976), 673-690.

[BH, 1974] R.L. Blair and A.W. Hager, 'Extensions of zero-sets and of real-valued functions', *Math. Z.* **136** (1974), 41-52.

[GJ, 1960] L. Gillman and M. Jerison, *Rings of continuous functions* (Van Nostrand, Princeton, N.J., 1960).

[I, 1967] T. Isiwata, Mappings and spaces, Pacific J. of Math., **20**, (1967), 455-480.

[L, 1977] R. Levy, Almost P-spaces, *Canad. J. Math.* **29** (1977) 284-288.

[M, 1970] J. Mack, 'Countable paracompactness and weak normality properties', *Trnas. Amer. Math. Soc.* **148** (1970), 265-272.

[Mo, 1970] W. Moran, 'Measure on metacompactness spaces,' *Proc. London Math. Soc.* **20** (1970), 507-526.

[S, 1975] J.O. Sawyer, Pseudocompactifications and Pseudocompact Spaces, Ph.D. Thesis, Virginia Polytechnic Institute and State University, (1975).

[Z, 1969] P. Zenor, 'A note on Z-mappings and WZ-mappings', *Proc. Amer. Math. Soc.* **23** (1969), 273-275.

The Quasi-F_κ cover of Compact Hausdorff Space and the κ-Ideal Completion of an Archimedean l-Group

Richard N. Ball
Department of Mathematics
Boise State University
Boise, Idaho 83725, U.S.A.

Anthony W. Hager
Department of Mathematics
Wesleyan University
Middletown, Connecticut 06457, U.S.A.

Charles W. Neville
Department of Mathematics
Central Connecticut State University
New Britain, Connecticut 06050, U.S.A.

It is a pleasure to dedicate this paper to Melvin Henriksen for his sixtieth birthday. We think it is very much in the spirit of some of Mel's work, and we hope he likes it.

1. INTRODUCTION

We prove the existence and uniqueness of an "overobject" of each sort mentioned in the title. Thus the results contained herein are both algebraic and topological in nature, and in approximately equal proportion. To so dichotomize them would be a mistake, however, since each viewpoint relies upon the other. Rather, the Yosida functor unites the two into a whole greater than the sum of its parts. We begin by summarizing the topological results.

We work in the category K of compact Hausdorff spaces with continuous maps; in particular, *all maps between spaces will be assumed to be continuous.* We assume the basic notation of Gillman and Jerison (1960). For an element $g \in C(X)$ we write $z(g)$ and $coz(g)$ for the zero set $g^{\leftarrow}\{0\}$ and cozero set $g^{\leftarrow}(\mathbb{R} \setminus \{0\})$, respectively. We use the symbol κ to represent either an infinite cardinal number or ∞. By a *κ-zero* set we mean any intersection of fewer than κ zero sets, and by a *κ-cozero* set we mean any union of fewer than κ cozero sets; thus any closed set is an ∞-zero set and any open set is an ∞-cozero set. Given an ℓ-group G of continuous extended real valued functions (see the discussion beginning Section 2) on $X \in$ K, we designate the lattice of κ-zero subsets generated by G as follows:

$$\mathcal{Z}^G_\kappa(X) = \{\cap_F z(f): F \subset G \text{ and } |F| < \kappa\}.$$

Here $|F|$ designates the cardinality of F. Of central importance is the image of $\mathcal{Z}^G_\kappa$ under the lattice homomorphism $Z \longmapsto clint\, Z$, namely

$$\mathfrak{Z}^G_\kappa(X) = \{clint\, Z: Z \in \mathcal{Z}^G_\kappa(X)\}.$$

(We concatenate topological operators, writing *clint* Z for the closure of the interior of Z, for instance.) The latter is a sublattice of the complete Boolean algebra $\mathcal{R}(X)$ of regular closed subsets of X, in which lattice, we remind the reader, $A \vee B = A \cup B$ and $A \wedge B = clint(A \cap B)$. In case G separates the points of X, and in particular if $G = C(X)$, we drop the superscript G, writing simply $\mathcal{Z}_\kappa(X)$ and $\mathfrak{Z}_\kappa(X)$.

The continuous maps of interest to us are the *κ-irreducible surjections*; such a surjection $\tau: Y \to X \in \mathrm{K}$ is defined by the property that for every $Z \in Z_\kappa(Y)$ there is some $W \in Z_\kappa(X)$ such that $int\ Z = int\ \tau^{\leftarrow}(W)$. Equivalently, τ is κ-irreducible if and only if the map $C \longmapsto \tau(C)$ is a lattice isomorphism from $\mathfrak{Z}_\kappa(Y)$ onto $\mathfrak{Z}_\kappa(X)$, in which case the inverse map is $D \longmapsto cl\ \tau^{\leftarrow}(int\ D)$. When these conditions obtain, we refer to Y as a *κ-irreducible preimage of* X. Observe that κ-irreducibility implies λ-irreducibility for $\kappa \leq \lambda$. This observation is valid even when $\lambda = \infty$, in which case ∞-irreducibility is equivalent to the usual notion of irreducibility found in the literature. In most cases we have chosen to use the latter term without the ∞.

Our major topological result is the identification of the projectives in the category of compact Hausdorff spaces and κ-irreducible maps; we term these objects *quasi-F_κ spaces.* When specialized to $\kappa = \infty$, these results recapture the classical development of extremally disconnected spaces of Gleason (1958), and when specialized to $\kappa = \aleph_1$ they reproduce the more recent development of quasi-F spaces. The results for intermediate cardinality are new, but they are not the only point of this paper. Most attempts to generalize the notion of extremal disconnectivity have been by means of various separation properties. But the present approach suggests that the quasi-F_κ property defined below may be a more natural and productive generalization.

The development of quasi-F spaces was initiated by Dashiell, Hager, and Henriksen (1980), and independently by Zakharov and Koldunov (1980). It has been carried forward by Henriksen, Vermeer, and Woods (1987, 1989). Neville (1979) extended these ideas, building on his work with Lloyd (1981), although Neville's notes enjoyed only limited circulation. We refer the reader to these papers for background and related results. Porter and Woods (1988) is an excellent general reference to these and related matters which has recently become available.

To state the result it is necessary to recall several definitions. A subspace $U \subset Y \in \mathrm{K}$ is said to be *κ-Lindelöf* provided that every open cover of U has a subcover of cardinality strictly less than κ. For example, κ-cozero subsets are κ-Lindelöf because they are unions of fewer than

κ compact sets. U is said to be *C^*embedded* provided that each bounded continuous function on U has a continuous extension to Y. Y is said to be an *F-space* if each cozero subset is C^*embedded in Y, and is said to be a *quasi-F space* if each dense cozero subset is C^*embedded in Y. F-spaces are discussed in 14.25-14.28 of Gillman and Jerison (1960). We shall call $Y \in \mathrm{K}$ *quasi-F_κ* provided that every dense κ-Lindelöf subspace is C^*embedded. Theorem 4.6 with $\kappa = \aleph_1$, combined with Dashiell, Hager, and Henriksen (1980), shows that the quasi-$F_{\aleph_1}$ spaces we treat here are precisely the quasi-F spaces in the literature. Finally, we say K morphisms $\varphi\colon Y \to X$ and $\tau\colon Z \to X \in \mathrm{K}$ are *equivalent over* X provided that there is a homeomorphism $\mu\colon Y \to Z$ such that

$$\begin{array}{ccc} Y & \xrightarrow{\ \mu\ } & Z \\ & {\scriptstyle\varphi}\searrow \quad \swarrow{\scriptstyle\tau} & \\ & X & \end{array}$$

commutes.

THEOREM 4.9. *For every $X \in K$ there is a κ-irreducible surjection $\varphi\colon Y \to X \in K$ with the following equivalent properties.*

(a) Y is quasi-F_κ.

(b) An irreducible surjection $\tau\colon Z \to X \in \mathrm{K}$ is κ-irreducible if and only if there is some surjection $\mu\colon Y \to Z \in K$ such that

$$\begin{array}{ccc} Y & \xrightarrow{\ \mu\ } & Z \\ & {\scriptstyle\varphi}\searrow \quad \swarrow{\scriptstyle\tau} & \\ & X & \end{array}$$

commutes.

φ is unique in the sense that it is equivalent over X to any other map with its properties.

We call the space Y of Theorem 4.9 the *quasi-F_κ cover* of X, written $QF_\kappa(X)$, and designate by φ_X the map φ of Theorem 4.9. Whereas Theorem 4.9 asserts that $QF_\kappa(X)$ is the maximal κ-irreducible preimage of X, Theorem 4.11 shows it to be also the minimal quasi-F_κ cover of X.

THEOREM 4.11. *Let* X, Y, *and* φ *have the meaning of Theorem 4.9. Then for any quasi-*F_κ *space* Z *and irreducible surjection* $\tau: Z \to X \in K$ *there is a surjection* $\mu: Z \to Y$ *such that*

$$\begin{array}{ccc} Y & \xleftarrow{\;\mu\;} & Z \\ & {\scriptstyle\varphi}\searrow \quad \swarrow{\scriptstyle\tau} & \\ & X & \end{array}$$

commutes.

Henriksen, Vermeer, and Woods (1989) propose a general scheme for generating covers of X by means of spaces of ultrafilters on sublattices of the regular closed algebra of X, and in particular they show in their 1987 paper that the quasi-F cover can be obtained in this way from $Z_{\aleph_1}(X)$. However, in Section 6 we show by example that $QF_\kappa(X)$ cannot in general be so obtained from $Z_\kappa(X)$, and characterize those spaces for which it can.

The domain of discourse for the algebraic investigation is the category **Arch** of archimedean lattice ordered groups (ℓ-groups) with ℓ-homomorphisms (maps which are simultaneously group and lattice homomorphisms). Given $G \in$ Arch, K is an *ℓ-subgroup* of G, written $K \leq G$, provided that K is both a subgroup and sublattice of G, and K is an *ideal* of G if K is an ℓ-subgroup which is *convex*, meaning $0 < g < k \in K$ implies $g \in K$. Ideals are precisely the kernels of ℓ-homomorphisms from G onto other ℓ-groups, though these other ℓ-groups need not be archimedean and so the maps are not generally **Arch** morphisms. The ideals upon which we fasten our attention, however, *are* kernels of **Arch** morphisms (Proposition 3.7). An ideal $K \leq G$ is a *κ-ideal* provided that for any subset $L \subset K$ of cardinality strictly less than κ and any $g \in G$ such that $\bigvee L = g$ we have $g \in K$. Thus an ∞-ideal contains the supremum of any subset which has one; such ideals are termed *closed* in the literature. A ***polar*** is a subset $X \subset G$ such that $X^{\perp\perp} = X$, where

$$X^{\perp} = \{g \in G: |g| \wedge |x| = 0 \text{ for all } x \in X\}.$$

The polars are precisely the ∞-ideals of an archimedean ℓ-group G; a proof of this fact can be found in Bigard, Keimel, and Wolfenstein (1977), an excellent general reference on ℓ-groups. κ-ideals are exactly the kernels of those ℓ-homomorphisms which are *κ-complete*, by which is meant those which preserve the existing suprema and infima of subsets of cardinality less than κ (Proposition 3.2). ∞-complete ℓ-homomorphisms are termed ***complete*** in the literature.

Here is the notion that drives the algebraic portion of the investigation: $G \leq H \in$ **Arch** is a *κ-ideal preserving extension* provided the κ-ideals of G and H are in one-to-one correspondence by intersection. The ∞-ideal preserving extensions are called a^*-extensions in the literature; they have been investigated in greater generality than we treat here by Bleier and Conrad (1973, 1975), Glass, Holland, and McCleary (1975), Ball (1975), and others. There is yet another term widely used in the literature whose meaning, for archimedean ℓ-groups, coincides with the notion of ∞-ideal preserving extension. An ℓ-group G is said to be *large* in an ℓ-supergroup H if every nontrivial ideal of H meets G nontrivially. The reasons for the coincidence of meanings are as follows. It is a folk theorem (Proposition 5.3 of Ball (1980)) that the polars of G and H are in one-to-one correspondence by intersection if and only if every nontrivial polar of H meets G nontrivially. Therefore every large embedding of archimedean ℓ-groups is ∞-ideal preserving. On the other hand, Bleier and Conrad (1975) proved that any ℓ-group is large in any a^*-extension, i.e. in any ∞-ideal preserving extension. Largeness is a useful condition in part because it implies that suprema and infima in G and H agree. Its recurrence in the algebraic material may be understood in light of the fact that, in the important weak unit case introduced in Section 2, it is dual to the irreducibility of the realizing continuous map (Hager and Robertson (1977)). Of the three terms available for this extension notion, we have chosen to use large most often.

Our major algebraic result is that every $G \in$ Arch has a κ-ideal preserving extension κG in which all other such extensions embed over G. (An embedding *over* G is one which is the identity on G.)

THEOREM 5.3. *For every $G \in$ **Arch** there is a κ-ideal preserving extension $G \leq H \in$ **Arch** with the following equivalent properties.*

(a) H has no proper κ-ideal preserving extensions.

(b) *An extension* $G \leq K \in$ ***Arch*** *is κ-ideal preserving if and only if there is an* Arch *injection* $\theta: K \longrightarrow H$ *over* G.

H is unique up to Arch *isomorphism over* G *with respect to its properties.*

We call the ℓ-group H of Theorem 5.3 the *κ-ideal completion of* G, and designate it by κG. We refer to an ℓ-group $G \in$ **Arch** for which $G = \kappa G$ as *κ-ideal complete.* Whereas Theorem 5.3 asserts that κG is the maximal κ-ideal preserving extension of G, Theorem 5.4 asserts that κG is also the smallest κ-ideal complete extension of G.

THEOREM 5.4. *Suppose* G *is a large ℓ-subgroup of the κ-ideal complete ℓ-group* $H \in$ Arch. *Then there is a unique* Arch *injection* $\psi: \kappa G \longrightarrow H$ *over* G.

Thus the $\kappa = \infty$ case of the two abovementioned theorems reproves the existence and uniqueness of the maximal a^*-extension of an archimedean ℓ-group, results due to Conrad (1971), and to Bleier and Conrad (1975). The $\kappa = \aleph_1$ case produces a completion which was also studied in the investigation of Ball and Hager (1989a). In this paper, which is closely related to the present one, it is shown that each object G in the weak unit subcategory **W** introduced in Section 2 admits a unique extension $G \leq aG \in$ **W** maximal with respect to the property that distinct kernels of **W** morphisms on aG trace distinctly on G. In fact, aG coincides with $\aleph_1 G$. The latter paper differs from the present one, however, in four respects. The extensions preserve κ-ideals rather than distinguish **W** kernels, the crucial idea of κ-sets is absent from Ball and Hager (1989a), the concept of the quasi-F_κ cover is apparently novel (though all the topological results contained herein were known to Neville (1979) ten years ago, and were recently rediscovered by another of us), and the algebraic results are extended from **W** to **Arch**. The completions which arise when κ is a cardinal between $\aleph_1$ and ∞ have not to our knowledge been studied, but they do not in themselves justify this work. Rather, we emphasize that our approach unites all mentioned completions.

2. THE YOSIDA REPRESENTATION

We work for the most part in **W**, the category whose objects are of the form (G,u), where G is an archimedean ℓ-group with designated *weak unit* u (meaning $0 < u \in G$, and $u^{\perp} = \{0\}$), and whose morphisms are the ℓ-homomorphisms that take the weak unit of the domain to the weak unit of the codomain. For example, when we say that H is a **W** extension of G, and write $G \leq H \in \mathbf{W}$, we mean that G is an ℓ-subgroup of the **W** object H having the same weak unit, i.e. that the inclusion map is a **W** morphism. With or without explicit mention, the weak unit is invariably taken to be 1, the constant function with value 1, whenever the **W** object consists of continuous functions on a space. To represent an abstract **W** object as an ℓ-group of continuous functions, however, it is necessary to allow the functions to take on the values $\pm\infty$ occasionally, as follows. The *extended real line* is the two point compactification of the real numbers obtained by adjoining points at $\pm\infty$, and a continuous extended real valued function g on a space Y is *almost finite* provided that $g^{\leftarrow}(\mathbb{R}) = \{y \in Y: g(y) \in \mathbb{R}\}$ is dense in Y. $D(Y)$ designates the set of such functions; it is a lattice under pointwise supremum and infimum operations, but may or may not be closed under the following addition operation. Given $f,g,h \in D(Y)$ we say $f + g = h$ provided that $f(y) + g(y) = h(y)$ for all $y \in f^{\leftarrow}(\mathbb{R}) \cap g^{\leftarrow}(\mathbb{R}) \cap h^{\leftarrow}(\mathbb{R})$. We shall say that G *is an ℓ-group in* $D(Y)$ whenever G is a subset of $D(Y)$ which is closed under the aforementioned group and lattice operations, and that G *is a* **W** *object in* $D(Y)$ if in addition G contains 1 as designated weak unit. We signify this relationship by writing $G \subset D(Y)$. We emphasize that $D(Y)$ is not an ℓ-group for arbitrary $Y \in \mathrm{K}$. In fact, the quasi-F condition is exactly what is required to extend, for $f,g \in D(Y)$, the pointwise sum $f(y) + g(y)$ from $g^{\leftarrow}(\mathbb{R}) \cap f^{\leftarrow}(\mathbb{R})$ to Y; that is, $D(Y)$ is an ℓ-group if and only if Y is quasi-F (Henriksen and Johnson (1961)).

The Yosida Representation asserts that every **W** object is **W** isomorphic to a **W** object in $D(Y)$ for some $Y \in \mathrm{K}$. We shall not prove this theorem here, but instead refer the interested reader to Bigard,

Keimel, and Wolfenstein (1975), to Luxemburg and Zaanen (1971), and to Hager and Robertson (1977).

THEOREM 2.1: THE YOSIDA REPRESENTATION OF **W** OBJECTS. *For any* $G \in \mathbf{W}$ *there is some* $Y \in \mathbf{K}$ *and* **W** *isomorphism* $\psi: G \to H \subset D(Y)$ *such that* H *separates the points of* Y. Y *and* ψ *are unique in the sense that for any other* $X \in \mathbf{K}$ *and* **W** *isomorphism* $\theta: G \to K \subset D(X)$ *with* K *separating the points of* X *there is a homeomorphism* $\tau: X \to Y$ *such that* $\psi(g)(\tau(x)) = \theta(g)(x)$ *for all* $g \in G$ *and* $x \in X$.

The space Y is called the *Yosida space* of G, and $H = \psi(G)$ is called the *Yosida representation of* G. Y can be realized as the set of *values* of the unit u, that is, the set of all convex ℓ-subgroups of G maximal with respect to omitting u, with the hull kernel topology. Furthermore,

$$\psi(g)(P) = \begin{cases} +\infty & \text{if } g \notin Q \text{ and } P + g > P \\ \theta(P+g) & \text{if } g \in Q \\ -\infty & \text{if } g \notin Q \text{ and } P + g < P \end{cases},$$

where P is a value of u, Q is the convex ℓ-subgroup generated by $P \cup \{u\}$, and θ is the unique ℓ-homomorphism from Q/P into $\mathbb{R}$ which takes u to 1. **Arch** objects can be represented as ℓ-groups in $D(Y)$ for $Y \in \mathbf{K}$, but the representation is less canonical than for **W** objects. $Y \in \mathbf{K}$ is said to be *extremally disconnected* if the closure of an open set is open. It can be shown that this property is equivalent to having every open subset C^*-embedded (1H6 of Gillman and Jerison (1960)), so that such spaces are quasi-F, hence $D(Y)$ is an ℓ-group. The next theorem is due to Bernau (1966), based on earlier work by Maeda and Ogasawara (1942).

THEOREM 2.2: THE REPRESENTATION OF **ARCH** OBJECTS. *For every* $G \in$ **Arch** *there is an extremally disconnected* $Y \in \mathbf{K}$ *and* **Arch** *isomorphism* $\psi: G \to H \subset D(Y)$ *such that* H *is large in* $D(Y)$. Y *and* ψ *are unique in the sense that for any extremally disconnected* $X \in \mathbf{K}$ *and*

Arch *isomorphism* $\theta: G \to K \subset D(X)$ *such that* K *is large in* $D(X)$ *there is a homeomorphism* $\tau: X \to Y$ *and weak unit* $f \in D(X)$ *such that* $\psi(g)(\tau(x)) = \theta(g)(x)f(x)$ *for all* $g \in G$ *and* $x \in X$.

Conrad (1971) characterized the extension $G \leq D(Y)$ of Theorem 2.2 as the maximal Arch extension in which G is large.

THEOREM 2.3. *Let* G, Y, *and* ψ *be as in Theorem 2.2. Then for any extension* $G \leq H \in$ Arch, G *is large in* H *if and only if there is an* Arch *injection* $\theta: H \to D(Y)$ *such that* $\theta(g) = \psi(g)$ *for all* $g \in G$. θ *is unique whenever it exists.*

The functoriality of the Yosida representation is the topic of the next result. Given an ℓ-group $G \subset D(X)$ and a function $\tau: Y \to X \in$ K, we shall say that τ *realizes* a mapping $\theta: G \to D(Y)$ provided that $\tau^{\leftarrow}g^{\leftarrow}(\mathbb{R})$ is dense in Y for all $g \in G$, in which case $\theta(g)(y) = g(\tau(Y))$ for all $g \in G$ and $y \in Y$. Observe that this condition is met by any irreducible surjection. If τ does realize such a mapping θ, then θ preserves the ℓ-group operations of $D(Y)$ and $D(X)$, so that if H is an ℓ-group in $D(X)$ such that $\theta(g) \in H$ for all $g \in G$ we say that τ realizes the ℓ-homomorphism $\theta: G \to H$. It is a remarkable fact that every W morphism is realized by some $\tau \in$ K (Hager and Robertson (1979)).

THEOREM 2.4: THE YOSIDA REPRESENTATION OF W MORPHISMS. *Suppose that* G *and* H *are* W *objects with respective Yosida spaces* X *and* Y. *Identify* G *and* H *with their Yosida representations. Then for every* $\theta: G \to H \in$ W *there is some* $\tau: Y \to X$ *which realizes it.* θ *is injective if and only if* τ *is surjective, and* τ *is injective whenever* θ *is surjective.*

If the elements of the Yosida spaces are realized as the values of the weak units, then the map τ of Theorem 2.4 is simply given by $\tau(P) = \theta^{\leftarrow}(P)$ for all $P \in Y$.

The kernels of **W** surjections are characterized by their zero sets. For any subset $K \subset D(Y)$, the *zero set* of K is $z(K) = \bigcap_K z(k)$, and for any subset $S \subset Y$ we let G_S designate $\{g \in G: z(g) \supset S\}$. We leave to the reader the proof of the following result, which is a rumination on the details of the Yosida Representation.

THEOREM 2.5. *Suppose that* $\theta: G \to H$ *is a* **W** *surjection with kernel* K, *let* Y *be the Yosida space of* G, *and identify* G *with its Yosida representation. Then the Yosida space of* H *can be taken to be* $S = z(K) \subset Y$, *and the Yosida representation of* H *accomplished by letting* $\theta(g)$ *be the restriction of* g *to* S. *Thus the inclusion map* $S \subset Y$ *realizes* θ. *Furthermore,* $K = G_S$.

3. κ-IRREDUCIBLE MAPS AND κ-SETS

In Theorem 3.12 we characterize κ-irreducible surjections in several ways. One of the most important conditions characterizing these maps makes reference to the complete Boolean algebra $\mathcal{P}$ of polars of $G \in$ Arch. If $\mathfrak{R}$ is a collection of polars, then the supremum and infimum of $\mathfrak{R}$ are given by

$$\bigvee \mathfrak{R} = (\bigcup \mathfrak{R})^{\perp\perp}, \text{ and}$$

$$\bigwedge \mathfrak{R} = \bigcap \mathfrak{R}.$$

Even though $D(X)$, $X \in$ K, is not generally an ℓ-group, we shall call an ℓ-group G *large* in $D(X)$ if $G \subset D(X)$ such that every strictly positive element of $D(X)$ has a positive multiple which exceeds ($\geq$) a strictly positive element of G. This terminology agrees with conventional usage in case $D(X)$ happens to be an ℓ-group. If G is a large ℓ-group in $D(X)$ for some $X \in$ K, then the map

$$P \longmapsto cl \bigcup\{coz(p): p \in P\}$$

with inverse

$$C \mapsto \{p: coz(p) \subset C\}$$

is a lattice isomorphism from $\mathscr{P}$ onto the Boolean algebra $\mathscr{P}(X)$ of regular closed subsets of X. It is to emphasize this association that we use the letter $\mathscr{P}$ for both algebras. If G is a large ℓ-subgroup of $H \in$ Arch, then the intersection map provides a Boolean isomorphism from the polars of H onto those of G. Consequently, we refrain from distinguishing these algebras notationally, using the letter $\mathscr{P}$ for both.

We employ conventional notation for ℓ-groups. In particular, for $g \in G \in$ Arch we use g^+ for $g \vee 0$ and $|g|$ for $g^+ \vee (-g)^+$. For example, $coz(f - g)^+$ denotes $\{y \in Y: f(y) > g(y)\}$ for elements $f,g \in D(Y)$. G^+ denotes $\{g \in G: g \geq 0\}$.

LEMMA 3.1. *For any subset* $K \subset G^+ \in$ Arch *and any element* $g \in G$,

$$g^{\perp\perp} \subset \vee_K k^{\perp\perp}$$

holds in $\mathscr{P}$ *if and only if*

$$g = \vee\{nk \wedge g: n \in N,\ k \in K\}$$

holds in G.

Proof. Represent G as a large ℓ-group in $D(X)$ for some $X \in$ K. Then $g^{\perp\perp} \subset \vee_K k^{\perp\perp}$ if and only if $\cup_K coz(k) \cap coz(g)$ is dense in $coz(g)$. But this is precisely the condition under which $g = \vee\{nk \wedge g: n \in N,\ k \in K\}$ would hold. Lemma 4.1 of Ball and Hager (1989b) is a generalization of this last assertion; its proof provides further details to the reader who wishes them.□

A principal polar is one of the form $g^{\perp\perp}$ for some $g \in G$. Since $g^{\perp\perp} \vee f^{\perp\perp} = (g \vee f)^{\perp\perp}$ and $g^{\perp\perp} \wedge f^{\perp\perp} = (g \wedge f)^{\perp\perp}$, these polars form a sublattice $\mathscr{Q}$ of $\mathscr{P}$. When applied to a subset $\mathscr{J} \subset \mathscr{Q}$, the term κ-ideal signifies the following two properties. For polars $P,Q \in \mathscr{Q}$, $P \subset Q \in \mathscr{J}$ implies $P \in \mathscr{J}$. And for any subset $\mathscr{R} \subset \mathscr{J}$ of cardinality strictly less than κ, $\vee\mathscr{R} = R \in \mathscr{Q}$ implies $R \in \mathscr{J}$.

PROPOSITION 3.2. *The following are equivalent for an ideal* K *of* $G \in$ Arch.

(a) K *is a* κ*-ideal.*

(b) The natural map $\theta\colon G \longrightarrow G/K$ *is* κ*-complete.*

(c) K *is of the form* $\bigcup\mathcal{J}$ *for some* κ*-ideal* $\mathcal{J}$ *in the lattice* $\mathcal{Q}$ *of principal polars.*

Proof. Suppose that K is a κ-ideal, and that $\bigvee F = g$ for some subset $F \subset G$ of cardinality less than κ. If it should happen that $\theta(f) \leq \theta(m)$ for some $m \in G$ and all $f \in F$, then

$$\theta((f - m)^+) = (\theta(f) - \theta(m)) \vee \theta(0) = 0$$

implies $(f - m)^+ \in K$ for all $f \in F$. But since

$$\bigvee_F (f - m)^+ = (\bigvee F - m)^+ = (g - m)^+,$$

we get $(g - m)^+ \in K$, or

$$\theta((g - m)^+) = (\theta(g) - \theta(m)) \vee \theta(0) = 0,$$

meaning that $\theta(g) \leq \theta(m)$. This completes the proof that (a) implies (b), and it is clear that (b) implies (a). Now assume that K satisfies (a) and let $\mathcal{J} = \{k^{\perp\perp}\colon k \in K\}$. Lemma 3.1 shows that $\mathcal{J}$ is a κ-ideal in $\mathcal{Q}$. To show that $K = \bigcup\mathcal{J}$, observe that $0 \leq k \in K$ and $0 \leq x \in k^{\perp\perp}$ imply $x^{\perp\perp} \subset k^{\perp\perp}$, so that $x = \bigvee_{\mathbb{N}}(nk \wedge x)$ by the Lemma 3.1, with the result that $x \in K$ by virtue of the fact that K is a κ-ideal. This proves that $k^{\perp\perp} \subset K$, so (c) holds. Finally, a third application of the lemma shows that (c) implies (a). □

The following easy consequence of Proposition 3.2, though well known for $\kappa = \aleph_1$, will be useful in what follows.

COROLLARY 3.3. *If* $g \in G \in$ Arch *then* $g^{\perp\perp}$ *is the smallest* κ*-ideal containing* g.

PROPOSITION 3.4. *If K is an ℓ-subgroup of an abelian ℓ-group G then the smallest κ-ideal of G containing K is*

$$\{g \in G: \vee_F(|g| \wedge f) = |g| \text{ for some } F \subset K^+ \text{ with } |F| < \kappa\}$$

Proof. The specified set, let us call it L, is clearly contained in any κ-ideal containing K. To show that it is a subgroup, consider two of its positive elements, say $g_i = \vee_{F_i}(g_i \wedge f)$ for subsets $F_i \subset K^+$ of cardinality less than κ. Then $g_1 + g_2$ is the supremum of elements of the form

$$(g_1 \wedge f_1) + (g_2 \wedge f_2) = (g_1 + g_2) \wedge (g_1 + f_2) \wedge (f_1 + g_2) \wedge (f_1 + f_2),$$

and since the latter is bounded above by $(g_1 + g_2) \wedge (f_1 + f_2)$,

$$g_1 + g_2 = \vee_F((g_1 + g_2) \wedge (f_1 + f_2)) \in L,$$

where F denotes $F_1 \cup F_2$. The rest of the argument consists of noting that $|g_1 \vee g_2|$, $|g_1 \wedge g_2|$, and $|g_1 + g_2|$ are all bounded above by $|g_1| + |g_2|$ in any abelian ℓ-group. □

COROLLARY 3.5. *If G is a large ℓ-subgroup of $H \in$ Arch and K is a κ-ideal of G then the smallest κ-ideal of H containing K meets G in K.*

Proof. Since G is large in H, suprema and infima in the two ℓ-groups agree. Thus if $g = \vee_F(g \wedge f)$ holds in G, then it holds in H also. □

COROLLARY 3.6. *An extension $G \leq H$ in Arch is κ-ideal preserving if and only if distinct κ-ideals of H trace distinctly on G.*

Proof. Suppose that distinct κ-ideals of H trace distinctly on G. Since polars are κ-ideals, it follows that distinct polars of H trace distinctly on G. As discussed in the introduction prior to Theorem 5.3, it follows that G is large in H. Corollary 3.5 then applies, and shows that the intersection map invariably takes the κ-ideals of H onto those of G. □

We now show that the natural map $\theta: G \to G/K$ induced by the κ-ideal K of $G \in$ Arch is not only an ℓ-homomorphism but is actually an Arch morphism. Moreover, $\theta \in$ W whenever $G \in$ W.

PROPOSITION 3.7. *Let* K *be a* κ*-ideal of* $G \in$ Arch *with* $\theta\colon G \longrightarrow G/K$ *the natural map. Then* $G/K \in$ Arch, *and if* $(G,u) \in$ W *and* K *is proper then* $(G/K,\theta(u)) \in$ W *and* $\theta \in$ W.

Proof. Consider $0 < f$, $g \in G$. If $K + nf \leq K + g$ for all $n \in N$, we have $K + ((nf \vee g) - g) = K$ or $(nf - g)^+ \in K$ for all $n \in N$. Now $\bigvee_N (f \wedge (nf - g)^+) = f$ in any archimedean ℓ-group, as can be readily verified in a representation of G. Then $f \in K$ because K is a κ-ideal, with the result that G/K is archimedean. Now suppose that K is proper and that u is a weak unit of G. Any $g \in .G^+$ satisfies $g = \bigvee_N (g \wedge nu)$; consequently, the propriety of K forces it to omit u. To see that $\theta(U)$ is a weak unit of G/K, observe that for $g \in G^+$, $\theta(g) \wedge \theta(u) = 0$ implies $g \wedge u \in K$, whereupon $g = \bigvee_N n(g \wedge u) \in K$, or $\theta(g) = 0$. □

The fact that κ-ideals are kernels of W morphisms when $G \in$ W implies by Theorem 2.5 that they are determined by their zero sets. This is the content of the next theorem, which is an important one for our purposes. Recall that, for a subset $S \subset Y \in$ K and an ℓ-group $G \subset D(Y)$, we denote by G_S the ℓ-subgroup $\{g \in G\colon S \subset z(g)\}$.

THEOREM 3.8. *The following are equivalent for a closed subspace* $S \subset Y \in$ K.

(a) $U \cap S$ *is dense in* S *whenever* U *is a dense* κ*-cozero set of* Y.

(b) $S \subset$ *clint* Z *whenever* S *is contained in the* κ*-zero set* Z.

(c) G_S *is a* κ*-ideal of* G *for every large* ℓ*-group* G *in* $D(Y)$.

Proof. Assume (a), consider a κ-zero set Z which contains S, and suppose for contradiction that there is some $s \in S \setminus$ *clint* Z. Since $\{s\}$ and *clint* Z are compact, they have disjoint cozero neighborhoods Q and R, respectively. Let U designate the dense κ-cozero set $R \cup (Y \setminus Z)$. But $U \cap S$ is not dense in S, for it misses the set $Q \cap S$, which is nonempty because it contains s. This contradiction proves (b). Now assume (b), let G be a large ℓ-group in $D(Y)$, suppose $\bigvee F = g$ for some $F \subset G_S^+$ such that $|F| < \kappa$, and let Z be the κ-zero set $z(F) = \bigcap_F z(f)$. Apparently $S \subset Z$, hence $S \subset$ *clint* Z. Observe that the largeness of G in $D(Y)$ implies that suprema and infima agree in the two structures, and that an element $h \in D(Y)$ is the supremum of a subset $K \subset D(Y)$ if and only if $\{y \in Y: h(y) = \bigvee_K k(y)\}$ is dense in Y (Proposition 4.1 of Ball and Hager (1989b)). Thus $\bigvee F = g$ implies $coz(g) \subset cl \bigcup_F coz(f)$, or $clint\, Z \subset z(g)$, meaning $S \subset z(g)$ and $g \in G_S$, which proves (c). Finally, assume (c) and consider a dense κ-cozero set $U \subset Y$. If it should happen that $U \cap S$ is not dense in S, then there is a cozero set $V \subset Y$ such that $S \cap U \cap V = \phi$ in spite of the fact that $S \cap V \neq \phi$. Now any κ-cozero set, and in particular U, is of the form $\bigcup_F coz(f)$ for some subset $F \subset C(Y)^+$ such that $|F| < \kappa$, and we may assume without loss of generality that V is $coz(g)$ for some $g \in C(Y)^+$. Then $C(Y)_S$ contains $\{nf \wedge g: n \in N, f \in F\}$ as a result of the fact that $coz(f \wedge g) \subset U \cap V \subset Y \setminus S$; furthermore, its supremum is g thanks to the density of U in Y. But $g \notin C(Y)_S$ because $coz(g) \cap S \neq \phi$, contrary to the assumption that $C(Y)_S$ is a κ-ideal. The proof is complete. □

Let us call any set satisfying the preceding proposition a *κ-set*. Thus an ∞-set is a simply a regular closed set. $\aleph_1$-sets have been studied by Veksler (1970, 1973) under the name P'-sets, and by Tzeng (1970), who called their defining property the D-restriction property. A *P_κ-set* is a closed set C such that the intersection of any collection of fewer than κ neighborhoods of C is itself a neighborhood of C. (See Gillman and Jerison (1960) for a fairly thorough treatment of the $\kappa = \aleph_1$ case.) A simple argument using Theorem 3.8 (a) reveals that these sets are among the κ-sets. In fact, Ball, Hager, and Macula (1989) show that the P_κ-sets are precisely the κ-sets in κ-disconnected spaces (i.e. spaces in which the closure of a κ-cozero set is open).

PROPOSITION 3.9. *Suppose that G is a* **W** *object in $D(Y)$ which separates the points of $Y \in$* K. *Then the map*

$$K \longmapsto z(K)$$

with inverse

$$S \longmapsto G_S$$

is an order reversing bijection from the set of κ-ideals of G onto the set of κ-sets of Y.

Proof. Suppose $S = z(K)$ for some κ-ideal K of G; we shall show that $G_S = K$. To that end consider $0 < g \in G_S$, and write $coz(g)$ as $\bigcup_N C_n$ for a suitable collection $\{C_n : n \in N\}$ of compact sets. Since each point of a particular C_n lies in $coz(k)$ for some $k \in K$, there is some $k_n \in K^+$ such that $C_n \subset coz(k_n)$. Then $g \in K$ because $g = \bigvee\{mk_n \wedge g : m, n \in N\}$ and K is a κ-ideal, thus proving $G_S = K$. The argument that (c) implies (a) in Theorem 3.8 then shows that S is a κ-set. If S is an arbitrary κ-set of Y then for any $y \in Y \setminus S$ a simple compactness argument produces $g \in G_S^+$ such that $y \in coz(g)$. Thus $z(G_S) = S$. □

Consider an irreducible surjection $\tau: Y \longrightarrow X \in \mathrm{K}$. We prove in the next proposition that $\tau(S)$ is a κ-set in X whenever S is a κ-set in Y. We shall term any $\tau \in \mathrm{K}$ *κ-set preserving* if the association

$$S \longmapsto \tau(S)$$

provides a one-to-one correspondence between the κ-sets of Y and those of X.

LEMMA 3.10. *A surjection $\tau: Y \longrightarrow X \in \mathrm{K}$ is κ-set preserving if and only if distinct κ-sets of Y have distinct images in X.*

Proof. Suppose that τ gives distinct images to distinct κ-sets of Y. In particular, τ distinguishes regular closed sets, from which it follows that τ is irreducible. We claim that $\tau(S)$ is a κ-set of X whenever S is a κ-set of Y. To see this, consider a dense κ-cozero $U \subset X$. Then the irreducibility of τ makes $\tau^{\leftarrow}(U)$ dense in Y, hence $\tau^{\leftarrow}(U) \cap S \neq \phi$, and therefore $U \cap \tau(S) \neq \phi$. That is, $\tau(S)$ is a κ-set in X. To show that τ is onto, consider a κ-set $S \subset X$, let G and H stand for $C(X)$ and $C(Y)$, respectively, and let $\theta: G \longrightarrow H$ be the ℓ-injection realized by τ. Then G_S is a κ-ideal of G by Theorem 3.8, and $\theta(G)$ is large in H because τ is irreducible, so by Corollary 3.5 there is a κ-ideal K of H such that $\theta(G_S) = K \cap \theta(G)$. $T = z(K)$ is a κ-set of Y according to Proposition 3.9; we claim that $\tau(T) = S$. If $y \in T$ but $\tau(y) \notin S$ then there is some $g \in G_S$ such that $g(\tau(y)) > 0$. But then $\theta(g)(y) = g(\tau(y)) > 0$, contrary to the assumption that $\theta(g) \in K = H_T$. This shows $\tau(T) \subset S$. On the other hand, if $\tau(T)$ is proper in S then the compactness of $\tau(T)$ implies the existence of some $g \in G^+ \setminus G_S$ such that $\tau(T) \subset z(g)$. But then $\theta(g) \in H_T = K$ in spite of the fact that $\theta(g) \notin \theta(G_S)$, contrary to the assumption that $\theta(G_S) = K \cap \theta(G)$. □

In order to more fully characterize the algebraic ramifications of κ-irreducibility, we need two more concepts. The first is a condition on the polar algebra $\mathscr{P}$ of $G \in$ **Arch**. We term a polar $P \in \mathscr{P}$ *κ-generated* provided that it is of the form $A^{\perp\perp}$ for some subset $A \subset G$ of cardinality strictly less than κ. We shall say of an extension $G \leq H \in$ **Arch** that *κ-generated polars in G and H correspond* provided that the intersection map provides a lattice isomorphism from the κ-generated polars of H onto those of G. This condition implies that for every $h \in H^+$ there is some $g \in G^+$ such that $g^{\perp\perp} \subset h^{\perp\perp}$, a consequence of which is that distinct closed ideals of H have distinct intersections with G. From this it follows (see the discussion prior to Theorems 5.3 in the Introduction) that G is large in H. The formulation of the definition of corresponding κ-generated polars which we find most useful is this: for every $h \in H$ there is a subset $A \subset G$ of cardinality less than κ such that $A^{\perp\perp} = h^{\perp\perp}$.

The second notion is a natural generalization to cardinal κ of the concept of sequential order density which was heavily used by Dashiell, Hager, and Henriksen (1980). We shall, however, refrain from following their example insofar as using κ-order convergence to construct a completion, though this can be done. (Ball (1980, 1984) discussed order convergence without the cardinality restriction and with an eye toward constructing completions.) Instead, we use the lattice theoretical nomenclature for the corresponding density notion. An ℓ-subgroup G is said to be *κ-join and meet dense* in $H \in$ Arch provided that for every $h \in H$ there are subsets $A, B \subset G$ of cardinality strictly less than κ such that $h = \bigvee A = \bigwedge B$. We often check the last condition by verifying that $A \leq h \leq B$ (i.e. $a \leq h \leq b$ for all $a \in A$ and $b \in B$) and that $\bigwedge(B - A) = 0$, where $B - A$ designates $\{b - a: a \in A,\ b \in B\}$. We willingly succumb to the temptation to abbreviate this terminology by saying that G is *κjamd* in H. The next result shows the close relationship between κ-join and meet density and corresponding κ-generated polars. Note that the convexification of G in H is all of H whenever G is κjamd in H.

PROPOSITION 3.11. *Suppose that $G \leq H \in \mathbf{W}$ and that G is divisible. Then κ-generated polars correspond in G and H if and only if G is κjamd in its convexification in H.*

Proof. Suppose first that G is κjamd in its convexification in H. Note that G is large in H. For given $h \in H^+$ find subsets $A,B \subset G^+$ of cardinality less than κ such that $h \wedge u = \bigvee A = \bigwedge B$, where u is the weak unit of G. Clearly $A^{\perp\perp} = (h \wedge u)^{\perp\perp} = h^{\perp\perp}$, thereby proving that κ-generated polars in G and H correspond.

Now suppose that κ-generated polars in G and H correspond, and assume without loss of generality that H is a large ℓ-group in $D(X)$ for some $X \in \mathrm{K}$. Consider $h \in H$ and $g \in G$ such that $0 \leq h \leq g$. Fix a rational number q such that $0 < q < 1$. Choose a specific subset $A(q) \subset G^+$ of cardinality less than κ such that $A(q)^{\perp\perp} = (h - qg)^{+\perp\perp}$, meaning that $\bigcup_{A(q)} coz(a)$ has the same closure as does $coz(h - qg)^+$. By replacing each element $a \in A(q)$ by all elements of the form $na \wedge qg$, we may assume that $a \leq h$ for all $a \in A(q)$, and that for every $y \in \bigcup_{A(q)} coz(a)$ there is some $a \in A(q)$ such that $a(z) = qg$ for all z in some neighborhood of y. Now repeat this argument with $g - h$ in place of h and $(1 - q)g$ in place of qg, thereby obtaining a set $B' \subset G^+$, and let $B(q) = \{g - b: b \in B'\}$. $B(q)$ is a set of cardinality less than κ with the following properties: $g \geq b \geq h$ for all $b \in B(q)$, $\bigcup_{B(q)} coz(g - b)$ has the same closure as does $coz(qg - h)^+$, and for every element y in $\bigcup_{B(q)} coz(g - b)$ there is some $b \in B(q)$ such that $b(z) = qg$ for all z in some neighborhood of y. Let $A = \bigcup\{A(q): 0 \leq q \leq 1\}$ and let $B = \bigcup\{B(q): 0 \leq q \leq 1\}$. We claim that $\bigwedge(B - A) = 0$. To establish this claim, suppose for contradiction that $0 < k \leq (B - A)$ for some $k \in H$. Then $k \leq g$, so there is some $r \in \mathbb{Q}$ and nonempty open $U \subset X$ such that $0 < r < 1$ and $rg(x) < k(x)$ for all $x \in U$. Because h is continuous, there are $q,s \in \mathbb{Q}$ and nonempty open $V \subset U$ with the following three properties: $0 < q < s < 1$, $s - q < r$, and $qg(x) < h(x) < sg(x)$ for all $x \in V$. Now $V \subset coz(h - qg)^+ \subset cl \bigcup_{A(q)} coz(a)$, so by assumption there is some $a \in A(q)$ such that $a(x) = qg$ for all x in some nonempty open subset $R \subset V$. Likewise there is some $b \in B(s)$ such that $b(x) = sg$ for all x in some nonempty open subset $T \subset R$. But then

$$b(x) - a(x) = sg(x) - qg(x) < rg(x) < k(x)$$

for all $x \in T$, the contradiction which proves the claim and the proposition. □

We arrive at last at the characterization of κ-irreducibility. The equivalence of (a) and (g) in this theorem is due to Dashiell (unpublished manuscript).

THEOREM 3.12. *The following are equivalent for a surjection* $\tau: Y \longrightarrow X \in K$.

(a) τ *is* κ*-irreducible.*

(b) *For distinct* $y_1, y_2 \in Y$ *there are* κ*-zero sets* $Z_1, Z_2 \subset X$ *such that* $y_i \in int\ \tau^{\leftarrow}(Z_i)$ *and* $int\ \tau^{\leftarrow}(Z_1 \cap Z_2) = \phi$.

(c) *For disjoint closed subsets* $C_1, C_2 \subset Y$ *there are* κ*-zero sets* $Z_1, Z_2 \subset X$ *such that* $C_i \subset int\ \tau^{\leftarrow}(Z_i)$ *and* $int\ \tau^{\leftarrow}(Z_1 \cap Z_2) = \phi$.

(d) τ *is* κ*-set preserving.*

(e) *Suppose that* G *is a* **W** *object in* $D(X)$ *which separates the points of* X*, that* H *is a large* ℓ*-group in* $D(Y)$*, and that* τ *realizes* $\theta: G \longrightarrow H$*. Then* H *is a* κ*-ideal preserving extension of* $\theta(G)$*.*

(f) *For* G *and* H *as in (e),* κ*-generated polars in* $\theta(G)$ *and* H *correspond.*

(g) *For* G *and* H *as in (e), if* $\theta(G)$ *is divisible then it is* κ*jamd in its convexification in* H*.*

(h) *For* G *and* H *as in (e) and for* $K = \theta(G)$, $\mathfrak{Z}^H_\kappa = \mathfrak{Z}^K_\kappa$.

Proof. Suppose (a) holds, consider distinct $y_1, y_2 \in Y$, and find disjoint zero set neighborhoods W_i of y_i (1.15 of Gillman and Jerison). Then use the κ-irreducibility of τ to find κ-zero sets $Z_i \subset X$ such that $int\ W_i = int\ \tau^{\leftarrow}(Z_i)$, thus establishing (b). (c) follows from (b) by an argument very similar to the familiar one used to show that a compact Hausdorff space is normal. Assume (c); to prove (d) we first show that τ is irreducible by showing that $\tau(C) \neq X$ for any proper closed subset $C \subset Y$. For we can let $C_1 = C$ and $C_2 = \{y\}$ for some $y \in Y \setminus C$ to get κ-zero sets $Z_i \subset X$ satisfying (c). But $\tau(C) \subset Z_1$, and if $\tau(C) = X$ then $\tau^{\leftarrow}(Z_1) = Y$, which makes it impossible for the sets $int\, \tau^{\leftarrow}(Z_i)$ to be nonempty and disjoint. This establishes the irreducibility of τ. Now consider distinct κ-sets $S_1, S_2 \subset Y$, say $s_2 \in S_2 \setminus S_1$. Let $C_1 = S_1$ and $C_2 = \{s_2\}$ to get κ-zero sets $Z_i \subset X$ satisfying (c). It follows that $int(Z_1 \cap Z_2) = \phi$ in X, so that $U = X \setminus (Z_1 \cap Z_2)$ is a dense κ-cozero set in X. It follows as a consequence of the irreducibility of τ that the κ-cozero set $V = \tau^{\leftarrow}(U)$ is dense in Y, and so $V \cap S_2$ must be dense in S_2. Since $Y \setminus S_1$ is an open set which meets S_2 (in s_2, for example) there must be some $s \in V \cap (S_2 \setminus S_1)$. Then $\tau(s) \in \tau(S_2) \subset Z_2$ and $\tau(s) \in U = (X \setminus Z_1) \cup (X \setminus Z_2)$ imply $\tau(s) \notin Z_1$, and since $\tau(S_1) \subset Z_1$, we get $\tau(s) \notin \tau(S_1)$. This proves that (d) holds.

Given G, H, and θ as in (e), let $K = \theta(G)$, and consider distinct κ-ideals $H_1, H_2 \subset H$. Then the corresponding κ-sets $T_i = z(H_i)$ are also distinct because $H_i = H_{T_i}$, and, assuming (d), have distinct images $S_i = \tau(T_i)$. Since G separates the points of X, we know by Proposition 3.9 that $z(G_{S_i}) = S_i$, so that the κ-ideals G_{S_i} are distinct. Let $\theta(G_{S_i}) = K_i$. Then one readily checks that $K \cap H_i = K_i$, which proves (e). To show (f) consider $h \in H$. The κ-ideal of H generated by h is $h^{\perp\perp}$, so that by (e) we must have that the κ-ideal L of H generated by $h^{\perp\perp} \cap K$ contains h. But $L = \bigcup\{A^{\perp\perp}\colon A \subset h^{\perp\perp} \cap K$ and $|A| < \kappa\}$ by Proposition 3.2. Therefore there is some $A \subset h^{\perp\perp} \cap K$ of cardinality less than κ such that $h^{\perp\perp} = A^{\perp\perp}$. (f) implies (g) by Proposition 3.10. To show (h) consider first a zero set $Z = z(h) \subset Y$. Assume without loss of generality that $h \in H^+$, and use (g) to find a subset $A \subset K^+$ of cardinality less than κ such that $\bigvee A = h$. Now $\bigcap_A z(a) \in Z_\kappa^K(Y)$ need not coincide with Z, but it must have the same interior (see the proof of Lemma 3.1), which is to say that $clint\, Z \in \mathfrak{Z}_\kappa^K(Y)$. It is then a short step to get $clint\, Z \in \mathfrak{Z}_\kappa^K(Y)$ for arbitrary $Z \in Z_\kappa^H(Y)$.

It remains is to prove (a) from (h). Take G and H to be $C(X)$ and $C(Y)$, respectively, and consider $Z \in \mathcal{Z}_\kappa^H(Y)$. By (h) there is a subset $A \subset K^+$ of cardinality less than κ such that $int\ Z = int \bigcap_A z(a)$. If we let $B = \{b \in G: \theta(g) \in A\}$ and $W = \bigcap_B z(b)$ we get

$$int\ Z = int \bigcap_B z(\theta(b)) = int \bigcap_B z(b\tau) = int \bigcap_B \tau^{\leftarrow}(z(b)) = int\ \tau^{\leftarrow}(W).$$

This completes the proof of (a) from (h), and of the theorem. □

4. THE QUASI-F_κ COVER OF A COMPACT HAUSDORFF SPACE

We show the existence and uniqueness of the minimal quasi-F_κ cover of an arbitrary $X \in \mathbf{K}$. We do this by taking the inverse limit of the κ-irreducible preimages of X. Thus this approach can be viewed as a particular instance of the general procedure of Hager (1989), which, in turn, is a development of ideas present in an earlier paper of Hager (1971), and and of those of Banaschewski (1968). In fact, this section could be embedded in Hager's recent paper, though we refrain from doing so in the interests of concision.

We begin with a full characterization of quasi-F_κ spaces (Theorem 4.6), and this requires the introduction of two completeness notions for $G \in$ **Arch**. Of these the first is *κ-repleteness*, which may be intuitively understood to require the incorporation into G of all those elements in some ℓ-supergroup in some $D(Y)$ which could possibly be incorporated without increasing $\mathcal{Z}_\kappa^G(Y)$. The formal definition requires the following construct. Suppose $Y \in \mathbf{K}$, and consider an ℓ-group $H \subset D(Y)$ with divisible ℓ-subgroup G. Let $H_\kappa(G)$ designate

$$\{h \in H: clintz(h - g)^+, clintz(g - h)^+ \in \mathcal{Z}_\kappa^G(Y) \text{ for all } g \in G\}.$$

We prove in Lemma 4.2 that $H_\kappa(G)$ is an ℓ-subgroup of H, but this requires a preliminary technical result.

LEMMA 4.1. *Suppose* H *is an* ℓ*-group in* $D(Y)$, $Y \in \mathrm{K}$, *with divisible* ℓ*-subgroup* G. *Then for any elements* $h_1, h_2 \in H$ *such that* $clintz(h_i) \in \mathfrak{Z}^G_\kappa(Y)$ *and for any* $g \in G$ *there is a subset* $A \subset G$ *of cardinality less than* κ *such that*

$$\cap\{z(h_1 - g_1)^+ \cup z(h_2 - g_2)^+ : g_1, g_2 \in A \text{ and } g_1 + g_2 = g\}$$

has the same interior as $z(h_1 + h_2 - g)^+$.

Proof. Let L designate the displayed set and let R designate $z(h_1 + h_2 - g)^+$. Note that $y \in R$ if and only if $h_1(y) + h_2(y) \leq g(y)$, with the result that either $h_1(y) \leq g_1(y)$ or $h_2(y) \leq g_2(y)$ for any g_1 and g_2 such that $g_1 + g_2 = g$. That is, $R \subset L$. It is the opposite containment which requires judicious choice of A. First use the assumption that $clintz(h_i) \in \mathfrak{Z}^G_\kappa(Y)$ to find subsets $A_i \subset G$ of cardinality less than κ such that $z(h_i)$ has the same interior as $\cap_{A_i} z(a)$. Then let A be the divisible subgroup of G generated by $A_1 \cup A_2 \cup \{g\}$. Now consider a nonempty open subset $U \subset L$, and suppose for contradiction that $V = U \setminus R \neq \phi$. Observe that $V \subset coz(h_1) \cup coz(h_2) \cup coz(g)$ because $h_1(y) + h_2(y) > g(y)$ for all $y \in V$. We treat the case in which $V \cap coz(h_1) \neq \phi$. In this case there must be some $a \in A_1$ such that $V \cap coz(a) \neq \phi$, say $y \in V \cap coz(a)$. Since $a(y) \neq 0$ there is some rational number q such that $h_1(y) > qa(y)$ and $h_2(y) > g(y) - qa(y)$. But by taking $g_1 = qa$ and $g_2 = g - qa$ we get $y \notin z(h_1 - g_1)^+ \cup z(h_2 - g_2)^+$, contrary to the assumption that $y \in V \subset L$. The case in which $V \cap coz(h_2) \neq \phi$ is handled analogously. If $y \in V \cap coz(g)$ then $g(y) \neq 0$ so there is some rational number q such that $h_1(y) > qg(y)$ and $h_2(y) > (1 - q)g(y)$, in which case g_1 can be taken to be qg. This completes the proof. □

LEMMA 4.2. *Let* Y, G, H, *and* $H_\kappa(G)$ *have the meaning above. Then* $H_\kappa(G)$ *is an* ℓ*-subgroup of* H.

Proof. $H_\kappa(G)$ clearly contains the negatives of its elements, so consider $h_1, h_2 \in H_\kappa(G)$ and $g \in G$. Now

$$z(h_1 \vee h_2 - g)^+ = z(h_1 - g)^+ \cap z(h_2 - g)^+ \text{ and}$$

$$z(g - h_1 \vee h_2)^+ = z(g - h_1)^+ \cup z(g - h_2)^+,$$

so that $clintz(h_1 \vee h_2 - g)^+$ and $clintz(g - h_1 \vee h_2)^+$ are the infimum and supremum in $\mathscr{P}(Y)$, respectively, of members of $\mathfrak{Z}_\kappa^G(Y)$, thus must themselves lie in $\mathfrak{Z}_\kappa^G(Y)$. The argument for $h_1 \wedge h_2$ is similar. Finally, $h_1 + h_2 \in H_\kappa(G)$ by the preceding lemma. □

In case $G \in \mathbf{W}$ the condition for membership in $H_\kappa(G)$ can be somewhat more simply put: $h \in H_\kappa(G)$ if and only if

$$clintz(h - q)^+, clintz(q - h)^+ \in \mathfrak{Z}_\kappa^G(Y)$$

for all rational numbers q, where q designates the function with constant value q. We leave the verification to the interested reader.

Here is the first of the two relevant completeness notions. An abstract **Arch** object G is *κ-replete* provided that, for any **Arch** isomorphism θ from G onto an ℓ-group K in $D(Y)$, $Y \in \mathbf{K}$, and for any ℓ-group H large in $D(Y)$ and containing K, we have

$$H_\kappa(K) = K.$$

PROPOSITION 4.3. *$G \leq H \in$ **Arch** is a κ-ideal preserving extension if and only if κ-generated polars in G and H correspond.*

Proof. Suppose that $G \leq H$ is κ-ideal preserving, and consider $h \in H$. The κ-ideal of H generated by h is $h^{\perp\perp}$ by Corollary 3.3, and if the κ-ideals are to correspond it must be true that the κ-ideal generated in H by $h^{\perp\perp} \cap G$ must contain h. But by Proposition 3.2 the latter κ-ideal is $\cup\{A^{\perp\perp}: A \subset h^{\perp\perp} \cap G, |A| < \kappa\}$. Thus $h^{\perp\perp} = A^{\perp\perp}$ for some subset $A \subset G$ of cardinality less than κ, which is to say that κ-generated polars in G and H correspond.

Now suppose that $G \leq H$ is *not* κ-ideal preserving, say H_1 and H_2 are κ-ideals of H such that there is some $h \in H_1^+ \setminus H_2$, but that $H_1 \cap G = H_2 \cap G$. We claim there can be no subset $A \subset G$ of cardinality less than κ such that $h^{\perp\perp} = A^{\perp\perp}$. For such a subset would be contained in $h^{\perp\perp}$ which is itself contained in H_1 by Corollary 3.3, and would therefore also lie within H_2. But Proposition 3.2 then asserts that *any*

κ-ideal of H containing A would also contain h, contrary to the assumption that $h \notin H_2$. This proves the claim and the lemma. □

A consequence of the following result is that $H_\kappa(G)$ does not depend on the particular representation chosen for H, so long as H is large in the $D(Y)$.

PROPOSITION 4.4. *Let H be an ℓ-group large in $D(Y)$ for some $Y \in \mathrm{K}$, and let G be a divisible ℓ-subgroup of H. Then $H_\kappa(G)$ is the largest κ-ideal preserving extension of G within H.*

Proof. Let K denote $H_\kappa(G)$. To show that K is a κ-ideal preserving extension of G we check that κ-generated polars in G and K correspond; for that purpose consider $k \in K^+$. Because $clintz(k)^+ \in \mathscr{Z}^G_\kappa(Y)$ there is a subset $A \subset G$ of cardinality less than κ such that $clintz(k) = clint \bigcap_A z(a)$. In view of the correspondence between polars and regular closed subsets of Y discussed at the beginning of Section 3, $k^{\perp\perp} = A^{\perp\perp}$ Thus K is a κ-ideal preserving extension of G.

Now suppose that L is an ℓ-subgroup of H which is a κ-ideal preserving extension of G, and consider $h \in L^+$ and $g \in G$. The κ-generated polars of G and L correspond by Lemma 4.3, so there is some subset $A \subset G^+$ of cardinality less than κ such that $A^{\perp\perp} = (h-g)^{+\perp\perp}$. Translating this by means of the aforementioned correspondence, we get $clintz(h-g)^+ = clint \bigcap_A z(a) \in \mathscr{Z}^G_\kappa(Y)$. This shows that $h \in K$. □

COROLLARY 4.5. *G is κ-ideal complete if and only if it is κ-replete.*

Proof. If G is κ-ideal complete then it follows directly from the previous proposition that G is κ-replete. Now suppose that G is κ-replete and consider a κ-ideal preserving extension H. By Theorem 2.2, H can be represented as a large ℓ-group in $D(Y)$ for some $Y \in \mathrm{K}$. But then $G = H$ by Proposition 4.4. Thus G is κ-ideal complete. □

The second relevant completeness concept is *κ-join and meet completeness,* or κjamd completeness for short. For $G \in$ Arch this means that for any nonempty subsets $A, B \subset G$ of cardinality less than κ such that

$A \leq B$ and $\wedge(B - A) = 0$ there is some $g \in G$ such that $A \leq g \leq B$. $\aleph_1$jamd completeness is termed sequential order completeness in the literature, and this idea was crucial to the analysis of quasi-F spaces by Dashiell, Hager, and Henriksen (1980). In fact, the equivalence of conditions (a) through (d) of the following theorem for $\kappa = \aleph_1$ can be found in this paper.

κjamd completeness and κ-repleteness are closely related, as indicated by Proposition 3.11. There is, however, an important distinction between them. Although both kinds of completeness imply that the underlying Yosida space X is quasi-F_κ (Theorem 4.6), κjamd completeness does not force large elements to be present in G. For example, if X is quasi-F_κ, then $C(X)$ is κjamd complete but not κ-replete, while $D(X)$ is both. And it is κ-repleteness which is the more useful notion in **Arch** (Section 5).

THEOREM 4.6. *The following are equivalent for* $X \in \mathrm{K}$.

(a) X *is quasi-*F_κ*, that is every dense* κ*-Lindelöf subspace of* X *is* C^**-embedded.*

(b) Every dense intersection of countably many κ*-cozero subsets of* X *is* C^**-embedded.*

(c) $C(X)$ *is* κ*jamd complete.*

(d) X *has no proper* κ*-irreducible preimage; that is, every* κ*-irreducible surjection onto* X *is a homeomorphism.*

(e) $D(X)$ *is a* κ*-ideal complete* ℓ*-group.*

(f) $D(X)$ *is a* κ*-replete* ℓ*-group.*

Proof. When $\aleph_0$ is replaced by κ, the elegant arguments (Lemma 9.7, Corollary 9.8, and Theorem 9.10(a)) of Comfort and Negrepontis (1975) show that a dense intersection of countably many κ-cozero sets is κ-Lindelöf, proving that (a) implies (b). Now assume (b), and consider subsets $A, B \subset C(X)$ of cardinality less than κ such that $\wedge(B - A) = 0$. For each rational $q \in \mathbb{Q}^+$ let $U_q = \cup\{coz(q + a - b)^+ : a \in A,\ b \in B\}$, a dense κ-cozero set, and let $U = \cap_{\mathbb{Q}^+} U_q$. It is easy to see that the function $h: U \longrightarrow \mathbb{R}$ defined by

$$h(x) = \vee_A a(x) = \wedge_B b(x)$$

is actually continuous on U. Because U is C^*-embedded in X, h has a continuous extension to X, which we also call h. Clearly $A \leq h \leq B$, proving (c). Assume (c) and consider a κ-irreducible surjection $\tau: Y \longrightarrow X \in$ K. Then τ realizes a **W** injection $\theta: C(X) \longrightarrow C(Y)$, and $\theta(C(X))$ is κjamd in $C(Y)$ by Theorem 3.12. θ is surjective since $C(X)$ is κjamd complete, hence τ is injective and a homeomorphism.

Henriksen, Vermeer, and Woods (1987) showed that there is a quasi-F space $QF(X) \in$ K and $\aleph_1$-irreducible surjection $\varphi_X: QF(X) \longrightarrow X \in$ K. (We use this notation for this paragraph of this proof only.) Since φ_X is also κ-irreducible, (d) implies that it is a homeomorphism. Consequently X is quasi-F and $D(X)$ is an ℓ-group in this case. Let $G = D(X)$; to prove G κ-ideal complete, consider a κ-ideal preserving extension H. As we have remarked, G is large in H, and so 1 is a weak unit of H as well as of G. That is, we regard H as a **W** extension of G. Identify H with its Yosida representation (Theorem 2.1) on its Yosida space Y, and let $\tau: Y \longrightarrow X$ be the surjection which realizes the embedding (Theorem 2.4). Then τ is κ-irreducible by Theorem 3.12, therefore onto by (d), hence a homeomorphism. That is, $G = H$, meaning that G is κ-ideal complete.

The equivalence of (e) and (f) is Proposition 4.5, so assume (f) to prove (a). Consider a dense κ-Lindelöf subspace $V \subset X$ and function $h \in C^*(V)$. Let G be $C(X)$, and fix $q \in \mathbb{Q}^+$. We claim that for each $x \in V$ there is some open subset $U(x,q) \subset X$ containing x for which there is some $g(x,q) \in G$ such that $g(x,q)(v) \leq h(v)$ for all $v \in V$, and such that $h(v) - g(x,q)(v) < q$ for all $v \in U(x,q) \cap V$. One might argue for the claim as follows. Assume for the sake of specificity that $h(x) > 0$. First use the continuity of h to find rationals t and r such that $0 < t < h(x) < r$ and $r - t < q$, and then use the boundedness of h to find a rational s such that $s < h(v)$ for all $v \in V$. Now an open subset of V containing is $coz((h - t) \wedge (r - h))^+$, so there is some $f \in G^+$ such that $x \in coz(f)$ and $coz(f) \cap V \subset coz((h - t) \wedge (r - h))^+$. Let n be a positive integer large enough that $s + nf(x) > t$. Then $U(x,q)$ can be taken to be $coz(s + nf - t)^+ \cap coz(f)$ and $g(x,q)$ can be taken to be $(s + nf) \wedge t$. With similar arguments for the other cases, this proves the claim. Now $\{U(x,q): x \in V\}$ is an open cover of the κ-Lindelöf subspace V and so has a subcover $\mathcal{U}(q)$ of cardinality less than κ. Let

$$A = \{g(x,q): U(x,q) \in \mathcal{U}(q) \text{ for some } q \in \mathbb{Q}^+\}.$$

Let $\varphi: Y \longrightarrow X$ be the canonical irreducible surjection from the Gleason cover (absolute) Y onto X, let H designate $D(Y)$, and let θ be the isomorphism from $D(X)$ onto $K \leq H$ realized by τ. We can (and do) consider h to be an element of H, since $\varphi^{\leftarrow}(V)$ is dense in the extremally disconnected space Y and is therefore C^*-embedded (1H6 of Gillman and Jerison (1960)). We claim that $h \in H_\kappa(K)$. To check the claim, consider arbitrary $k \in K$. Since $\theta(a) \leq h$ for all $a \in A$, it follows that

$$U = coz(h - k)^+ \supset \bigcup_A coz(\theta(a) - k)^+ = R,$$

and since $h(v) = \bigvee_A a(v)$ for all $v \in V$, it follows that $cl\, U = cl\, R$. This shows that $clintz(h - k)^+ \in \mathcal{Z}_\kappa^K(Y)$, and an analogous argument shows that $clintz(k - h)^+ \in \mathcal{Z}_\kappa^K(Y)$, proving the claim. We can finally invoke (f) to assert $h \in K$, say $h = \theta(g)$ for some $g \in D(X)$. Then $g \in C(X)$ since h is bounded, and g clearly agrees with h on V. We conclude that V is C^*-embedded in X, proving (a). □

In order to perform the inverse limit construction of the quasi-F_κ cover of $X \in \mathrm{K}$, it is first necessary to define the system whose inverse limit we seek. Theorem 3.12 (b) shows that for any κ-irreducible surjection $\varphi_A: A \longrightarrow X \in \mathrm{K}$ the cardinality of A is limited by 2^λ, where $\lambda = |Z_\kappa(X)|$. Therefore it is possible to find a *set* $\mathcal{C}$ of such surjections such that an arbitrary irreducible surjection $\tau: Z \longrightarrow X \in \mathrm{K}$ is equivalent over X to exactly one member of $\mathcal{C}$. Let $\mathcal{A} = \{A: \varphi_A \in \mathcal{C}\}$, partially ordered in the usual way (Porter and Woods (1988)): $A \geq B$ whenever there is some surjection $\varphi_A^B: A \longrightarrow B$ such that $\varphi_B \varphi_A^B = \varphi_A$, in which case the φ_A^B is unique. We next need to show that $\mathcal{C}$ is upward directed in this partial order.

LEMMA 4.7. *Given κ-irreducible surjections α and β, there are surjections γ and δ such that $\alpha\delta = \beta\gamma$ is κ-irreducible.*

$$\begin{array}{ccc} A & \xrightarrow{\gamma} & Z \\ \delta\big\downarrow & & \big\downarrow\beta \\ Y & \xrightarrow{\alpha} & X. \end{array}$$

Proof. Let $B = \{(y,z): \alpha(y) = \beta(z)\} \subset Y \times Z$, and let γ and δ be the appropriate projection maps. Note that $B \in \mathrm{K}$, that $\alpha\delta = \beta\gamma$ on B, but that, in general, this composition is not irreducible. Using a trick that Gleason (1958) described as being "well known", we apply Zorn's Lemma to find a minimal element A among the closed subspaces of B which map onto X under $\alpha\delta = \beta\gamma$. From the irreducibility of $\alpha\delta = \beta\gamma$ on A follows the irreducibility of γ and δ. To check that the composition map is κ-irreducible, we use Theorem 3.12 (b). If (y_1,z_1) and (y_2,z_2) are distinct points of A, say $y_1 \neq y_2$, then the κ-irreducibility of α yields κ-zero sets $W_1, W_2 \subset X$ with $y_i \in int\, \alpha^{\leftarrow}(W_i)$ and $int\, \alpha^{\leftarrow}(W_1 \cap W_2)$ empty. Then $(y_i,z_i) \in int\, (\alpha\delta)^{\leftarrow}(W_i)$, and, by virtue of the irreducibility of δ,

$$\phi = int\, \delta^{\leftarrow}\alpha^{\leftarrow}(W_1) \cap int\, \delta^{\leftarrow}\alpha^{\leftarrow}(W_2) = int\, (\alpha\delta)^{\leftarrow}(W_1 \cap W_2) = \phi,$$

which proves $\alpha\delta$ κ-irreducible. □

LEMMA 4.8. *Suppose that $\{\varphi^B_A: A \to B: A,B \in \mathscr{A}\}$ is an upward directed system of maps in* K *with inverse limit Y having projections $\{\varphi^A: Y \to A: A \in \mathscr{A}\}$. If all the bonding maps φ^B_A are irreducible, then so are the projections φ^A.*

Proof. Suppose the projection φ^A maps the closed subset $S \subset Y$ onto A. Now $\{\varphi^{B\leftarrow}(T)$: closed $T \subset B \in \mathscr{A}\}$ is a basis for the closed sets of Y, so that, for the purpose of testing the propriety of S in Y, we may assume $S = \varphi^{B\leftarrow}(T)$ for some closed subset T of some $B \in \mathscr{A}$. Since $\mathscr{A}$ is upward directed, we may assume that there is an irreducible map $\varphi^A_B: B \to A$. But since $\varphi^A_B\varphi^B = \varphi_A$, φ^A_B must take T onto A. It follows from the irreducibility of φ^A_B that $T = B$, and therefore that $S = Y$. □

THEOREM 4.9. *For every* $X \in \mathrm{K}$ *there is a* κ*-irreducible surjection* $\varphi\colon Y \dashrightarrow X \in \mathrm{K}$ *with the following equivalent properties.*

(a) Y *is quasi-*F_{κ}.

(b) *A surjection* $\tau\colon Z \dashrightarrow X \in \mathrm{K}$ *is* κ*-irreducible if and only if there is some surjection* $\mu\colon Y \dashrightarrow Z \in K$ *such that*

$$\begin{array}{ccc} Y & \xrightarrow{\ \mu\ } & Z \\ & \varphi\searrow \quad \swarrow\tau & \\ & X & \end{array}$$

commutes.

φ *is unique in the sense that it is equivalent over* X *to any other map with its properties.*

Proof. Let φ_A, $\mathscr{C}$, $\mathscr{A}$, φ^B_A, φ^A, and Y have the meanings above. Let $\varphi = \varphi_A \varphi^A$ for any $A \in \mathscr{A}$; φ is independent of the choice of A. To show that φ is κ-irreducible consider $y_1 \neq y_2 \in Y$. Then $\varphi^A(y_1) \neq \varphi^A(y_2)$ for some $A \in \mathscr{A}$. The κ-irreducibility of φ_A then produces κ-zero sets $Z_1, Z_2 \subset X$ such that $\varphi^A(y_i) \in int\ \varphi_A^{\leftarrow}(Z_i)$ and $int\ \varphi_A^{\leftarrow}(Z_1 \cap Z_2) = \phi$. But then $y_i \in int\ \varphi^{\leftarrow}(Z_i)$, and, by virtue of the irreducibility of φ^A guaranteed by the previous lemma,

$$\phi = int\ \varphi^{A\leftarrow}\varphi_A^{\leftarrow}(Z_1) \cap int\ \varphi^{A\leftarrow}\varphi_A^{\leftarrow}(Z_2) = int\ \varphi^{\leftarrow}(Z_1 \cap Z_2),$$

proving φ κ-irreducible. This shows that φ is equivalent over X to some member of $\mathscr{C}$, and that we can take Y to be the greatest element of $\mathscr{A}$. Thus Y satisfies Theorem 4.6 (d), and so is quasi-F_{κ}. We leave the demonstration of the equivalence of (a) and (b) and of the uniqueness of φ to the reader. □

It is not necessary to use all κ-irreducible preimages of X in the direct limit to form $QF_{\kappa}(X)$. It is sufficient to use only those of the form βF, where F is a countable intersection of dense κ-cozero sets with Stone-Čech compactification βF. We elaborate.

Let $\mathscr{F}$ be the filter base of all countable intersections of dense κ-cozero subsets of X. For each $F \in \mathscr{F}$ let $\varphi_F\colon \beta F \longrightarrow X$ be the Stone-Čech extension of the inclusion $F \subset X$. For $F, M \in \mathscr{F}$ such that $F \subset M$, let $\varphi^M_F\colon \beta F \longrightarrow \beta M$ be the Stone-Čech extension of that inclusion. Note that $\varphi_M \varphi^M_F = \varphi_F$. Since $\mathscr{F}$ is downward directed by inclusion,

$\{\beta V: V \in \mathcal{F}\}$ with bonding maps $\{\varphi_F^M: F,M \in \mathcal{F}\}$ forms an upward directed system in K. Let Z be the inverse limit of this system with projections $\{\varphi^F: Z \to \beta F: F \in \mathcal{F}\}$, and let $\varphi: Z \to X$ be $\varphi_F\varphi^F$ for any $F \in \mathcal{F}$. Note that φ is independent of the choice of F.

PROPOSITION 4.10. *In the notation above,* $Z = QF_\kappa(X)$.

Proof. Fix $F \in \mathcal{F}$. We first claim that $\varphi_F: \beta F \to X$ is κ-irreducible by showing that $\theta(C(X))$ is κjamd in $C^*(V) = C(\beta F)$, where $\theta: C(X) \to C(Z)$ is the **W** injection realized by φ_F, and then appealing to Theorem 3.12. Consider $h \in C^*(V)^+$. The first paragraph of the argument that (f) implies (a) in Theorem 4.6 shows that there is some subset $A \subset C(X)^+$ of cardinality less than κ such that $\bigvee_A \theta(a(v)) = h(v)$ for all $v \in V$; a similar line of reasoning produces $B \subset C(X)^+$ of cardinality less than κ such that $\bigwedge_B \theta(b(v)) = h(v)$ for all $v \in V$. This proves the first claim. From the κ-irreducibility of the φ_F's follows the κ-irreducibility of the bonding maps φ_F^M; the first paragraph of the proof of Theorem 4.9 then shows that the κ-irreducibility of φ follows from these two facts.

It remains to show that Z is quasi-F_κ by showing that an arbitrary κ-Lindelöf subset $U \subset Z$ is C^*-embedded. For that purpose consider $h \in C^*(U) = C(\beta U)$. Let $\tau: \beta U \to Z$ be the Stone-Čech extension of the inclusion $U \subset Z$, and let $\psi: C(Z) \to C(\beta U)$ be the **W** injection realized by τ. Let $\theta: C(X) \to C(Z)$ be the **W** injection realized by φ. In the previous paragraph we established the κ-irreducibilty of φ, and the first part of this argument also shows that τ is κ-irreducible. Therefore $\varphi\tau$ is κ-irreducible, with the result that $\psi\theta(C(X))$ is κjamd in $C(\beta U)$. Therefore we can find sets $A,B \subset C(X)$ of cardinality less than κ such that $\bigvee\psi\theta(A) = h = \bigwedge\psi\theta(B)$. For each $q \in \mathbb{Q}^+$ let

$$F(q) = \bigcup\{coz(q + a - b)^+: a \in A, b \in B\},$$

and let $\bigcap_{\mathbb{Q}^+}F(q) = F \in \mathcal{F}$. Therefore we have a projection $\varphi^F: Z \to \beta F$, and $h\varphi_F$ can be seen to be the continuous extension of h to Z. □

THEOREM 4.11. *Let X, Y, and φ have the meaning of Theorem 4.9. Then for any quasi-F_κ space Z and irreducible surjection $\tau: Z \to X$ there is a surjection $\mu: Z \to Y$ such that*

$$\begin{array}{ccc} Y & \xleftarrow{\;\mu\;} & Z \\ & {\scriptstyle\varphi}\searrow \quad \swarrow{\scriptstyle\tau} & \\ & X & \end{array}$$

commutes.

Proof. Let $\nu: T \to X$ be the canonical irreducible surjection from the Gleason cover (absolute) T onto X. By the projectivity of Z in K there are $\alpha,\beta \in$ K such that

$$\begin{array}{ccc} T & \xrightarrow{\;\beta\;} & Z \\ {\scriptstyle\alpha}\downarrow & \searrow{\scriptstyle\nu} & \downarrow{\scriptstyle\tau} \\ Y & \xrightarrow[\;\varphi\;]{} & X \end{array}$$

commutes; these maps realize W injections

$$\begin{array}{ccc} C(T) & \xleftarrow{\;\rho\;} & C(Z) \\ {\scriptstyle\lambda}\uparrow & & \uparrow{\scriptstyle\psi} \\ C(Y) & \xleftarrow[\;\theta\;]{} & C(X) \end{array}$$

which also commute. The irreducibility of φ, τ, and ν imply that of α and β, and therefore also the largeness of the W injections. The claim is that $\lambda(C(Y)) \subset \rho(C(Z))$. To settle the claim consider $h \in C(Y)^+$. The κ-irreducibility of φ implies that $\theta(C(X))$ is κjamd in $C(Y)$, so there are subsets $A, B \subset C(X)$ of cardinality less than κ such that $\wedge(B - A) = 0$ and $\theta(A) \le h \le \theta(B)$, where $\theta(A)$ designates $\{\theta(a): a \in A\}$ and likewise for $\theta(B)$. Now $C(Z)$ is κjamd complete by virtue of the assumption that Z is quasi-F_κ, and $\wedge(\psi(B) - \psi(A)) = 0$ by virtue of the largeness of $\psi(C(X))$ in $C(Z)$. Consequently there is some $k \in C(Z)$ satisfying $\psi(A) \le k \le \psi(B)$. Since both $\lambda(h)$ and $\rho(k)$ are trapped between $\lambda\theta(A) = \rho\psi(A)$ and $\lambda\theta(B) = \rho\psi(B)$, they must coincide. This proves the claim. By following λ with the inverse of ρ we get a W injection $\eta: C(Y) \to C(Z)$ such that $\eta\theta = \psi$. The desired $\mu: Z \to Y \in$ K is the map realizing η. □

5. THE κ-IDEAL COMPLETION OF AN ARCH OBJECT

We begin by obtaining the maximal κ-ideal preserving extension of a W object. This is done by simply reversing the arrows of Section 4 by means of the Yosida Functor.

THEOREM 5.1. *For every* $G \in \mathbf{W}$ *there is a* κ*-ideal preserving extension* $G \leq H \in \mathbf{W}$ *with the following equivalent properties.*

(a) H *is* κ*-replete.*

(b) H *is* κ*-ideal complete.*

(c) *An extension* $G \leq K \in \mathbf{W}$ *is* κ*-ideal preserving if and only if there is a* $\mathbf{W}$ *injection* $\theta\colon K \longrightarrow H$ *over* G.

(d) H *is of the form* $D(Y)$ *for some quasi-*F_κ $Y \in \mathbf{K}$.

H is unique up to $\mathbf{W}$ *isomorphism over* G *with respect to its properties.*

Proof. We argue first for the existence of an extension $G \leq H$ satisfying (d). Let X be the Yosida space of G, let $\psi\colon G \longrightarrow L \subset D(X)$ be the Yosida representation of G (Theorem 2.1), let $\varphi_X\colon Y \longrightarrow X$ be the κ-irreducible surjection from $Y = QF_\kappa(X)$ onto X (Theorem 4.8), let H be $D(Y)$, and let $\xi\colon L \longrightarrow H$ be the $\mathbf{W}$ injection realized by φ_X (Theorem 2.4). Identify G with $\xi\psi(G)$. Then the extension $G \leq H$ is κ-ideal preserving by Theorem 3.12 (e).

Now suppose only that $G \leq H$ is a κ-ideal preserving $\mathbf{W}$ extension satisfying (d). To prove (c) consider a κ-ideal preserving extension $G \leq K \in \mathbf{W}$. Let X, Y, and T be the Yosida spaces of G, H, and K, respectively, and identify the ℓ-groups with their Yosida representations. Then these extensions are realized by κ-irreducible surjections φ and τ, so that by Theorem 4.11 there is a surjection μ making

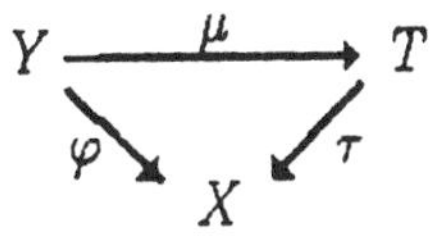

commute. Then μ realizes a $\mathbf{W}$ injection $\theta\colon K \longrightarrow H$ over G.

Assuming $G \leq H$ satisfies (c), consider a κ-ideal preserving extension $H \leq K$. Then K is also a κ-ideal preserving extension of G, and thus there is some $\mathbf{W}$ injection $\theta\colon K \longrightarrow H$ over G. But the restriction of θ to H agrees with the identity map on G, and so *is* the identity map by the uniqueness clause of Theorem 2.3. We conclude that $H = K$, proving (b). The equivalence of (a) and (b) is Proposition 4.3. To prove (d) from (b) it is sufficient to observe that the existence argument above establishes that H has a κ-ideal preserving extension of the form required in (d), and so by (b) must coincide with it. □

We shall refer to the **W** object H of Theorem 5.1 as the *κ-ideal completion* of G, and write it κG.

THEOREM 5.2. *If G is a large* **W** *subobject of the κ-ideal complete* **W** *object H, then there is a* **W** *injection $\theta\colon \kappa G \longrightarrow H$ over G.*
Proof. Apply the Yosida functor to Theorem 4.11. □

Now we construct the κ-ideal completion of an **Arch** object G. In this broader context we lack the fruitful topological counterpoint provided by the Yosida representation; we nevertheless succeed in this construction by fastening on the idea of corresponding κ-generated polars.

THEOREM 5.3. *For every $G \in$* **Arch** *there is a κ-ideal preserving extension $G \leq L \in$* **Arch** *with the following equivalent properties.*

(a) L is κ-replete.

(b) L is κ-ideal complete.

(c) An extension $G \leq K \in$ **Arch** *is κ-ideal preserving if and only if there is an* **Arch** *injection $\theta\colon K \longrightarrow L$ over G.*

L is unique up to **Arch** *isomorphism over G with respect to its properties.*

Proof. We first show the existence of an extension satisfying (c). Let $Y \in \mathbf{K}$ be the extremally disconnected space for which there is some **Arch** injection $\theta\colon G \longrightarrow D(Y) = H$ such that $\theta(G) = M$ is large in H (Theorem 2.2). Since $\{h \in D(Y)\colon nh \in M$ for some $n \in \mathbb{Z}\}$ is apparently a κ-ideal preserving extension of M, we may as well assume that M is divisible. Let $L = H_{\kappa}(M)$, a κ-ideal preserving extension of M in H (Proposition 4.4). To prove (c) consider now an arbitrary κ-ideal preserving extension $G \leq K$. By Theorem 2.3 there is some **Arch** isomorphism ψ from K onto some ℓ-group $F \subset H$ which agrees with θ on G. But since L is the largest κ-ideal preserving extension of M in H, ψ takes K into L.

The equivalence of (a) and (b) is Corollary 4.5. To show that (c) implies (a) consider a κ-ideal preserving extension $L \leq K$, and let $\theta: K \longrightarrow L$ be the injection given by (c). The restriction of θ to L agrees with the identity map on G, and so must *be* the identity map on all of H by the uniqueness clause of Theorem 2.3. From this it follows that $L = K$, meaning that (a) holds. Now suppose that (a) holds for L, and let L' be the κ-ideal preserving extension of G discussed above. Then by property (c) of L' there is some **Arch** injection $\theta: L \longrightarrow L'$ over G. But since L' is clearly a κ-ideal preserving extension of $\theta(L)$, property (a) of L implies that θ is an isomorphism. That is, L has property (c). The argument for the uniqueness of L is similarly straightforward. □

We shall also refer to the extension L of Theorem 5.3 as κG, and to any **Arch** object G for which $G = \kappa G$ as *κ-ideal complete.* As in the case of **W**, κG is not only the maximal κ-ideal preserving extension of G, it is also the minimal κ-ideal complete extension of G.

THEOREM 5.4. *Suppose G is a large ℓ-subgroup of the κ-ideal complete ℓ-group $K \in$* **Arch**. *Then there is an* **Arch** *injection $\psi: \kappa G \longrightarrow K$ over G.*

Proof. Let $Y \in K$ be the extremally disconnected space and θ an **Arch** injection from H into $D(Y)$ such that $\theta(G)$ is large in $D(Y)$ (Theorem 2.3). Let M, N, and H designate $\theta(G)$, $\theta(H)$, and $D(Y)$, respectively. We know from the proof of Theorem 5.4 that $\kappa G = H_\kappa(M)$ and that $K = \kappa K = H_\kappa(N)$. But one sees directly from the definitions that $H_\kappa(M) \subset H_\kappa(N)$. □

Having treated separately the κ-ideal completions of **W** objects and of **Arch** objects, we seek now to compare them.

PROPOSITION 5.5. *The κ-ideal completions of $G \in$* **W** *in* **Arch** *and in* **W** *coincide.*

Proof. Consider $(G,u) \in$ **W**. Forget the weak unit u, and consider a κ-ideal preserving extension $G \leq H \in$ **Arch**. Since G is large in H, u is a weak unit of H also, so $(G,u) \leq (H,u)$ is a κ-ideal preserving extension in **W**. It follows that the maximal κ-ideal preserving extensions in **W** and in **Arch** coincide. □

The preceding proposition justifies using the same notation κG for both completions. However, if $G \in$ Arch then κG need *not* have a weak unit. A necessary condition that κG have a weak unit is the existence of some subset $A \subset G$ of cardinality less than κ such that $A^{\perp\perp} = G$. For if $u \in \kappa G$ is a weak unit then, since κ-generated polars in G and κG correspond, there must be a subset $A \subset G$ of cardinality less than κ such that $A^{\perp\perp} = u^{\perp\perp} = \kappa G$ when polars are computed in κG, from which it follows that $A^{\perp\perp} = G$ when polars are computed in G. It is easy to find examples which violate this condition. But the larger question remains: characterize those $G \in$ Arch such that κG has a weak unit.

6. THE QUASI-F_κ COVER OF Y VERSUS THE SPACE OF MAXIMAL FILTERS ON $\mathcal{Z}_\kappa(Y)$.

Henriksen, Vermeer, and Woods (1987) construct the quasi-F cover of $X \in$ K as the space of maximal filters on $\mathcal{Z}_{\aleph_1}(X)$ (which, recall, is $\mathcal{Z}^G_{\aleph_1}(X)$ for $G = C(X)$), and then generalize this construction in their 1989 paper to a usefully broad class of other sublattices of the algebra $\mathcal{R}(X)$ of regular closed subsets of X. It develops that $QF_\kappa(X)$ can be realized as the space of all maximal ideals on $\mathcal{Z}_\kappa(X)$ in many but not all instances. We characterize those instances, and give an example in which $QF_\kappa(X)$ cannot be so realized.

We abbreviate $\mathcal{Z}_\kappa(X)$ to $\mathcal{Z}_\kappa$ wherever the context makes this unambiguous. A *filter* on $\mathcal{Z}_\kappa$ is a subset $t \subset \mathcal{Z}_\kappa$ with the following three properties: $C_1 \supset C_2 \in t$ implies $C_1 \in t$, $C_1, C_2 \in t$ imply $C_1 \wedge C_2 \in t$, and $\phi \notin t$. Remember that suprema and infima are those in $\mathcal{R}(X)$: $C_1 \vee C_2 = C_1 \cup C_2$, and $C_1 \wedge C_2 = clint(C_1 \cap C_2)$. Filters maximal in the containment order exist by Zorn's Lemma, and are characterized by the property that for every $C \notin t$ there is some $D \in t$ such that $C \wedge D = \phi$. Let

$T_\kappa(X)$ represent

$$\{t\colon t \text{ is a maximal filter on } \mathcal{Z}_\kappa\},$$

topologized by using sets of the form

$$M(C) = \{t \in T_\kappa(X)\colon C \in t\},\ C \in \mathcal{Z}_\kappa,$$

as a base for the closed sets. Observe that

$$M(C_1 \vee C_2) = M(C_1) \cup M(C_2), \text{ and}$$

$$M(C_1 \wedge C_2) = M(C_1) \cap M(C_2).$$

We abbreviate $T_\kappa(X)$ by T_κ whenever it is unambiguous to do so. The canonical map $\varphi\colon T_\kappa \to X$, defined by the formula $\varphi(t) = x$ if and only if $\cap t = \{x\}$, can readily be seen to be a continuous surjection. It is straightforward to show that T_κ is a compact space; the crucial issue addressed in Theorem 6.2 is whether it is Hausdorff.

Recall that a polar P of $G \in$ Arch is said to be κ-generated if it is of the form $A^{\perp\perp}$ for some subset $A \subset G$ of cardinality less than κ. We shall say of G that *κ-generated polars can be disjointified* provided that for every pair of κ-generated polars P_1 and P_2 such that $P_1 \vee P_2 = G$ there are disjoint κ-generated polars Q_1 and Q_2 (that is, $Q_1 \wedge Q_2 = 0$) such that $P_1 \vee Q_2 = Q_1 \vee P_2 = G$. Keep in mind that the suprema and infima are those in the algebra $\mathcal{P}$ of polars: $P_1 \wedge P_2 = P_1 \cap P_2$, and $P_1 \vee P_2 = (P_1 \cup P_2)^{\perp\perp}$.

PROPOSITION 6.1. *$\aleph_1$-generated polars of any $G \in$ W can always be disjointified.*

Proof. Consider $\aleph_1$-generated polars P_i of G, say $P_i = A_i^{\perp\perp}$ for countable subsets $A_i \subset G$. Let X be the Yosida space of G, and identify G with its Yosida representation. Then $U_i = \cup_{A_i} coz(a)$ is a cozero subset of X, and so is of the form $coz(a_i)$ for some $a_i \in C(X)^+$. Let $b_1 = a_1 - (a_1 \wedge a_2)$ and $b_2 = a_2 - (a_1 \wedge a_2)$, so that $b_1 \wedge b_2 = 0$ and $coz(b_1) \cup coz(a_2) = coz(a_1) \cup coz(b_2) = coz(a_1) \cup coz(a_2)$. Now every polar principal in $C(X)$ is a countable supremum of polars principal in G. (Given $g \in G^+$, note that for each positive integer n, $coz(g - 1/n)^+$ is completely separated from $z(g - 1/(n+1))^+$, with the result that there is some $g_n \in G$ such that $coz(g_n)$ contains the former set and misses the latter. Then $coz(g) = \bigcup_N coz(g_n)$, so $g^{\perp\perp} = \bigcup\{g_n : n \in N\}$.) Therefore the polars $Q_i = b_i^{\perp\perp}$ disjointify the P_i's. □

The equivalence of conditions (b) through (e) of the following theorem was proved by Henriksen, Vermeer, and Woods (1987) for the case $\kappa = \aleph_1$. Condition (e), in particular, received heavy use in that paper as the working definition of the quasi-F property. It is precisely the utility of this characterization which places upon us the obligation of determining the relevance of the analogous condition for quasi-F_κ spaces.

THEOREM 6.2. *The following are equivalent for* $G \in \mathbf{W}$ *with Yosida space* X.

(a) *κ-generated polars of G can be disjointified.*
(b) *For any disjoint $C_1, C_2 \in \mathcal{Z}_\kappa$ there are $D_1, D_2 \in \mathcal{Z}_\kappa$ such that $D_1 \vee D_2 = G$ and $C_1 \wedge D_2 = D_1 \wedge C_2 = 0$.*
(c) *T_κ is Hausdorff.*
(d) *$QF_\kappa(X)$ is homeomorphic to T_κ.*
(e) *For any pair $D_1, D_2 \in \mathcal{Z}_\kappa(QF_\kappa(X))$, $D_1 \wedge D_1 = \phi$ implies $D_1 \cap D_2 = \phi$.*
(f) *κ-generated polars of κG can be disjointified.*

Proof. Identify G with its Yosida representation. The equivalence of (a) and (b) is clear in light of the association of each κ-generated polar $A^{\perp\perp}$, $|A| < \kappa$, with $clint \bigcap_A z(a) \in \mathcal{Z}_\kappa$. Now assume (b), and to prove (c) consider distinct points $t_1, t_2 \in T_\kappa$. By the maximality of the t_i's there are disjoint $C_1, C_2 \in \mathcal{Z}_\kappa$ such that $C_i \in t_i$. Find $D_i \in \mathcal{Z}_\kappa$ satisfying (b). Then the basic open sets $U_i = T_\kappa \setminus M(D_i)$ are disjoint, and $t_i \in U_i$, proving (c).

Suppose that T_κ is Hausdorff. We shall show that $\varphi: T_\kappa \longrightarrow X$ is equivalent to $\varphi_X: QF(X) \longrightarrow X$ over X. To show φ κ-irreducible consider distinct $t_1, t_2 \in T_\kappa$. By the normality of T_κ one may find disjoint closed neighborhoods N_i of t_i. By appealing to the compactness of T_κ one may then locate disjoint basic closed sets $M(C_i) \supset N_i$. Then $t_i \in$

$int\, M(C_i) \subset int\, \varphi^{\leftarrow}(C_i)$, and $\phi = M(C_1) \cap M(C_2) = M(C_1 \wedge C_2)$ implies that $C_1 \wedge C_2 = \phi$, hence $int\, \varphi^{\leftarrow}(C_1 \wedge C_2) = \phi$, proving φ κ-irreducible by Theorem 3.12 (b). To complete the demonstration that φ is equivalent to φ_X over X we show T_K has property (b) of Theorem 4.9. Consider a κ-irreducible surjection $\tau: Z \longrightarrow X \in$ K. For $t \in T_K$ define

$$\mu(t) = \bigcap\{cl\, \tau^{\leftarrow}(int\, C): C \in t\}.$$

Because the map $D \longmapsto \tau(D)$ is a lattice isomorphism from $\mathcal{Z}_K(Z)$ onto $\mathcal{Z}_K(X)$ with inverse $C \longmapsto cl\, \tau^{\leftarrow}(int\, C)$, $\{cl\, \tau^{\leftarrow}(int\, C): C \in t\}$ is a maximal filter on $\mathcal{Z}_K(Z)$ whose intersection is therefore a singleton. And $\tau\mu = \varphi$ because

$$\tau(\mu(t)) \subset \bigcap\{\tau(cl\, \tau^{\leftarrow}(int\, C)): C \in t\} = \bigcap_t C = \{\varphi(t)\}.$$

μ is surjective, for given $z \in Z$ then $\{\tau(C): z \in int\, C\}$ is a filter on $\mathcal{Z}_K(X)$, and any maximal filter $t \in T_K$ which refines it must map to z under μ. To see that μ is continuous, observe that $\mathcal{Z}_K(Z)$ is a basis for the closed sets of Z, and that for any $C \in \mathcal{Z}_K(Z)$ we have $\mu^{\leftarrow}(C) = \varphi^{\leftarrow}(\tau(C))$, a closed subset of T_K. This completes the proof that φ is equivalent to φ_X over X, and thus also the proof of (d).

Suppose that (d) holds. Then T_K is Hausdorff, and by the above argument φ is equivalent to φ_X over X. We claim that $M(C) = cl\, \varphi^{\leftarrow}(int\, C)$ for all $C \in \mathcal{Z}_K(X)$. To verify the claim consider first $t \in \varphi^{\leftarrow}(int\, C)$. If $t \notin M(C)$ it is only because $D \in t$ for some $D \in \mathcal{Z}_K(X)$ such that $C \wedge D = \phi$. But that implies that $\varphi(t) \in D = clint\, D$, a set disjoint from $int\, C$. Therefore $t \in M(C)$, $\varphi^{\leftarrow}(int\, C) \subset M(C)$, and $cl\, \varphi^{\leftarrow}(int\, C) \subset M(C)$. On the other hand consider a basic open set $T_K \setminus M(D)$ containing $t \in M(C)$. There must be some $E \in \mathcal{Z}_K(X)$ such that $E \cap D = \phi$ and $E \in t$. It follows from $C \cap E \in t$ that $\phi \neq C \cap E = clint(C \cap E)$. Let $x \in int(C \cap E)$. Then $\phi \neq \varphi^{\leftarrow}(x) \subset M(E) \subset T_K \setminus M(D)$, which shows that $t \in cl\, \varphi^{\leftarrow}(int\, C)$ and completes the proof of the claim. To prove (e) consider $D_1, D_2 \in \mathcal{Z}_K(T_K)$ such that $\phi = D_1 \wedge D_2$ and let $C_i = \varphi(D_i)$. Then $C_i \in \mathcal{Z}_K(X)$ by the κ-irreducibility of φ, and $D_i = M(C_i)$ by the claim above. Now $D_1 \wedge D_2 = \phi$ makes $C_1 \wedge C_2 =$

ϕ, hence $D_1 \cap D_2 = M(C_1) \cap M(C_2) = M(C_1 \wedge C_2) = \phi$.

To show that (e) implies (f) consider κ-generated polars P_1 and P_2 of κG such that $P_1 \vee P_2 = \kappa G$. Let $P_i = A_i^{\perp\perp}$ for subsets $A_i \subset \kappa G$ of cardinality less than κ. Let Y designate $QF_\kappa(X)$, and identify κG with $D(Y)$. Let $C_i = clint \cap_{A_i} z(a) \in \mathfrak{Z}_\kappa(QF_\kappa(X))$. Since $C_1 \wedge C_2 = \phi$ as a result of the assumption that $P_1 \vee P_2 = \kappa G$, we get $C_1 \cap C_2 = \phi$ from (e). This condition asserts the existence for each $y \in Y$ of some $h \in \kappa G$ such that $y \in coz(h)$ and such that either $C_1 \subset z(h)$ or $C_2 \subset z(y)$, that is, such that $h \in P_1 \cup P_2$. The compactness of Y then produces $h_i \in P_i$ such that $z(h_1) \cap z(h_2) = \phi$. Let $Q_i = h_i^{\perp\perp}$. These principal polars satisfy $Q_i \subset P_i$ and $Q_1 \vee Q_2 = \kappa G$. By Proposition 6.1 there are disjoint $\aleph_1$ generated (actually principal) polars R_i such that $R_1 \vee Q_2 = Q_1 \vee R_2 = \kappa G$. But since $Q_i \subset P_i$, $R_1 \vee P_2 = P_1 \vee R_2 = \kappa G$, which proves (f). Finally, the implication from (f) to (a) is clear since κ-generated polars in G and κG correspond by intersection. This completes the proof. □

Here is an example of a **W** object G which violates the conditions of Theorem 6.2. Assume $\kappa \geq \aleph_2$, let S_1, S_2, and T be disjoint sets such that $\aleph_1 \leq |S_1| = |S_2| < \kappa = |T|$, let X designate the collection of non-empty finite subsets of $S_1 \cup S_2$, and let Y designate the Cartesian product $X \times T$ with the discrete topology. G will be the ℓ-subgroup of $C(Y)$ generated by the following functions. For each $s \in S_1 \cup S_2$ let $\mathfrak{s} \in C(Y)$ be defined by

$$\mathfrak{s}(x,t) = \begin{cases} 1 & \text{if } s \in x \\ 0 & \text{if } s \notin x \end{cases}$$

for $x \in X$ and $t \in T$. For each $y \in Y$ let

$$\chi_y(z) = \begin{cases} 1 & \text{if } y = z \\ 0 & \text{if not} \end{cases}$$

for all $z \in Y$. Then G is the ℓ-subgroup of $C(Y)$ generated by

$$\{\chi_y : y \in Y\} \cup \{\mathfrak{s} : s \in S_1 \cup S_2\} \cup \{1\}.$$

For any subset $P \subset G$ let $coz(P)$ stand for $\cup_P coz(p)$. Because G contains all integer valued elements of $C(Y)$ with finite cozero sets, the correspondence $P \longmapsto coz(P)$ is a Boolean isomorphism from the polar algebra of G onto the power set of Y. Let A_i be $\{\hat{s}: s \in S_i\}$. Observe that each $A_i^{\perp\perp}$ is a κ-generated polar, and that $A_1^{\perp\perp} \vee A_2^{\perp\perp} = G$ because

$$coz(A_1) \cup coz(A_2) = \{(x,t): x \cap S_1 \neq \phi\} \cup \{(x,t): x \cap S_2 \neq \phi\} = Y.$$

We claim these polars cannot be disjointified. For if B_1 and B_2 are subsets of G such that $B_1^{\perp\perp} \vee A_2^{\perp\perp} = A_1^{\perp\perp} \vee B_2^{\perp\perp} = G$ then

$$coz(B_1) \supset \{(\{s\},t): s \in S_1,\ t \in T\}, \text{ and}$$

$$coz(B_2) \supset \{(\{s\},t): s \in S_2,\ t \in T\}.$$

Therefore it is enough to establish the following claim. For any subsets $B_1, B_2 \subset G^+$ of cardinality less than κ satisfying the conditions set out above, it must be true that $coz(B_1) \cap coz(B_2) \neq \phi$.

Consider two such subsets B_1 and B_2. For $x \in X$ let $\{x\} \times T = \{(x,t): t \in T\}$. Note that for any $x \in X$ and $g \in G$, $coz(g) \cap \{x\} \times T$ is either finite or cofinite in T, since this is true of the generators. The subclaim is that for every $s \in S_1$ there is some $f_s \in G^+$ such that $\{s\} \times T \setminus coz(f_s)$ and $coz(f_s) \setminus coz(B_1)$ are both finite. First fix $s \in S_1$ to prove this subclaim. Since $|B_1| < \kappa = |T|$, there must be some element $b \in B_1$ such that $coz(b) \cap \{s\} \times T$ is cofinite in T. Now b may be written $\vee_I \wedge_J h_{ij}^+$, where each h_{ij} is of the form $\sum_{k=1}^{m} n_k g_k$ for integers n_k and generators g_k. Therefore there must be some $i \in I$ such that $coz(\wedge_J h_{ij}^+) \cap \{s\} \times T$ is cofinite in T. Replace b by $\wedge_J h_{ij}^+$, and suppress the index i, writing $b = \wedge_J h_j^+$. For the sake of specificity let $h_j = \sum_{k=1}^{m} n_{jk} g_k$, where we may as well assume that g_1 is $\hat{s}$, that g_2 is 1, and that $g_i \neq g_j$ whenever $i \neq j$. For subsequent use we denote by $S(s)$ the finite set $\{r \in S_1 \cup S_2: \hat{r} = g_k \text{ for some } k,\ 3 \leq k \leq m\}$. Then a little reflection on the definitions shows that $f_s = (\hat{s} - 2\sum_{k=3}^{m} g_k)^+$ has the properties desired for the subclaim. Likewise, for every $s \in S_2$ there is some $f_s \in G^+$ such that $\{s\} \times T \setminus coz(f_s)$ and $coz(f_s) \setminus coz(B_2)$ are both finite.

We prove the claim by producing $s_i \in S_i$ such that $s_i \notin S(s_j)$, $i \neq j$. For if this can be done then the definition of f_{s_i} shows that $\{s_1, s_2\} \times T \setminus (coz(f_{s_1}) \cap coz(f_{s_2}))$ is finite, and consequently that $\{s_1, s_2\} \times T \setminus (coz(B_1) \cap coz(B_2))$ is also finite, with the result that $coz(B_1) \cap coz(B_2)$ is nonempty.

Suppose no such s_1 and s_2 exist. Select a countable subset of S_1, $\{s_{1,n}: n \in N\}$. For each $n \in N$ let $S_{2,n} = \{s_2 \in S_2: s_{1,n} \in S(s_2)\}$. The hypothesis implies that $S_{2,n}$ is a cofinite subset of S_2. Then $\bigcap_N S_{2,n}$ cannot be empty since S_2 is uncountable. But if $s_2 \in \bigcap_N S_{2,n}$ then $s_{1,n} \in S(s_2)$ for all $n \in N$, contrary to the finiteness of $S(s_2)$. We can only conclude that elements s_1 and s_2 exist as required. This completes the example.

REFERENCES

R.N.Ball, *Convergence and Cauchy structures on lattice ordered groups*, Trans.Am.Math.Soc. 259(1980), 357-392.

R.N.Ball, *Distributive Cauchy lattices*, Alg.Univ. 18(1984), 134-174.

R.N.Ball, *Full convex ℓ-subgroups and the existence of a^*closures of lattice ordered groups*, Pacific J.Math. 61(1975), 7-16.

S.J.Bernau, *Unique representation of archimedean lattice groups and normal archimedean lattice rings*, Proc.London Math.Soc.(3)16(1966), 107-130.

R.N.Ball and A.W.Hager, *Archimedean kernel distinguishing extensions of archimedean ℓ-groups with weak unit*, to appear in Indian J. Math.,(1989a).

R.N.Ball and A.W.Hager, *Epimorphisms in archimedean ℓ-groups and vector lattices with weak unit (and Baire functions)*, to appear in J.Austral.Math.Soc.,(1989b).

B.Banaschewski, *Projective covers in categories of topological spaces and topological algebras*, General Topology and its relation to modern analysis and algebra, Proceedings of the Kanpur Topological Conference, (1968).

A.Bigard, K.Keimel, and S.Wolfenstein, *Groupes et Anneaux Reticules*, Springer Lecture Notes 608, Berlin-Heidelberg-New York, 1977.

R.D.Bleier and P.F.Conrad, *a^*closures of lattice ordered groups*, Trans.Amer.Math.Soc.209(1975), 367-387.

R.D.Bleier and P.F.Conrad, *The lattice of closed ideals and a^*extensions of an abelian ℓ-group,* Pacific J.Math.47(1973), 329-340.

R.D.Byrd, *Archimedean closures in lattice ordered groups,* Canad.J. Math.21(1969), 1004-1011.

W.W.Comfort and S.Negrepontis, *Continuous Pseudometrics,* Lecture Notes in Pure and Applied Mathematics 14, Marcel Dekker, Inc., New York, 1975.

P.F.Conrad, *Archimedean extensions of lattice-ordered groups,* J. Indian Math.Soc.30(1966), 131-160.

P.F.Conrad, *The essential closure of an archimedean lattice ordered group*, Proc.London Math.Soc.38(1971), 151-160.

F.Dashiell, private communication.

F.Dashiell, A.W.Hager, and M.Henriksen, *Order-Cauchy completions of rings and vector lattices of continuous functions*, Can.J.Math. 32(1980), 657-685.

N.J.Fine, L.Gillman, and J.Lambek, *Rings of quotients of rings of functions*, McGill University Press, Montreal, 1965.

A.M.W.Glass, W.C.Holland, and S.H.McCleary, *a^*closures of completely distributive lattice-ordered groups*, Pacific J.Math.59(1975), 43-67; Correction, ibid., 61(1975), 66.

L.Gillman and M.Jerison, *Rings of Continuous Functions*, Van Nostrand Co., 1960, reprinted as Springer-Verlag Graduate Text 43, Berlin Heidelberg New York, 1976.

A.M.Gleason, *Projective topological spaces*, Illinois J.Math.2(1958), 482-489.

A.W.Hager, *Minimal covers of topological spaces,* Proc.1987 Northeast Topology Conference, Annals N.Y.Acad.Sciences, to appear (1989).

A.W.Hager, *The projective resolution of a compact space*, Proc.Amer. Math.Soc.28(1971), 262-266.

A.W.Hager and L.C.Robertson, *Representing and ringifying a Riesz space*, Symposia Math.21(1977), 411-431.

A.W.Hager and L.C.Robertson, *On the embedding into a ring of an archimedean ℓ-group*, Can.J.Math.31(1979), 1-8.

M.Henriksen and D.G.Johnson, *On the structure of a class of archimedean lattice -ordered algebras,* Fund.Math.50(1961), 73-94.

M.Henriksen, J.Vermeer, and R.G.Woods, *Quasi-F covers of Tychonoff spaces*, Trans.Am.Math.Soc.303(1987), 779-803.

M.Henriksen, J.Vermeer, and R.G.Woods, *Wallman covers,* to appear in Diss.Math.(1989).

H.Herrlich and G.Strecker, *Category Theory,* Allyn and Bacon Co., Boston, 1973.

W.Luxemburg and A.Zaanen, *Riesz Spaces,* Vol.I, North Holland Co., Amsterdam, 1971.

F.Maeda and T.Ogasawara, *Representations of vector lattices,* J. Science Hiroshima Univ.(A)12(1942), 17-35.

C.W.Neville, *Quasi-$F_{\aleph}$ spaces as projectives,* unpublished manuscript (1979).

C.W.Neville and S.P.Lloyd, *ℵ-projective spaces,* Illinois J.Math.25 (1981), 159-168.

J.R.Porter and R.G.Woods, *Extensions and Absolutes of Hausdorff Spaces,* Springer-Verlag, New York Heidelberg Berlin, 1988.

Z.Tzeng, *Extended real valued functions and the projective resolution of a compacxt Hausdorff space,* Thesis, Wesleyan University, Middletown, Ct., (1970).

A.I.Veksler, *P'-points, P'-sets, P'-spaces. A new class of order-continuous measures and functionals,* Soviet Math.Dokl.14(1973), 1445-1450.

A.I.Veksler, *P-sets in topological spaces,* Soviet Math.Dokl.11(1970), 953-956.

V.K.Zakharov and A.V.Koldunov, *The sequential absolute and its characterizations,* Soviet Math.Dokl.22(1980), 70-74.

On the Classification of Minimally Free Rings of Continuous Functions

Paul Bankston
Marquette University
and
Robert A. McCoy
Virginia Polytechnic Institute and State University

Abstract

A universal algebra A is *minimally free* if there is a subset X of A such that every function from X into A extends uniquely to an endomorphism on A. If κ is a cardinal number, A is *κ-free* if the set X above can be chosen to have cardinality κ. (0-free is the same as (endomorphism-) rigid.) For a topological space $\mathcal{X}$, we let $C(\mathcal{X})$ be the unital ring of continuous real-valued functions on $\mathcal{X}$. We are interested in the problem of classifying the κ-free rings of the form $C(\mathcal{X})$. In particular we prove that $\mathbf{R} = C(\{\text{point}\})$ is the only 0-free such ring; and for $1 \leq \kappa \leq \omega, C(\mathbf{R}^\kappa)$ and $\mathbf{R}^c = C(\{\text{discrete space of cardinality=continuum}\})$ are the only κ-free such rings.

Subject Classification: 08A35, 54A10 54C40

0. Introduction

This article partially answers a question posed in [**1**], whether one can classify (effectively list) all κ-free commutative unital rings $C(\mathcal{X})$ of continuous real-valued functions (Question 3.14 of [**1**]). A fair amount of progress was made on the question by the first author at the time [**3**] was written, but still a satisfactory classifaction theorem was lacking. Now, at least, we can classify the κ-free rings $C(\mathcal{X})$ for countable κ : $\mathbf{R} = C(\{\text{point}\})$ is the only 0-free such ring; and for $1 \leq \kappa \leq \omega, C(\mathbf{R}^\kappa)$ and $\mathbf{R}^c = C(\{\text{discrete space of cardinality=continuum}\})$ are the only κ-free such rings.

We follow the notation and terminology of [1]. Let A be a universal algebra of arbitrary type. A is *minimally free* if there is a subset X of A, called a ***pseudobasis***, such that every mapping of X into A extends uniquely to an endomorphism on A. If κ is a cardinal number, A is *κ-free* if a pseudobasis of cardinality κ can be found. (Our terminology "κ-free" inadvertently clashes with that of Eklof [**6**], et al. Unfortunately, we were unaware of their usage when we wrote [**1**], and offer our apologies.) In [**1**], [**2**] and the Kříž-Pultr paper [**10**], there are theorems that describe how badly a pseudobasis can fail to generate the algebra. Indeed, there are varieties of algebras (e.g., the commutative unital rings) that have arbitrarily large κ-free algebras for any fixed κ (see [**2**], [**10**] for details).

Any pseudobasis is "Marczewski independent" (see Głazek [**8**]): every mapping from X into A extends (uniquely) to a homomorphism from the subalgebra generated by X. Furthermore, if the pseudobasis actually generates A, then X is a free basis for A relative to the smallest variety containing A. (Hall [**9**] introduces the notion "relatively free group" in this way. See also Neumann's book [**11**].)

Here we concentrate on minimally free rings. In our usage, "ring" always means "commutative unital ring", and homomorphisms preserve the unity element. We let RCF be the class of rings $C(\mathcal{X})$ of continuous real-valued functions where $\mathcal{X}$ is a topological space. It is well known [**7**] that $RCF = \{C(\mathcal{X}) : \mathcal{X}$ is realcompact and Tichonov $\}$. Moreover, by the duality theorem of Gel'fand-Kolmogorov, $C(\cdot)$ is a contravariant equivalence of categories; so two realcompact Tichonov spaces $\mathcal{X}$ and $\mathcal{Y}$ are homeomorphic if and only if the rings $C(\mathcal{X})$ and $C(\mathcal{Y})$ are isomorphic.

As mentioned above, there are arbitrarily large κ-free rings for every κ. However, RCF contains very few of them [**1**], and no κ-free ring in RCF is generated by any of its pseudobases (as we prove in 0.3 below). In brief, let $\mathbf{R}$ be the set of real numbers; and for each cardinal κ, let $\mathcal{U}^\kappa$ be the κ-fold Tichonov product of the usual topology $\mathcal{U}$ on $\mathbf{R}$. We are interested in enrichments of this topology on $\mathbf{R}^\kappa$. If $\mathcal{X} = \langle X, \mathcal{T}\rangle$ and $\mathcal{X}' = \langle X', \mathcal{T}'\rangle$ are any topological spaces, with $f : \mathcal{X} \to \mathcal{X}'$ continuous, define f to be a *coreflection map* if for each continuous $g : \mathcal{X} \to \mathcal{X}'$ there is a unique continuous $h : \mathcal{X} \to \mathcal{X}$ such that $g = f \circ h$. Coreflection maps are discussed more fully in [**1**]; easy to verify is the following.

0.1 PROPOSITION. *(Proposition 3.3 of [**1**]) Coreflection maps with nonempty domains are continuous bijections.*

Clearly every map with empty domain is coreflective; however if $f : \langle X, \mathcal{T}\rangle \to \langle X', \mathcal{T}'\rangle$ is a coreflection map with X nonempty, we may view X and X' as the same set with $\mathcal{T}$ an enrichment of $\mathcal{T}'(\mathcal{T} \supseteq \mathcal{T}')$ satisfying the condition that any $\langle \mathcal{T}, \mathcal{T}'\rangle$-continuous map (i.e., a map that pulls $\mathcal{T}'$-open sets back to $\mathcal{T}$-open sets) is also a $\langle \mathcal{T}, \mathcal{T}\rangle$-continuous map. Under these circumstances we call $\mathcal{T}$ a *coreflective enrichment* of $\mathcal{T}'$. Examples of this phenomenon arise most naturally in connection with topological coreflective functors (e.g., discretization, k-modification, G_δ-modification, to name a few).

The only $A \in RCF$ that is simultaneously 0-free (i.e., (endomorphism-) rigid) and κ-free for some $\kappa > 0$ is clearly the degenerate ring $C(\emptyset)$ (in which case $\kappa = 1$). This said, we henceforth consider only the nondegenerate case.

0.2 THEOREM. *(Theorem 3.10 of [1]) $A \in RCF$ is κ-free if and only if there is a coreflective enrichment $\mathcal{T}$ of $\mathcal{U}^\kappa$ such that $\langle \mathbf{R}^\kappa, \mathcal{T} \rangle$ is a realcompact Tichonov space and $A \cong C(\langle \mathbf{R}^\kappa, \mathcal{T} \rangle)$. A pseudobasis for $C(\langle \mathbf{R}^\kappa, \mathcal{T} \rangle)$ is the set Π of projection maps from $\mathbf{R}^\kappa$ to $\mathbf{R}$.*

For any set X, let $\mathcal{D}(X)$ be the discrete space with point set X.

0.3 COROLLARY.

(i) *The only 0-free ring in RCF is $\mathbf{R} = C(\{point\})$.*
(ii) *If $\kappa > 0$, the cardinality of any κ-free ring in RCF lies between $c \cdot \kappa^\omega$ and $\exp^2(\omega \cdot \kappa)$.*
(iii) *If $\kappa > 0, C(\langle \mathbf{R}^\kappa, \mathcal{U}^\kappa \rangle)$ is κ-free; and if in addition $\kappa <$ the first measurable cardinal, then $C(\mathcal{D}(\mathbf{R}^\kappa))$ is κ-free as well.*
(iv) *No minimally free ring in RCF is generated by any pseudobasis. (Thus no member of RCF is relatively free.)*

PROOF: Clauses (i)-(iii) are treated in [**1**]. As for clause (iv), assume $A \in RCF$ is minimally free (and nondegenerate), and let $X \subseteq A$ be a pseudobasis of cardinality κ. Let $\mathcal{T}$ be a topology on $\mathbf{R}^\kappa$ prescribed by 0.2, with $f : A \to C(\langle \mathbf{R}^\kappa, \mathcal{T} \rangle)$ an isomorphism. The set Π of projection maps is a pseudobasis for $C(\langle \mathbf{R}^\kappa, \mathcal{T} \rangle)$, of cardinality κ; and, given a bijection g between $f[X]$ and Π, there is a unique automorphism on $C(\langle \mathbf{R}^\kappa, \mathcal{T} \rangle)$ extending g. Thus there is an isomorphism from A to $C(\langle \mathbf{R}^\kappa, \mathcal{T} \rangle)$ taking X to Π. Clearly Π does not generate $C(\langle \mathbf{R}^\kappa, \mathcal{T} \rangle)$; consequently X does not generate A. ∎

The classification theorem, which we prove in the next section, is the following.

0.4 THEOREM. *Let $1 \leq \kappa \leq \omega$. Then the only κ-free rings in RCF are $C(\langle \mathbf{R}^\kappa, \mathcal{U}^\kappa \rangle)$ and $C(\mathcal{D}(\mathbf{R}))$.*

1. Proof of the Classification Theorem

In view of 0.2, all we need show is that, for $1 \leq \kappa \leq \omega$, the κ-fold Tichonov power $\mathcal{U}^\kappa$ of the usual topology $\mathcal{U}$ on $\mathbf{R}$ has no proper nondiscrete coreflective enrichments that are realcompact and Tichonov. As it turns out, we can do much better than this.

The notion "coreflective enrichment" has two successive generalizations, "C-enrichment" and "H-enrichment", which are defined as follows: Let $\mathcal{T}$ and $\mathcal{T}'$ be two topologies on a set X with $\mathcal{T} \supseteq \mathcal{T}'$. $\mathcal{T}$ is a *C-enrichment* (resp. *H-enrichment*) of $\mathcal{T}'$ if whenever $f : X \to X$ is a $\langle \mathcal{T}', \mathcal{T}' \rangle$-continuous map (resp. a $\langle \mathcal{T}', \mathcal{T}' \rangle$-homeomorphism), f is also a $\langle \mathcal{T}, \mathcal{T} \rangle$-continuous map (resp. a $\langle \mathcal{T}, \mathcal{T} \rangle$-homeomorphism). The study of H-enrichments is interesting in itself, as part of the theory of bitopological spaces, and is taken up in another paper [4]. What we need in order to finish the proof of 0.4 is the following topological result.

1.1 THEOREM. *Let X be a normed linear space over the real field. Then any proper C-enrichment of the norm-topology is discrete.*

PROOF: Let $X = \langle X, +, 0, \|\cdot\| \rangle$ be a normed linear space, and denote the norm-topology by $\mathcal{E}$. Let $\mathcal{T}$ be any proper C-enrichment of $\mathcal{E}$. Because $\langle X, \mathcal{E} \rangle$ is point-homogeneous and $\mathcal{T}$ is an H-enrichment of $\mathcal{E}$, $\langle X, \mathcal{T} \rangle$ is also point-homogeneous. Thus it suffices to prove that 0 is a $\mathcal{T}$-isolated point. Again using homogeneity and the fact that $\mathcal{T} \neq \mathcal{E}$, there exists a sequence $\langle x_n \rangle = \langle x_0, x_1, \cdots \rangle$ of distinct points of X such that $\langle x_n \rangle$ $\mathcal{E}$-converges to 0 and the set $S = \{x_n : n < \omega\}$ is $\mathcal{T}$-closed in X. Let $U_0 = X$; and, for $n > 0$, let $U_n = \{x : \|x\| < 1/n\}$ be the open $\frac{1}{n}$-ball about 0. For $n = 1, 2, \cdots$ let A_n be the "annulus" $U_{n-1} \backslash U_n$. Then $X = (\bigcup_{n=1}^{\infty} A_n) \cup \{0\}$. For each $x \neq 0$, let R_x be the ray $\{tx : t \geq 0\}$. For $n = 1, 2, \cdots$, define $\ell_n : A_n \to (0,1), m_n : A_n \to (0,1]$, and, for $n \geq 2, M_n : A_n \to (1,2]$ to be $\ell_n(x) = 1/((n+1)\|x\|), m_n(x) = 1/(n\|x\|)$, and $M_n(x) = 1/((n-1)\|x\|)$. Then $R_x \bigcap A_n = \{tx : m_n(x) \leq t < M_n(x)\}, n \geq 2$, and for all $n, R_x \cap A_{n+1} = \{tx : \ell_n(x) \leq t < m_n(x)\}$. The functions ℓ_n, m_n and M_n are clearly continuous where A_n has the inherited norm-topology.

Define $f : X \to X$ and $g : X \to X$ as follows:

$$f(x) = \begin{cases} \frac{1-m_{2n}(x)}{M_{2n}(x)-m_{2n}(x)} x_n + \frac{M_{2n}(x)-1}{M_{2n}(x)-m_{2n}(x)} x_{n+1}, & \text{if } x \in A_{2n} \\ x_n, & \text{if } x \in A_{2n-1} \\ 0, & \text{if } x = 0 \end{cases}$$

$$g(x) = \begin{cases} \left(\frac{1-m_n(x)}{M_n(x)-m_n(x)} m_n(x) + \frac{M_n(x)-1}{M_n(x)-m_n(x)} \ell_n(x) \right) x, & \text{if } x \in A_n, n \geq 2 \\ \ell_1(x)x, & \text{if } x \in A_1 \\ 0, & \text{if } x = 0 \end{cases}$$

Now f and g are $\mathcal{E}$-continuous, hence $\mathcal{T}$-continuous. Since $f^{-1}[S] = \bigcup_{n=1}^{\infty} A_{2n-1}$, we know $\bigcup_{n=1}^{\infty} A_{2n-1}$ is $\mathcal{T}$-closed. But $g^{-1}\left[\bigcup_{n=1}^{\infty} A_{2n-1}\right] = \bigcup_{n=1}^{\infty} g^{-1}\left[A_{2n-1}\right] = \bigcup_{n=1}^{\infty} A_{2n}$, so $\bigcup_{n=1}^{\infty} A_{2n}$ is also $\mathcal{T}$-closed. Therefore $\bigcup_{n=1}^{\infty} A_n$ is $\mathcal{T}$-closed, hence $\{0\}$ is $\mathcal{T}$-open. ∎

1.2 COROLLARY. *Let $1 \leq \kappa \lesssim \omega$. Then the only proper C-enrichment of $\mathcal{U}^{\kappa}$ is discrete.*

PROOF: For finite κ, we can apply 1.1 directly: the usual euclidean norm gives rise to the euclidean topology. In the infinite case, apply 1.1 to the Banach space ℓ^2 of square-summable real sequences, with the obvious norm.

Then use the celebrated result of Anderson [**12**] that ℓ^2 and $\langle \mathbf{R}^\omega, \mathcal{U}^\omega \rangle$ are homeomorphic. (The topological vector space $\langle \mathbf{R}^\omega, \mathcal{U}^\omega \rangle$ is well known not to be normable.) ∎

1.3 REMARK. *1.1 actually holds for any metrizable locally convex topological vector space. The proof is similar to the above, but less concrete (see [4]). Thus Anderson's theorem can be avoided in 1.2.*

2. Rings in RCF that are κ-free for uncountable κ.

When we pass from countable κ to uncountable κ, the problem of classifying the κ-free rings in RCF seems to become impossibly difficult. By 0.3(iii) there always exists a κ-free ring in RCF, namely $C(\langle \mathbf{R}^\kappa, \mathcal{U}^\kappa \rangle)$. Moreover, this ring has the attractive property of being *uniquely* κ-free; i.e., it is not λ-free for any $\lambda \neq \kappa$ (Theorem 3.18 of [**1**]). (Note that $C(\mathcal{D}(\mathbf{R}))$ is κ-free for all $1 \leq \kappa \leq \omega$.) Recall that κ is *Ulam-measurable* if there exists a countably complete nonprincipal ultrafilter on a set of cardinality κ. Let μ be the smallest Ulam-measurable cardinal (should one exist). Then κ is Ulam-measurable if and only if $\kappa \geq \mu$. When $\kappa < \mu$ we also know from 0.3 (iii) that $C(\mathcal{D}(\mathbf{R}^\kappa))$ is κ-free. (Whether it is uniquely κ-free depends on obvious cardinality issues.) However, for $\kappa \geq \mu, C(\mathcal{D}(\mathbf{R}^\kappa))$ is not minimally free (Theorem 3.12 in [**1**]).

By work of Comfort-Retta [**5**] and Williams [**13**], one can always find at least three κ-free rings in RCF for uncountable κ. In order to state the result of theirs that we need, we adopt the following notation. For any topology $\mathcal{T}$ on a set X, and any $\lambda \geq \omega_1$, let $(\mathcal{T})_\lambda$ be the topology basically generated by intersections of fewer than λ open sets from $\mathcal{T}$. Clearly $(\mathcal{T})_\lambda$ is a coreflective enrichment of $\mathcal{T}$.

2.1 THEOREM. *(i) (Comfort-Retta) Let $\mathcal{T}$ be a realcompact Tichonov topology on a set X, with $\mathcal{T}'$ any Tichonov topology satisfying $\mathcal{T} \subseteq \mathcal{T}' \subseteq (\mathcal{T})_{\omega_1}$. Then $\mathcal{T}'$ is also realcompact.*

(ii) (Comfort-Retta) Let $\mathcal{T}$ be a realcompact Tichonov topology. Then $(\mathcal{T})_\mu$ is also realcompact Tichonov.

(iii) (Williams) Let $\mathcal{T}$ be a realcompact Tichonov topology, $\alpha < \mu$ a cardinal, and $\mathcal{T}'$ any Tichonov topology satisfying $\mathcal{T} \subseteq \mathcal{T}' \subseteq (\mathcal{T})_{\alpha^+}$ and $\mathcal{T}' = (\mathcal{T}')_{cf(\alpha)}$ (where α^+ is the cardinal successor of α and $cf(\alpha)$ is the cofinality of α). Then $\mathcal{T}'$ is also realcompact.

With the aid of 2.1, we immediately have the following.

2.2 THEOREM. *(Theorem 3.15 of [**1**]) Let κ be a cardinal, and α any cardinal $\leq \mu$. Then $C(\langle \mathbf{R}^\kappa, (\mathcal{U}^\kappa)_\alpha \rangle)$ is κ-free.*

2.3 COROLLARY. *Let κ be uncountable. Then $C(\langle \mathbf{R}^\kappa, \mathcal{U}^\kappa \rangle)$, $C(\langle \mathbf{R}^\kappa, (\mathcal{U}^\kappa)_{\omega_1} \rangle)$, and $C(\langle \mathbf{R}^\kappa, \mathcal{U}^\kappa)_\mu \rangle)$ are three (isomorphically) distinct κ-free rings in RCF.*

Regarding the question of which κ-free rings in RCF are uniquely κ-free, other than $C(\langle \mathbf{R}^\kappa, \mathcal{U}^\kappa \rangle)$, we have the following result.

2.4 THEOREM. *(i) Assume κ is a regular cardinal that is not Ulam-measurable, and assume further that either: (a) $\kappa^+ = \exp(\kappa)$; or (b) $\kappa = \sup\{\exp(\alpha) : \alpha < \kappa\}$. Then $C(\langle \mathbf{R}^\kappa, (\mathcal{U}^\kappa)_\kappa \rangle)$ is uniquely κ-free.*

(ii) $C(\langle \mathbf{R}^\mu, (\mathcal{U}^\mu)_\mu \rangle)$ is uniquely μ-free.

PROOF: (i) Note that $\kappa = \omega$ is the only cardinal satisfying the hypotheses absolutely; that case being covered already. (The case $\kappa = \omega_1$ relies on either $\omega_2 = \exp(\omega_1)$ or $\omega_1 = c = \exp(\omega)$ holding.) So assume κ is uncountable and let $\mathcal{X} = \langle \mathbf{R}^\kappa, (\mathcal{U}^\kappa)_\kappa \rangle$. In Williams' theorem 2.1(iii), set $\mathcal{T} = \mathcal{U}^\kappa$ and $\mathcal{T}' = (\mathcal{U}^\kappa)_\kappa$. Then we may conclude that $(\mathcal{U}^\kappa)_\kappa$ is a realcompact Tichonov coreflective enrichment of $\mathcal{U}^\kappa$; whence $C(\mathcal{X})$ is κ-free. Suppose $C(\mathcal{X})$ is λ-free, and let $f : \mathcal{X} \to \langle \mathbf{R}^\lambda, \mathcal{U}^\lambda \rangle$ be a coreflective map. f is a continuous bijection by 0.1, and the weight $w(\langle \mathbf{R}^\lambda, \mathcal{U}^\lambda \rangle)$ of $\langle \mathbf{R}^\lambda, \mathcal{U}^\lambda \rangle$ is $\lambda \cdot \omega$. Since κ is uncountable regular, $(\mathcal{U}^\kappa)_\kappa$ is nondiscrete; and the intersection of $\lambda \cdot \omega$ open sets in $\mathcal{X}$ cannot be a singleton if $\lambda < \kappa$. Thus $\lambda \geq \kappa$. Suppose (a) holds. Then $\kappa^+ = \exp(\kappa) = \exp(\lambda) > \lambda$, so $\lambda = \kappa$. Now suppose (b) holds. It is easy to construct a discrete subset D of $\langle \mathbf{R}^\lambda, \mathcal{U}^\lambda \rangle$ of cardinality λ; thus $f^{-1}[D]$ is a discrete subset of $\mathcal{X}$, also of cardinality λ. This forces $w(\mathcal{X}) \geq \lambda$. On the other hand, we have $w(\mathcal{X}) \leq$ the number of intersections of $< \kappa$ open subsets of $\langle \mathbf{R}^\kappa, \mathcal{U}^\kappa \rangle \leq$ the number of intersections of $< \kappa$ open sets from $\mathcal{B}$, where $\mathcal{B}$ is an open basis of $\langle \mathbf{R}^\kappa, \mathcal{U}^\kappa \rangle$ of cardinality κ. This number is in turn $\leq \sup\{\exp(\alpha) : \alpha < \kappa\}$, since κ is a regular cardinal. In short $\lambda \leq w(\mathcal{X}) \leq \kappa$; whence $\kappa = \lambda$.

To prove (ii), we use the Comfort-Retta theorem 2.1(ii) to infer that $(\mathcal{U}^\mu)_\mu$ is realcompact. Thus $C(\langle \mathbf{R}^\mu, (\mathcal{U}^\mu)_\mu \rangle)$ is μ-free. Since measurable cardinals are strongly inaccessible, condition (b) holds, and we can argue as above. ∎

Theorem 1.1 can also be put to use in helping us get a better idea of what kinds of topologies $\mathcal{T} \supseteq \mathcal{U}^\kappa$ make it possible for $C(\langle \mathbf{R}^\kappa, \mathcal{T} \rangle)$ to be κ-free, even when κ is uncountable. For each $J \subseteq \kappa$, let $A_J \subseteq \mathbf{R}^\kappa$ be the "flat" $\{x \in \mathbf{R}^\kappa : x_\xi = 0 \text{ for } \xi \notin J\}$. Let $\pi_J : \mathbf{R}^\kappa \to \mathbf{R}^\kappa$ be the projection map onto A_J. For any topology $\mathcal{T}$ on $\mathbf{R}^\kappa$, let $\mathcal{T}|J$ be the restriction of $\mathcal{T}$ to A_J. The following small collection of facts is easy to prove.

2.5 PROPOSITION. *(i) If $\mathcal{T}$ is an H-enrichment (resp. C-enrichment, coreflective enrichment) of $\mathcal{U}^\kappa$, then $\mathcal{T}|J$ is an H-enrichment (resp. C-enrichment, coreflective enrichment) of $\mathcal{U}^\kappa|J$.*

(ii) If $\mathcal{T}$ is an H-enrichment of $\mathcal{U}^\kappa$, then all straight lines in $\mathbf{R}^\kappa$ (viewed as a vector space) are equivalent via $\langle \mathcal{T}, \mathcal{T} \rangle$-homeomorphisms.

(iii) If $\mathcal{T}$ is a C-enrichment of $\mathcal{U}^\kappa$, then each map π_J is $\langle\mathcal{T},\mathcal{T}\rangle$-continuous. Moreover, if $U \in \mathcal{T}|\xi(= \mathcal{T}|\{\xi\})$, then $\pi_\xi^{-1}[U] \in \mathcal{T}$; whence $\mathcal{T}$ is an enrichment of the product topology $\Pi_{\xi<\kappa}\mathcal{T}|\xi$.

2.6 THEOREM. *Let $\mathcal{T}$ be a C-enrichment of $\mathcal{U}^\kappa$. Then either: (a) $\mathcal{T}$ is a pathwise connected topology; or (b) $\mathcal{T} \supseteq (\mathcal{U}^\kappa)_{\omega_1}$.*

PROOF: **First Proof.** Let $\xi < \kappa$. By 2.5 (i), $\mathcal{T}|\xi$ is a C-enrichment of $\mathcal{U}^\kappa|\xi$ on the ξ-axis A_ξ. ($\langle A_\xi,\mathcal{U}^\kappa|\xi\rangle$ is naturally homeomorphic to $\langle\mathbf{R},\mathcal{U}\rangle$.) By 2.5 (ii), all the subspaces $\langle A_\xi,\mathcal{T}|\xi\rangle$ are homeomorphic; and by 1.1 they are either all euclidean lines or all discrete. Using 2.5 (ii) again, we know that if all the axes are euclidean, then all straight lines in $\langle\mathbf{R}^\kappa,\mathcal{T}\rangle$ are euclidean; hence $\mathcal{T}$ is a pathwise connected topology.

Suppose all the axes are discrete. For each countable $J \subseteq \kappa$, then, $\mathcal{T}|J$ is a proper C-enrichment of $\mathcal{U}^\kappa|J$ on A_J; so, by 1.2 (which relies on Anderson's theorem: $\langle\mathbf{R}^\omega,\mathcal{U}^\omega\rangle \simeq \ell^2$), $\mathcal{T}|J$ is discrete. ($\langle A_J,\mathcal{U}^\kappa|J\rangle$ is naturally homeomorphic to $\langle\mathbf{R}^\omega,\mathcal{U}^\omega\rangle$.) Because π_J is $\langle\mathcal{T},\mathcal{T}\rangle$-continuous by 2.5 (iii), every set of the form $\Pi_{\xi<\kappa}U_\xi$, where U_ξ is a singleton if $\xi \in J$ and $U_\xi = \mathbf{R}$ if $\xi \notin J$, is $\mathcal{T}$-open. This implies $(\mathcal{U}^\kappa)_{\omega_1} \subseteq \mathcal{T}$. ∎

Second Proof. Repeat the argument in the first paragraph of the proof above. We alter the second paragraph so as not to rely on Anderson's theorem (in 1.2).

Suppose all the axes are discrete, and let $J = \{\xi_n : n < \omega\}$, where $\xi_0 < \xi_1 < \cdots$. Define $f : \mathbf{R}^\kappa \to \mathbf{R}^\kappa$ to be the function taking $\langle x_\xi\rangle$ to $\langle \sum_{n<\omega} \frac{1}{2^n}\frac{|x_{\xi_n}|}{1+|x_{\xi_n}|}, 0, 0, \cdots\rangle \in A_0$. Then f is $\langle\mathcal{U}^\kappa,\mathcal{U}^\kappa\rangle$-continuous, hence $\langle\mathcal{T},\mathcal{T}\rangle$-continuous. Moreover, if $U \in \mathcal{T}$ is such that $U \cap A_0 = \{\langle 0,0,\cdots\rangle\}$, then $f^{-1}[U] = \Pi_{\xi<\kappa}U_\xi$, where $U_\xi = \{0\}$ if $\xi \in J$ and $U_\xi = \mathbf{R}$ if $\xi \notin J$. By homogeneity, this implies $(\mathcal{U}^\kappa)_{\omega_1} \subseteq \mathcal{T}$. ∎

2.7 REMARKS. *(i) In 2.6, conclusion (a) implies that $C(\langle\mathbf{R}^\kappa,\mathcal{T}\rangle)$ is a connected ring; i.e., there are no nontrivial idempotents. When κ is countable, there is the much stronger conclusion that $\mathcal{T} = \mathcal{U}^\kappa$. However, for uncountable κ, there is at least one other topology that could possibly work, namely the k-modification of $\mathcal{U}^\kappa$. Recall that a space $\langle X,\mathcal{T}\rangle$ is a k-space if for any $C \subseteq X$, C is $\mathcal{T}$-closed if and only if $C \cap K$ is closed in K for every compact subspace K of $\langle X,\mathcal{T}\rangle$. The k-modification $k(\mathcal{T})$ of $\mathcal{T}$ is that topology on X, whose closed sets are precisely those sets whose intersections with the compact subspaces of $\langle X,\mathcal{T}\rangle$ are closed in those subspaces (see [**12**]). It is well known that $k(\mathcal{T})$ is always a coreflective enrichment of $\mathcal{T}$, and that $\langle\mathbf{R}^\kappa,\mathcal{U}^\kappa\rangle$ is not a k-space for $\kappa \geq \omega_1$. Also, since the compact subspaces of $\langle X,k(\mathcal{T})\rangle$ are exactly those of $\langle X,\mathcal{T}\rangle$. the modified topology is pathwise connected whenever the old topology is. Thus $k(\mathcal{U}^\kappa)$ is a proper C-enrichment of $\mathcal{U}^\kappa$ for which conclusion (a) obtains. Unfortunately we do not know whether this* *topology is realcompact, and hence do not know whether $C(\langle \mathbf{R}^\kappa, k(\mathcal{U}^\kappa)\rangle)$ is κ-free.*

*(ii) If conclusion (b) holds in 2.6, then $\langle \mathbf{R}^\kappa, \mathcal{T}\rangle$ is a space with the following properties. (1) The space is totally disconnected; (2) the only compact subspaces are the finite ones; (3) the only convergent sequences are eventually constant; (4) the only first countable subspaces are discrete; and (5) all countable subsets are closed. We do not know whether $\mathcal{T}$ is itself closed under countable intersections (i.e., a P-space topology). If it were, and Tichonov as well, we could conclude that $C(\langle \mathbf{R}^\kappa, \mathcal{T}\rangle)$ is a (von Neumann) regular ring (see Exercise 4J in [**7**]).*

REFERENCES

1. P. Bankston and R. Schutt, *On minimally free algebras*, Can. J. Math. **37** (1985), 963-978.
2. P. Bankston, *A note on large minimally free algebras*, Algebra Universalis (to appear).
3. P. Bankston, *Characterizing Minimally Free Rings of Continuous Functions*, Marquette University Technical Report #278 (February, 1988).
4. P. Bankston and R. A. McCoy, *H-enrichments of topologies*, (in preparation).
5. W. W. Comfort and T. Retta, *Generalized perfect maps and a theorem of I. Juhász*, in "Rings of Continuous Functions" (Proc. A.M.S. Special Session, Cincinnati, Ohio, 1982, C. E. Aull, ed.) Marcel Dekker, New York, 1985, pp. 79-102..
6. P. Eklof, *Infinitary equivalence of abelian groups*, Fund. Math. **81** (1974), 305-314.
7. L. Gillman and M. Jerison, "Rings of Continuous Functions," Van Nostrand, Princeton, 1960.
8. K. Głazek, *Some old and new problems in the independence theory*, Coll. Math. **42** (1979), 127-189.
9. P. Hall, *The splitting properties of relatively free groups*, Proc. London Math. Soc. **3** (1954), 343-356.
10. I. Kříž and A. Pultr, *Large k-free algebras*, Algebra Universalis **21** (1985), 46-53.
11. H. Neumann, "Varieties of Groups," Ergebnisse der Math. und ihrer Grenzgebiete, Band 37, Springer-Verlag, Berlin-New York, 1967.
12. S. Willard, "General Topology," Addison Wesley, Reading, MA., 1970.
13. S. W. Williams, *More realcompact spaces*, in "Rings of Continuous Functions" (Proc. A.M.S. Special Session, Cincinnati, Ohio, 1982; C. E. Aull, ed.) Marcel Dekker,New York, 1985, pp. 289-300.

Department of Mathematics, Statistics and Computer Science, Milwaukee, WI 53233 (first author)
Department of Mathematics, Blacksburg, VA 24061 (second author)

Sums of Ultrafilters and the Rudin-Keisler and Rudin-Frolik Orders

Andreas Blass*, Department of Mathematics, University of Michigan, Ann Arbor, MI 48109, USA

Neil Hindman*, Department of Mathematics, Howard University, Washington, DC 20059, USA

I. INTRODUCTION

We take the Stone-Čech compactification βN of the set N of positive integers to be the set of ultrafilters on N, the points of N being identified with the principal ultrafilters. Given p and q in βN, define as usual their sum p + q by agreeing that a subset A of N is in p + q if and only if $\{x \in N: A - x \in p\} \in q$. (Here $A - x = \{y \in N: y + x \in A\}$.)

An alternative characterization due to van Douwen [5] utilizes the notion "q-lim". In any Hausdorff space X, given a sequence $\langle x_n \rangle_{n \in N}$ in X and $q \in \beta N$, one has $q\text{-}\lim_{n \in N} x_n = z$ if and only if $\{n \in N: x_n \in U\} \in q$ whenever U is a neighborhood of z. Observe that z is a limit (or cluster) point of the sequence $\langle x_n \rangle_{n \in N}$ if and only if there is some $q \in \beta N \backslash N$ with

---

The authors gratefully acknowledge support received from the NSF via grants DMS-8501752 and DMR-8801988 (for the first author) and DMS 8520873 (for the second author).

$q\text{-}\lim_{n\in N} x_n = z$. Using this notion one first defines for $p \in \beta N \setminus N$ and $m \in N$, $p + m = p\text{-}\lim_{n\in N} (n + m)$. One then defines for $p \in \beta N$ and $q + \beta N \setminus N$, $p + q = q\text{-}\lim_{n\in N} (p + n)$.

It is clear that members of p + q look like members of p. Indeed, every member of p + q translates to a member of p.

Two measures of the complexity of the structures of ultrafilters are the Rudin-Keisler and the Rudin-Frolik orders of βN. We denote these orders by $\leq$ and $\preceq$ respectively and define them as follows. (See [4] for considerable information about these orders including the fact that $p \preceq q$ implies $p \leq q$.) Given p and q in βN, we say $p \leq q$ if and only if there is a function $f\colon N \to N$ such that $f^\beta(q) = p$. (Here $f^\beta\colon \beta N \to \beta N$ is the unique continuous extension of f.) We say $p \preceq q$ provided there is a one-to-one function $g\colon N \to \beta N$ with $\{g(n)\colon n \in N\}$ discrete and $g^\beta(p) = q$. We say $p < q$ (respectively $p \prec q$) when $p \leq q$ (respectively $p \preceq q$) and p and q are not type equivalent. (That is, there is no permutation π of N with $\pi^\beta(p) = q$.)

Since larger members with respect to either of these orders have more structure it is not surprising, in view of the discussion above, that often one has $p \leq p + q$. It is in fact more surprising that often this inequality fails. What seems more interesting is that we often get $q \prec p + q$.

We establish in Section 2 a long list of equivalent conditions on p which guarantee that all q satisfy $q \prec p + q$. We also determine certain concrete situations in which these equivalent conditions do and do not hold. In this section we utilize the results about sums to show that the minimal ideal of $(\beta N,+)$ is cofinal upward in the Rudin-Keisler order.

In Section 3 we show that it is consistent that one always has $q \prec p + q$ but p does not satisfy the sufficiency conditions from Section 2.

For $A \subseteq N$ we write $\bar{A} = \{p \in \beta N\colon A \in p\}$. It is a fact that $\bar{A} = clA$. Of more interest is that $\{\bar{A}\colon A \subseteq N\}$ forms a basis for the open sets of βN. We write $N^* = \beta N \setminus N$. Given $p \in \beta N$, the function $\lambda_p\colon \beta N \to \beta N$ is defined by $\lambda_p(q) = p + q$ and is always continuous.

II. DISCRETE SETS AND THE RUDIN-KEISLER AND RUDIN-FROLIK ORDERS.

The simplest of our list of equivalent conditions is the discreteness of $\{p + n: n \in N\}$. We recall that if $\langle a_n \rangle_{n \in N}$ is a one-to-one sequence in βN and $\{a_n: n \in N\}$ is discrete, then it is strongly discrete. That is, there exist $A_n \in a_n$ such that $A_n \cap A_m = \emptyset$ whenever $n \neq m$. (The proof of this assertion is an easy exercise.)

2.1 <u>THEOREM</u>. Let $p \in N^*$ and let D be an infinite subset of N. Statements (a) through (h) are equivalent and imply (i) which implies (j).

(a) $\{p + n: n \in D\}$ is discrete.

(b) For each $q \in \bar{D}\backslash N$, $p < p + q$.

(c) For each $q \in \bar{D}\backslash N$, $p \approx p + q$.

(d) For each $a \in D$, $p \notin p + \overline{D - a}$.

(e) The restriction of λ_p to $\bar{D}$ is one-to-one.

(f) There is a one-to-one function $f: N \to \beta N$ such that $\{f(n): n \in N\}$ is discrete and for all $q \in \bar{D}$, $f^{\beta}(q) = p + q$.

(g) There is a function $g: N \to N$ such that for all $q \in \bar{D}$, $g^{\beta}(p + q) = q$.

(h) There is a function $h: N \to N$ such that for all $q \in \bar{D}$, $h^{\beta}(p + q) = p$.

(i) For all $q \in \bar{D}$, $q \lessdot p + q$.

(j) For all $q \in \bar{D}$, $q < p + q$.

Proof. We establish the following pattern of implications.

$$\begin{array}{ccccccccccc}
(b) & \Longleftarrow & (a) & \Longleftarrow & (h) & \Longleftarrow & (g) & & & & \\
\Downarrow & & \Uparrow & \searrow & & & \Uparrow & & & & \\
(c) & \Longrightarrow & (d) & & (e) & \Longrightarrow & (f) & \Longrightarrow & (i) & \Longrightarrow & (j)
\end{array}$$

(a) $\Longrightarrow$ (b). Observe that $p + n \neq p + m$ whenever $n \neq m$. (The congruence class mod $(m + n)$ which is in $p + n$ can't be in $p + m$.) For each $n \in D$ pick $B_n \in p + n$ such that $B_n \cap B_m = \emptyset$ whenever $n \neq m$. Define $h: N \to N$ by $h(x) = x - n$ if $x \in B_n$ and $x > n$ and $h(x) = 1$ (or whatever one feels like) otherwise. Let $q \in \overline{D}\backslash N$. We show first that $h^\beta(p + q) = p$. Let $A \in p$. We show $h^{-1}[A] \in p + q$ by showing $D \subseteq \{n: h^{-1}[A] - n \in p\}$. Let $n \in D$. Then $B_n \in p + n$ so $B_n - n \in p$. We show $(B_n - n) \cap A \subseteq h^{-1}[A] - n$. Let $x \in (B_n - n) \cap A$. Then $x + n \in B_n$ and $x + n > n$ so $h(x + n) = x \in A$ so $x + n \in h^{-1}[A]$ as required. Thus $p \leq p + q$.

Suppose $p \approx p + q$. Then $h^\beta(p + q) \approx p + q$ so by [4, Theorem 9.2] there is a member C of $p + q$ such that $h|_C$ is one-to-one. Then $D \cap \{n \in N: C - n \in p\} \in q$ and is hence infinite. Pick $n \neq m$ in D such that $C - n \in p$ and $C - m \in p$. Pick $a \in (C - n) \cap (B_n - n) \cap (C - m) \cap (B_m - m)$. Now $a + n \in B_n$ and $a + n > n$ so $h(a + n) = a$. Similarly $h(a + m) = a$. But $a + n \in C$ and $a + m \in C$ so $a + n = a + m$, a contradiction.

That (b) $\Rightarrow$ (c) is trivial.

(c) $\Rightarrow$ (d). Let $a \in D$ and suppose we have $q \in \overline{D - a}$ with $p = p + q$. Since $p = p + q$, $q \notin N$ so $q + a \notin N$. Thus $q + a \in \overline{D}\backslash N$. Thus $p \approx p + (q + a)$. But trivially $p \approx p + a$ so $p \approx p + a = p + q + a$, a contradiction.

(d) $\Rightarrow$ (a). Let $a \in D$. We need to show $p + a \notin cl\{p + b: b \in D\backslash\{a\}\} = cl(\lambda_p[D\backslash\{a\}]) = \lambda_p[cl(D\backslash\{a\})] = p + \overline{D\backslash\{a\}}$. Suppose instead we have $q \in \overline{D\backslash\{a\}}$ with $p + a = p + q$. Since, as already observed, $p + b \neq p + a$ for any $b \in N\backslash\{a\}$, we must have $q \in \overline{D}\backslash N$. Consequently $q - a = \{A - a: A \in q\}$ is a non-principal ultrafilter and $p = p + (q - a)$, a contradiction since $q - a \in \overline{D - a}$.

(a) $\Rightarrow$ (e). This is an immediate consequence of the observation of Rudin [14, p. 354] that if f is a one-to-one function from a countable discrete space D to βN with discrete range, then $f^\beta: \beta D \to \beta N$ is one-to-one.

(e) $\Rightarrow$ (f). We first note that we can pick a one-to-one sequence $\langle x_m \rangle_{m=1}^{\infty}$ in N such that for each $n \in N$, $\{x_m - n: m \in N\backslash D\} \notin p$. In fact either $x_m = 2^{2m}$ or $x_m = 2^{2m+1}$ will do the job.

Now define $f: N \to \beta N$ by $f(n) = p + n$ if $n \in D$ and $f(n) = x_n$ if $n \in N\backslash D$. Then f is one-to-one and, since f and λ_p agree on D, $f^{\beta}(q) = \lambda_p(q) = p + q$ whenever $q \in \bar{D}$. If $n \in N\backslash D$, then $\{x_n\}$ is a neighborhood of f(n) missing $f[N\backslash\{n\}]$. Finally suppose we have $n \in D$ with $f(n) \in cl f[N\backslash\{n\}] = cl f[N\backslash D] \cup cl f[D\backslash\{n\}]$. Now $f(n) \notin cl f[N\backslash D] = cl\{x_m: m \in N\backslash D\}$ by the choice of the x_m's. Thus $f(n) \in cl f[D\backslash\{n\}]$ so there is some $q \in \beta(D\backslash\{n\}) = \overline{D\backslash\{n\}}$ with $f(n) = q\text{-}\lim_{m \in D\backslash\{n\}} f(m)$. But this says $p + n = q\text{-}\lim_{m \in D\backslash\{n\}} (p + m) = p + q$ so that λ_p is not one-to-one on $\bar{D}$.

(f) $\Rightarrow$ (g). For each $n \in N$ pick $B_n \in f(n)$ such that $B_n \cap B_m \neq \emptyset$ whenever $n \neq m$. Define $g[B_n] = \{n\}$ (and define g at will off $\cup_{n \in N} B_n$). Given $n \in N$, $g^{\beta}(f(n)) = n$ so $g^{\beta} \circ f$ is the identity on N so $(g^{\beta} \circ f)^{\beta}$ is the identity on βN. Since $g^{\beta} \circ f^{\beta}$ is a continuous extension of $g^{\beta} \circ f$, we have $g^{\beta} \circ f^{\beta} = (g^{\beta} \circ f)^{\beta}$. Thus for all $q \in \beta N$, $g^{\beta}(f^{\beta}(q)) = q$. If $q \in \bar{D}$ we then have $q = g^{\beta}(f^{\beta}(q)) = g^{\beta}(p + q)$.

(g) $\Rightarrow$ (h). Given $x \in N$, if $x > g(x)$ define $h(x) = x - g(x)$ and define h(x) at will otherwise. Let $q \in \bar{D}$. To see that $h^{\beta}(p + q) = p$, let $A \in p$. We claim $D \subseteq \{n: h^{-1}[A] - n \in p\}$ so that $h^{-1}[A] \in p + q$. To prove the claim let $n \in D$. Then $g^{\beta}(p + n) = n$ so $g^{-1}[\{n\}] \in p + n$ so $g^{-1}[\{n\}] - n \in p$. We claim $A \cap (g^{-1}[\{n\}] - n) \subseteq h^{-1}[A] - n$ so that $h^{-1}[A] - n \in p$. Let $x \in A \cap (g^{-1}[\{n\}] - n)$. Then $g(x + n) = n$ so $x + n > g(x + n)$ so $h(x + n) = x \in A$. Thus $x \in h^{-1}[A] - n$ as required.

(h) $\Rightarrow$ (a). For each $n \in D$, let $B_n = \{x \in N: x = h(x) + n\}$. Then trivially $B_n \cap B_m = \emptyset$ for $n \neq m$. We claim each $B_n \in p + n$ so that $\{p + n: n \in D\}$ is discrete.

Let $n \in D$ and suppose $B_n \notin p + n$. Define $f: N \to N$ by $f(x) = h(x) + n$ if $x \notin B_n$ and $f(x) = x + 1$ if $x \in B_n$. Then for all $x \in N$, $f(x) \neq x$. Then by [4, Lemma 9.1], due to Katětov [12], there exists a partition $\{A_0, A_1, A_2\}$

of N so that $A_i \cap f[A_i] = \emptyset$ for all $i \in \{0,1,2\}$. Pick $i \in \{0,1,2\}$ such that $A_i \in p + n$. Then $A_i \backslash B_n \in p + n$ so, since $h^\beta(p + n) = p$, $h[A_i \backslash B_n] \in p$. Then $h[A_i \backslash B_n] + n \in p + n$ so pick $x \in A_i \cap (h[A_i \backslash B_n] + n)$ and pick $y \in A_i \backslash B_n$ such that $x = h(y) + n$. Since $y \notin B_n$, $x = f(y)$ so $x \in A_i \cap f[A_i]$, a contradiction.

(f) $\Rightarrow$ (i). We have immediately that for all $q \in \bar{D}$, $q \preceq p + q$. Let $q \in \bar{D}$ and suppose $q \approx p + q$. Pick for each $n \in N$, $B_n \in f(n)$ such that $B_n \cap B_m = \emptyset$ when $n \neq m$ and define $g: N \to N$ by $g[B_n] = \{n\}$ (and define g at will elsewhere). Then as shown in the proof of "(f) $\Rightarrow$ (g)" we have $g^\beta(p + q) = q$. Thus $g^\beta(p + q) \approx p + q$ so by [4, Lemma 9.2], pick $E \in p + q$ such that $g|_E$ is one-to-one. Pick $n \in D \cap \{x \in N: E - x \in p\}$. Then $E \in p + n$ and, since $n \in D$, $f(n) = p + n$ so $B_n \in p + n$. Since g is one-to-one on E and constant on B_n we have $|E \cap B_n| = 1$. But this contradicts the fact that $p + n$ is non-principal.

That (i) $\Rightarrow$ (j) is trivial. □

The concept of the tensor product of two ultrafilters, defined by $p \otimes q = \{A \subseteq N \times N: \{n \in N: \{m \in N: (n,m) \in A\} \in p\} \in q\}$, lies behind several of the equivalences in Theorem 2.1. In fact, as the referee pointed out, parts (a) through (h) are equivalent to the following statement relating sums to tensor products: For each $q \in \bar{D} \backslash N$, $p + q \approx q \otimes p$.

As we shall see in Section 3, it is at least consistent that (i) does not imply the other equivalent statements (and so neither does (j)). It is worth noting that if all $q \in \bar{D}$ have $q \preceq p + q$ <u>by means of the same function</u>, we have one of the equivalent statements, namely (f). Likewise, if all $q \in \bar{D}$ have $q \leq p + q$ by means of the same function we have statement (g), another of the equivalent statements.

We also remark that, if $D = N$, statement (d) takes the simpler form "$p \notin p + \beta N$".

We now note a simple situation in which all the statements in Theorem 2.1 hold.

2.2 THEOREM. Let $\langle x_n \rangle_{n=1}^{\infty}$ be a sequence in N such that $\lim_{n\to\infty} (x_{n+1} - x_n) = \infty$ and let $C = \{x_n: n \in N\}$. For all $p \in \overline{C}\backslash N$, $\{p + n: n \in N\}$ is discrete.

Proof. For each $n \in N$, $C + n \in p + n$. Also, since $\lim_{n\to\infty} (x_{n+1} - x_n) = \infty$ we have $(C + n) \cap (C + m)$ is finite whenever $n \neq m$. Given n, let $B_n = (C + n)\backslash(\cup_{m=1}^{n-1} C + m)$. Then $B_n \in p + n$ and $B_n \cap B_m = \emptyset$ for $n \neq m$. □

As the referee has pointed out, if $p \in N^*$ is a weak P-point, then $p + n$ is also a weak P-point for all $n \in N$ so that $\{p + n: n \in N\}$ is discrete.

The semigroup $(\beta N,+)$, like any compact left topological semigroup, has a smallest two sided ideal [15] which we will denote by **M**. Theorem 2.2 refers to ultrafilters p living on thin sets. On the other hand (see [9, Theorem 3.8]) if $p \in \mathbf{M}$, in fact if $p \in cl\mathbf{M}$, then every member of p is piecewise syndetic -- the opposite of thin. As a consequence, the following is perhaps not surprising.

2.3 THEOREM. Let $p \in \mathbf{M}$. Then there exist q and $r \in \beta N$ such that $q \nleq p + q$ and $r \nleq r + p$. In particular, $\{p + n: n \in N\}$ is not discrete.

Proof. **M** is the union of all of the minimal left ideals of βN and also is the union of all of the minimal right ideals of βN [15]. Distinct minimal left ideals are disjoint (trivially) as are distinct minimal right ideals. βN has 2^c minimal right ideals [3] and 2^c minimal left ideals [1]. (or see [11].) We will show that given any minimal left and right ideals L and R there exist $q \in R$ and $r \in L$ with $p + q = p$ and $r + p = p$. Consequently $|\{q \in \beta N: p + q = p\}| = |\{r \in \beta N: r + p = p\}| = 2^c$. Since $|\{q \in \beta N: q \leq p\}| \leq c$ one must have q and r with $q \nleq p = p + q$ and $r \nleq p = r + p$.

Now $p \in T$ where T is a minimal right ideal and $p + R \subseteq T + R \subseteq T$ while $(p + R) + \beta N \subseteq p + (R + \beta N) \subseteq p + R$. Thus $p + R$ is a right ideal contained in T so $p + R = T$. Since $p \in T$, there is some $q \in R$ with $p + q = p$. Likewise there is $r \in L$ with $r + p = p$. □

As we have already mentioned, if $p \in cl\mathbf{M}$, then every member is piecewise syndetic. That is, each member A has some k such that $\bigcup_{t=1}^{k} (A - t)$ has arbitrarily long blocks. However members p of $\mathbf{M}$ may have members with density zero. The following result shows that all members of p may be large in both of these senses, yet $\{p + n: n \in N\}$ may be discrete.

2.4 <u>THEOREM</u>. There exists $p \in cl\mathbf{M}$ such that every member of p has positive upper density and $\{p + n: n \in N\}$ is discrete.

Proof. Let $\Delta = \{p \in \beta N$: each $A \in p$ has positive upper density$\}$. By [5, Theorem 7.4] (or see [7]) Δ is a right ideal of $(\beta N,+)$ and of $(\beta N,\cdot)$. Since Δ is a right ideal of $(\beta N,+)$, $\Delta \cap \mathbf{M} \neq \emptyset$. By [9, Theorem 3.9], $cl\mathbf{M}$ is a right ideal of $(\beta N,\cdot)$ so $\Delta \cap cl\mathbf{M}$ is a right ideal of $(\beta N,\cdot)$ hence meets the minimal ideal K of $(\beta N,\cdot)$. Pick $p \in \Delta \cap cl\mathbf{M} \cap K$. Given any $n \in N$, $\overline{Nn}$ is an ideal of $(\beta N,\cdot)$ so $K \subseteq \overline{Nn}$. Since $p \in K$ there is some q with $p \cdot q = p$. (To see this let R be a minimal right ideal of $(\beta N,\cdot)$ with $p \in R$. Then $p \cdot R$ is a right ideal and $p \cdot R \subseteq R$ so $p \cdot R = R$.) Thus by [8, Theorem 5.3] $p \notin N^* + N^*$. In particular $p \notin p + N^*$ so statement (d) of Theorem 2.1 holds. □

We see next that $\mathbf{M}$ has members with arbitrarily complex structure.

2.5 <u>THEOREM</u>. Let L be a minimal left ideal of βN. Then L is cofinal upward in the Rudin-Keisler order.

Proof. Let $r \in N^*$ and let $f: N \to N$ be an increasing function with $\lim_{n\to\infty} (f(n + 1) - f(n)) = \infty$. Let $p = f^{\beta}(r)$. Then $p \approx r$ and, by Theorem 2.2, $\{p + n: n \in N\}$ is discrete. Pick $q \in L$. Then by Theorem 2.1(b), $p < p + q \in L$. □

As a consequence of the following theorem we have that no member of $\mathbf{M}$ is Rudin-Keisler minimal. ($N^* + N^*$ is an ideal so $\mathbf{M} \subseteq N^* + N^*$.)

2.6 THEOREM. Let $p,q \in N^*$. Then $p + q$ is not minimal in the Rudin-Keisler order.

Proof. We have $p + q = q\text{-}\lim_{n \in N} (p + n)$ while $p + q = p + n$ for at most one n. Thus $p + q$ is not a weak P-point hence not a P-point while Rudin-Keisler minimal ultrafilters are selective, hence P-points. (See [4, Theorem 9.6].) □

III. CONSISTENCY RESULTS.

We show here that it is consistent that statement (i) of Theorem 2.1 does not imply statements (a) through (h).

3.1 THEOREM. Each of the following statements implies the one following it.

(a) There exists $p \in N^*$ such that $|\{q \in N^*: p + q = p\}| = 1$.

(b) There exists $p \in N^*$ such that $\{p + n: n \in N\}$ is not discrete but for all $q \in \beta N$ $q \prec p + q$.

(c) There exists $p \in N^*$ such that $\{p + n: n \in N\}$ is not discrete but for all $q \in \beta N$, $q < p + q$.

(d) There exists $p \in N^*$ such that $0 < |\{q \in N^*: p + q = p\}| < \omega$.

Proof. (a) $\Rightarrow$ (b). Let $\{s\} = \{q \in N^*: p + q = p\}$. Then $p + (s + s) = (p + s) + s = p + s = p$ so $s + s = s$. Further

(*) If $s + t = s$, then $p + t = p + s + t = p + s = p$ so $t = s$.

Pick by Theorems 2.2 and 2.1, $r \in N^*$ such that for all $q \in \beta N$, $q \prec r + q$. Let $v = r + s$. We claim v is as required for (b). First $v + s = r + s + s = r + s = v$ so $v \in v + \beta N$ so, by Theorem 2.1, $\{v + n: n \in N\}$ is not discrete.

Now let $q \in \beta N$. If $q \in N$, the conclusion is trivial so assume $q \in N^*$. Assume first that $s + q = q$. Then $q \prec r + q = r + s + q = v + q$ as required. Thus we assume $s + q \neq q$. Pick $A \in q \backslash (s + q)$ and let

$D = A \cap \{x \in N: (N\backslash A) - x \in s\}$. Then $D \in q$ and for all $d \in D$, $A - d \notin s$ so $D - d \notin s$. That is for all $d \in D$, $s \notin \overline{D - d}$. We claim that for all $d \in D$, $s \notin s + (\overline{D - d})$. Indeed suppose we have $t \in \overline{D - d}$ with $s = s + t$. Then by (*) we have $s = t$ so $s \in \overline{D - d}$, a contradiction. Therefore, by Theorem 2.1 we have for all $t \in \overline{D}$, $t \prec s + t$. In particular, since $q \in \overline{D}$, we have $q \prec s + q \prec r + (s + q) = (r + s) + q = v + q$, as required.

That (b) $\Rightarrow$ (c) is trivial.

(c) $\Rightarrow$ (d). Let $T = \{q \in N^*: p + q = p\}$. By Theorem 2.1 $p \in p + \beta N$ so $T \neq \emptyset$. (Recall no $n \in N$ has $p + n = p$.) Now suppose T is infinite. Since $T = \lambda_p^{-1}[\{p\}]$, we have T is infinite and compact, hence contains a copy of βN. (This fact is attributed to Cech in [6].) But then $|T| = 2^c$ so there is some $q \in T$ with $q \nleq p = p + q$, a contradiction. □

We do not know whether (d) implies (a) in Theorem 3.1. We do however have the following (which is completely trivial if $k = 1$). (The term "left-zero" is multiplicative terminology.)

3.2 <u>THEOREM</u>. Let $k = \min(\{|\{q \in N^*: p + q = p\}|: p \in N^*\}\backslash\{0\})$ and assume $k < \omega$. Let $p \in N^*$ with $|\{q \in N^*: p + q = p\}| = k$. Then $D = \{q \in N^*: p + q = p\}$ is a left-zero semigroup. That is, for all $q,r \in D$ $q + r = q$.

Proof. Let $q,r \in D$ and suppose $q + r \neq q$. Now if $t \in N^*$ and $q + t = q$, then $p = p + q = p + (q + t) = (p + q) + t = p + t$, so $t \in D$. Since $q + r \neq q$ we have then $\{t \in N^*: q + t = q\} \subsetneq D$ so, by the minimality of k, $\{t \in N^*: q + t = q\} = \emptyset$. But then $q \notin q + \beta N$ so by Theorem 2.1 λ_q is one-to-one. But also $\lambda_q: D \to D\backslash\{q\}$ while D is finite, a contradiction. □

Statement (b) in Theorem 3.1 is precisely the assertion that in Theorem 2.1, (i) does not imply (a). We show that (b) is consistent by showing that statement (a) follows from the existence of a strongly summable ultrafilter. (A strongly summable ultrafilter is one with a basis of sets of the form FS(A), where $FS(A) = \{\Sigma F$: F is a finite non-empty subset of A.}.) It was shown in [10] that Martin's Axiom implies the existence of strongly summable ultrafilters and in [13] and, independently, in [2] that the existence of strongly summable ultrafilters implies the existence of P-points in N^*, hence cannot be shown in ZFC.

3.3 <u>THEOREM</u>. Assume $p \in N^*$ is a strongly summable ultrafilter. Then $|\{q \in N^*: p + q = p\}| = 1$.

Proof. By [10, Theorem 2.3(a)], $p + p = p$, so it suffices to let $q \neq p$ and show $p + q \neq p$. To this end let $q \neq p$ and suppose $p + q = p$. Pick $A \in p\backslash q$. By the proof of Lemma 1A of [2] pick a sequence $\langle y_n \rangle_{n=1}^{\infty}$ in N such that for each n, $y_{n+1} > 2\Sigma_{k=1}^{n} y_k$ and so that if $B = \{y_n: n \in N\}$ then $FS(B) \in p$ and $FS(B) \subseteq A$.

We now claim that $\{x \in N: FS(B) - x \in p\} \subseteq FS(B)$. (In fact, equality holds, but we don't care about that.) To this end, let $x \in N$ with $FS(B) - x \in p$. Then $(FS(B) - x) \cap FS(B) \in p$. Pick C with $FS(C) \in p$ and $FS(C) \subseteq (FS(B) - x) \cap FS(B)$. Pick $a,b \in C$ with $a \neq b$ and pick finite $F,G \subseteq N$ with $a = \Sigma_{n\in F} y_n$ and $b = \Sigma_{n\in G} y_n$. By [2, Lemma 1D], $F \cap G = \emptyset$. Now $x + a \in FS(B)$ and $x + b \in FS(B)$ so pick finite $H,K \subseteq N$ with $x + a = \Sigma_{n\in H} y_n$ and $x + b = \Sigma_{n\in K} y_n$. Then $x = \Sigma_{n\in H} y_n - \Sigma_{n\in F} y_n$ and $x = \Sigma_{n\in K} y_n - \Sigma_{n\in G} y_n$ so $\Sigma_{n\in H} y_n + \Sigma_{n\in G} y_n = \Sigma_{n\in K} y_n + \Sigma_{n\in F} y_n$. Thus $\Sigma_{n\in H\Delta G} y_n + 2\Sigma_{n\in H\cap G} y_n = \Sigma_{n\in K\Delta F} y_n + 2\Sigma_{n\in K\cap F} y_n$ so by [2, Lemma 1C] $H\Delta G = K\Delta F$ and $H \cap G = K \cap F$.

We now claim $F \subseteq H$. To see this, let $z \in F$. Now $F \cap G = \emptyset$ so $z \notin G$ so $z \notin H \cap G = K \cap F$ and since $z \in F$ we infer $z \notin K$. Thus $z \in K\Delta F$ so $z \in H\Delta G$ and hence $z \in H$. Since $F \subseteq H$, we have $x = \Sigma_{n\in H} y_n - \Sigma_{n\in F} y_n = \Sigma_{n\in H\backslash F} y_n$, so $x \in FS(B)$ as required.

Since $FS(B) \subseteq A$ we have $\{x \in N: FS(B) - x \in p\} \notin q$ so $FS(B) \notin p + q = p$, a contradiction. □

3.4 <u>COROLLARY</u>. It is consistent that statement (i) of Theorem 3.1 does not imply statements (a) through (h).

Proof. Theorems 3.1 and 3.3. □

REFERENCES

1. Baker, J. and Milnes, P. (1977). The ideal structure of the Stone-Cech compactification of a group, Math. Proc. Cambridge Phil. Soc. 82: 401-409.
2. Blass, A. and Hindman, N. (1987). On strongly summable ultrafilters and union ultrafilters, Trans. Amer. Math. Soc. 304: 83-99.
3. Chou, C. (1969). On the size of the set of left invariant means on a semigroup. Proc. Amer. Math. Soc. 23: 199-205.
4. Comfort, W. and Negrepontis, S. (1974), The Theory of Ultrafilters. Springer-Verlag, Berlin.
5. van Douwen, E., The Cech-Stone compactification of a discrete groupoid, Topology and its Applications, to appear.
6. Gillman, L., and Jerison, M., (1960). Rings of Continuous Functions. Van Nostrand, Princeton.
7. Hindman, N. (1979). Ultrafilters and combinatorial number theory, in Number Theory Carbondale 1979, ed. M. Nathanson, Lecture Notes in Math, 751: 119-184.
8. Hindman, N. (1980). Sums equal to products in βN, Semigroup Forum 21: 221-255.
9. Hindman, N. (1982). Minimal ideals and cancellation in βN, Semigroup forum 25: 291-310.
10. Hindman, N (1987). Summable ultrafilters and finite sums, in Logic and Combinatorics, ed. S. Simpson, Contemporary Mathematics 65: 263-274.
11. Hindman, N., and Pym, J. (1984). Free groups and semigroups in βN, Semigroup Forum 30: 177-193.
12. Katětov, M. (1967). A theorem on mappings, Comment. Math. Univ. Carolinae 8: 431-433.
13. Matet, P. (1988). Some filters of partitions, J. Symbolic Logic 53: 540-553.
14. Rudin, M. (1971). Partial orders of the types in βN, Trans. Amer. Math. Soc. 155: 350-362.
15. Ruppert, W. (1973). Rechstopologische Halbgruppen. J Reine Angew. Math. 261: 123-133.

A Countably Compact, Countably Tight, Non-Sequential Space

ALAN DOW[1] Department of Mathematics, York University, North York, M3J 1P3 Canada

1 Introduction

Z. Balogh [2] showed that the Proper Forcing Axiom, PFA, implies that each compact space of countable tightness is sequential. The main portion of the proof is an investigation of countably compact subsets of spaces with countable tightness. Hence it is natural to consider the situation in which the compactness condition is weakened to countable compactness. Indeed, if a space X is not sequential the either it contains a countably compact subset Y which is not closed or it contains a completely divergent sequence. A sequence is said to be completely divergent if it has no converging subsequence. In this paper we shall construct, assuming $\underline{b} = \underline{c}$ (which follows from PFA) , a countably compact space of countable tightness which contains a completely divergent sequence. The question of whether or not PFA implies countably compact sets are closed in spaces of countable tightness remains open and has been asked in [3] .

Recall that $\underline{b}$ is the minimum cardinality of a set $B \subset {}^{\omega}\omega$ such that there is no $g \in {}^{\omega}\omega)$ such that $f <^* g$ for all $f \in B$, where as usual $<^*$ is the mod finite ordering.

[1]Research supported by NSERC of Canada

2 A Preliminary Construction

We shall be interested in spaces which contain ω as a dense discrete subset. When we write $\omega \cup X$ it will generally be understood that $X \cap \omega = \emptyset$ and frequently that $\omega \cup X$ has a topology in which ω is a dense open discrete set.

Definition 2.1 *Suppose that $\mathcal{F} \subset \mathcal{P}(\omega)$ and $\omega \cup X$ is a space.*
$\omega \cup X$ is an <u>$\mathcal{F}$-space</u> if each $x \in X$ has a neighbourhood U such that $U \cap \omega \subset^* F$ $(\forall F \in \mathcal{F})$.
$\omega \cup X$ is <u>$\mathcal{F}$-sequential</u> if for each $I \subset \omega$ such that $I \subset^* F$ $(\forall F \in \mathcal{F})$ there is an $x \in X$ such that some subsequence of I converges to x.
$CO(X, \mathcal{F}) = \{U \subset \omega : (\forall\ F \in \mathcal{F}) U \subset^* F$ and $\overline{U}$ is compact open in $X\}$.
$\omega \cup X$ is <u>ω--compact</u> if X is countably compact.*

Lemma 2.2 *If $\underline{b} = \underline{c}$ and $|\mathcal{F}| < \underline{c}$ then there is an $\mathcal{F}$-sequential $\mathcal{F}$-space which is locally compact, locally countable and which is ω-*-compact.*

PROOF. A standard Ostaszewski-type induction.

Observation 2.3 *With $\omega \cup X$ as in Lemma 2, let $\mathcal{B}$ be the Boolean subalgebra of $\mathcal{P}(\omega)$ generated by $CO(X, \mathcal{F})$. Note that the Stone space of $\mathcal{B}$ is just the one point compactification of $\omega \cup X$.*

Definition 2.4 *Let $\omega \cup X$ be a 0-dimensional locally compact $\mathcal{F}$-sequential $\mathcal{F}$-space and let $\mathcal{A}$ be a maximal subset of $\{ A \subset \omega : (\exists x \in X)\ A$ converges to $x \}$ such that the members of $\mathcal{A}$ are almost disjoint.*

Define $\mathcal{B}(X, \mathcal{F}, \mathcal{A})$ to be the Boolean subalgebra of $\mathcal{P}(\omega)$ which is generated by $\mathcal{A} \cup CO(X, \mathcal{F})$.
For each $x \in X$, let $\mathcal{A}_x = \{A \in \mathcal{A} : A$ converges to $x\}$

$\omega \cup \mathcal{A} \cup X$ embeds canonically into the Stone space $S(\mathcal{B}(X, \mathcal{F}, \mathcal{A}))$ as follows:

1. $A \in \mathcal{A}$ is identified with
   the ultrafilter generated by $\{A - n : n \in \omega\}$;

2. $x \in X$ is identified with the ultrafilter which has, as a base,

$$\{U - [\cup \mathcal{A}' \cup n] : U \in CO(X, \mathcal{F}) \ , \ x \in \overline{U} \ , \mathcal{A}' \in [\mathcal{A}_x]^{<\omega} \ , \ n \in \omega\}$$

With these identifications note that, for each $U \in CO(X, \mathcal{F})$

$$(\omega \cup \mathcal{A} \cup X) \cap \tilde{U} = cl_X U \cup \bigcup \{\mathcal{A}_y : y \in cl_X U\} \ ;$$

where, as usual, $\tilde{U}$ is the set of ultrafilters in $S(\mathcal{B}(X, \mathcal{F}, \mathcal{A}))$ which include U .

Observation 2.5 *With $S(\mathcal{B}(X, \mathcal{F}, \mathcal{A}))$ defined as above:*

1. *X (with its original topology) is a subspace;*
2. *for each $x \in X$, $x \in \overline{\mathcal{A}'}$ for all infinite $\mathcal{A}' \subset \mathcal{A}_x$;*
3. *if $x \in X$ and $\mathcal{A}' \subset \mathcal{A} - \mathcal{A}_x$ then*

$$x \in \overline{\mathcal{A}'} \iff x \in cl_X\{y : \mathcal{A}' \cap \mathcal{A}_y \neq \emptyset\} \ . \square$$

An immediate consequence of Observation 5 is that we have constructed a new $\mathcal{F}$-sequential $\mathcal{F}$-space .

Observation 2.6 *When given the topology induced by $S(\mathcal{B}(X, \mathcal{F}, \mathcal{A}))$ $\omega \cup \mathcal{A} \cup X$ is a locally compact 0-dimensional $\mathcal{F}$-sequential ω-*-compact $\mathcal{F}$-space . Furthermore $\omega \cup \mathcal{A} \cup X$ is sequential iff $\omega \cup X$ is sequential .*

3 The Branch Space

In [5] , Simon constructs a compactification of ω in which ω is a completely divergent sequence and the remainder is sequentially compact. Later Nyikos, Pelant and Simon [4], give a more constructive definition of such a space using a *tree which is dense in* $\mathcal{P}(\omega)$. The advantage of the second construction is that it is very easy to visualize the space and to see exactly at which points the space fails to have countable tightness. Our construction is a modification of this *branch space* ; in fact our space will be a subspace of a preimage of this space. Let us new recall the definition of the *branch space* .

It follows immediately from the results in [1] that there is a (necessarily) countably complete subtree $T \subset {}^{<\underline{c}}\underline{c}$ and an isomorphic embedding of T^+ (the successor nodes of T) into $\mathcal{P}(\omega)$, where $\mathcal{P}(\omega)$ is ordered by reverse-inclusion mod finite, ${}^*\supset$. The minimum height of such a tree is denoted by $\underline{h}$ and it is proven in [1] that $\underline{h} \le \underline{b}$. They also show that the distributivity degree of such a tree (when turned upside down) is exactly $\underline{h}$.

Let us now fix a tree $T \subset {}^{<\underline{h}}\underline{c}$ such that there is an embedding of T^+ into $\mathcal{P}(\omega)$ as described above. We may assume that T is full-branching, that is, $t^\frown\alpha \in T$ for each $\alpha \in \underline{c}$ and $t \in T$. For each $t \in T^+$ let A_t be the image of t under this embedding. Note that for each $\alpha < \underline{h}$ the set $\{A_t : t \in T_{\alpha+1}\}$ is a maximal almost disjoint family on ω and that for each infinite $I \subset \omega$, there is a $t \in T^+$ such that $A_t \subset^* I$.

The construction of the *branch space* is very simple. Let $\mathcal{B}_T$ be the subalgebra of $\mathcal{P}(\omega)$ generated by $\{A_t : t \in T^+\}$. The *branch space* is just the Stone space of $\mathcal{B}_T$. However the beauty of this space is that it has the following description. Let

$$bT = \{t \in {}^{\le\underline{h}}\underline{c} : s \in T \text{ for each } s < t\} \ .$$

It turns out that $S(\mathcal{B}_T)$ can be viewed as just a rather natural topology on $\omega \cup bT$ where each $t \in bT$ is identified with the ultrafilter generated by

$$\left\{A_s - \left[\bigcup_{r\in F} A_r \cup n\right] : s \in T^+ \ , \ s \le t \ , \ n \in \omega \ , \text{and} \quad F \in [T \cap \{t^\frown\xi : \xi \in \underline{c}\}]^{<\omega}\right\} \ .$$

Furthermore, if $t \in T^+$ and $\widetilde{A_t} \subset S(\mathcal{B}_T)$ is the clopen set of ultrafilters including A_t then under our identification

$$\widetilde{A_t} = A_t \cup \{s \in bT : t \le s\} \ .$$

It then follows that a neighbourhood base for $t \in bT$ is

$$\left\{\widetilde{A_s} - \left[\bigcup_{r\in F} \widetilde{A_r} \cup n\right] : s \in T^+ \ , \ s \le t \ , \ n \in \omega \ , \text{and} \quad F \in [T \cap \{t^\frown\xi : \xi \in \underline{c}\}]^{<\omega}\right\} \ .$$

Since bT is just a subtree of ${}^{\le\underline{h}}\underline{c}$, there is no harm in assuming that, in fact, $T = bT$. There would be no change in the above discussion if we had begun by assuming that $T \subset {}^{\le\underline{h}}\underline{c}$ rather than ${}^{<\underline{h}}\underline{c}$ and also T^+ would be unchanged.

Proposition 3.1 *ω is a completely divergent sequence in $\omega \cup T$.*

PROOF. Let I be any infinite subset of ω and choose $t \in T^+$ so that $A_t - I$ is finite. Note that since T is full-branching, $A_t - [\bigcup_{r \in F} A_r]$ is infinite for each finite subset F of $\{t^\frown\xi : \xi \in \underset{\sim}{c}\}$. Therefore it follows that $I \cap \widetilde{A}_t - \left[\bigcup_{r \in F} \widetilde{A}_r\right]$ is infinite for each finite subset F of $\{t^\frown\xi : \xi \in \underset{\sim}{c}\}$; hence t is a limit point of I . But now if r is any immediate successor of t in T, then $\widetilde{A}_r$ is a clopen set not containing t but which does meet I in an infinite set (since $A_r \subset^* I$). ⊣

For each $t \in T^+$ let $t^\uparrow = \widetilde{A}_t - \omega$. More generally let

$$t^\uparrow = \bigcap_{\substack{s \le t \\ s \in T^+}} \widetilde{A}_s = \{r \in T : t \le r\}$$

Also for $t \in T$ let $o(t) = \alpha$ and $cf(t) = cf(\alpha)$ where α is the ordinal such that $t \in T_\alpha$. If $cf(t) \ge \omega$ we shall call t a limit node and if $cf(t) > \omega$ we shall say that t is a node of uncountable cofinality. It will be useful for us to observe the following facts.

Observation 3.2 *For each $t \in T$*

1. *$t^\uparrow$ is closed in $\omega \cup T$;*
2. *t is in the closure of $Y \subset t^\uparrow - \{t\}$ iff there are infinitely many immediate successors r of t such that Y contains of point from $r^\uparrow$ (in which case this subset of Y converges to t);*
3. *t is the only point in $t^\uparrow$ which may be in the closure of $T - t^\uparrow$ (and it __is__ exactly when it is a limit node);*
4. *if $cf(t) = \omega$ then t has a countable neighbourhood base in the subspace $\{t\} \cup T - t^\uparrow$;*
5. *if $cf(t) > \omega$ then $T - t^\uparrow$ is $\aleph_0$-bounded .□*

Corollary 3.3 *The only points of $\omega \cup T$ at which it does not have countable tightness are those t with $cf(t) > \omega$. Furthermore such t are not in the closure of any countable subsets of $T - t^\uparrow$.*

Of course the next result was known to the authors of [4]

Corollary 3.4 *If $\underline{h} = \omega_1$ there is a countably compact space with countable tightness containing a completely divergent sequence.*

PROOF. Simply remove from $\omega \cup T$ all nodes of uncountable cofinality. Since $\underline{h} = \omega_1$, none of these nodes have any successors hence it is easily checked that the space remains countably compact.

4 The Branch Space Modification

Roughly speaking, in this section we paste the spaces constructed in §1 into the correct positions in the branch space from §2 . Recalling that our goal is to construct a countably compact space of countable tightness it seems that we should remove the nodes of uncountable cofinality and *glue* in their places some points to ensure (at least) that all countable sets of immediate successors have limit points. We can regard t as the collection of subsets of ω , $\{A_s : s < T \, , \; s \in T^+\}$. We have a t-sequential t-space, say $\omega \cup X_t$, as constructed in §1 and can take a suitable $\mathcal{A}_t$ to obtain $\omega \cup \mathcal{A}_t \cup X_t$. If we were fortunate enough to have that $\mathcal{A}_t = \{A_r : r \text{ is an immediate successor of } t\}$, then we would indeed just glue X_t in place. Since this is unlikely to be the case however, we show how to fix it up.

Fix the tree $T \subset {}^{<\underline{h}}\underline{c}$ and the embedding into $\mathcal{P}(\omega)$ as chosen in §2 .

Definition 4.1 *Assume that $\underline{b} = \underline{c}$. For each node, $t \in T$, of uncountable cofinality choose (by §1 Lemma 2) a space $\omega \cup X_t$ which is a locally countable, locally compact, t-sequential ω-*-compact t-space .*

Definition 4.2 *Choose a set $\mathcal{A}_t$ as in §1 Definition 4 so that*

1. *$\mathcal{A}_t$ is an almost disjoint family of infinite subsets of ω*
2. *each $A \in \mathcal{A}_t$ converges to some $x \in X_t$;*
3. *$\mathcal{A}_t$ is maximal with respect to the first two properties;*
4. *$\mathcal{A}_t$ is a subset of $\{A_s : s < t \, , \; s \in T^+\}$*

It should cause no confusion if we let the context decide whether or not the symbol "$\mathcal{A}_t$" refers to a subset of $\mathcal{P}(\omega)$ (as it is defined to be) or to the subset of T of which it is the image, i.e. $\{s \in T^+ : A_s \in \mathcal{A}_t\}$.

We shall now choose a dense subtree, $\hat{T}$, of T . We shall choose the levels of $\hat{T}$ according to the following induction. More precisely we define recursively subsets of T which can be shown (by induction) to be the levels of a subtree of T .

Definition 4.3 *For each $\alpha \leq \omega_1$ let $\hat{T}_\alpha = T_\alpha$.*

Suppose $\alpha < \underline{h}$ and that we have chosen $\hat{T}_\beta$ for each $\beta < \alpha$.

If α is a limit ordinal then we define $\hat{T}_\alpha$ to be
$\{t \in T : (\forall \beta < \alpha)(\forall s < t)(\exists \gamma)(\exists r)(r \in \hat{T}_\gamma,\ s < r < t \text{ and } \beta < \gamma < \alpha)\}$

If $\alpha = \beta + 1$ and $cf(\beta) \leq \omega$ then we define $\hat{T}_\alpha$ to be the set of all nodes in T which are the immediate successors of some node in $\hat{T}_\beta$.

Finally suppose that $\alpha = \beta + 1$ and $cf(\beta) > \omega$. By the inductive assumption that $\hat{T}_\gamma$ is the "γ-th" level of a subtree of T and the definition of $\hat{T}_\beta$, it follows that each member of $\hat{T}_\beta$ is a node of uncountable cofinality in T . So we shall define

$$\hat{T}_\alpha = \bigcup \left\{ \mathcal{A}_t : t \in \hat{T}_\beta \right\} . \square$$

Perhaps the only non-trivial aspect to the above induction is the claim that $\hat{T} = \bigcup_{\alpha < \underline{h}} \hat{T}_\alpha$ is a *dense* subtree. Equivalently the implicit assertion that for each $\alpha < \underline{h}$ the antichain $\hat{T}_\alpha$ is a *maximal* antichain of T . To see this, recall that the distributivity degree of T is $\underline{h}$, and simply observe that for each $\alpha < \underline{h}$ $\hat{T}_\alpha$ is indeed an antichain which is maximal in the set $\{s \in T : (\forall \beta < \alpha)(\exists t \in \hat{T}_\beta) t < s\}$.

Let us make several observations about $\hat{T}$.

1. a node $t \in \hat{T}$ has uncountable cofinality in $\hat{T}$ iff t had uncountable cofinality in T;
2. if $t \in \hat{T}$ has uncountable cofinality then $\mathcal{A}_t$ is the set of immediate successors of t in $\hat{T}$;
3. $\hat{T}^+ = T^+ \cap \hat{T}$
4. if $\mathcal{B}_{\hat{T}}$ and $S(\mathcal{B}_{\hat{T}})$ are defined as in §2 then ω is still a completely divergent sequence in $S(\mathcal{B}_{\hat{T}})$.

Definition 4.4 *Let $\mathcal{B}$ be the subalgebra of $\mathcal{P}(\omega)$ generated by*

$$\{A_t : t \in T^+ \cap \hat{T}\} \cup \bigcup \left\{ CO(X_t, t) : t \in \cup\{\hat{T}_\alpha : cf(\alpha) > \omega\} \right\} .$$

Recall that $CO(X_t,t) = \{U \subset \omega : cl_{\omega\cup X_t} U \text{ is compact open in } \omega \cup X_t\}$ and note that since $\omega \cup X_t$ is a t-space each member of $CO(X_t,t)$ is almost contained in each A_s for $s < t$ in T^+. Further recall the definition of $\mathcal{B}(X_t, t, \mathcal{A}_t)$ as given in §1 Definition 2. For convenience, let us assume that for any subset Y of $\omega \cup \mathcal{A}_t \cup X_t$ $cl_{X_t} Y$ denotes the closure of Y in $\omega \cup \mathcal{A}_t \cup X_t$ as a subspace of $S(\mathcal{B}(X_t, t, \mathcal{A}_t))$

Just as we did for the branch space we can identify the members of $S(\mathcal{B})$ in a rather natural fashion. Let t be an arbitrary member of $\widehat{T}$ and let $\alpha \leq \underline{h}$ be such that $t \in \widehat{T}_\alpha$.

1. if $cf(\alpha) < \omega$ then let u_t be the ultrafilter generated by the filter base

$$\left\{ A_t - \left[\bigcup_{\xi \in F} [A_{\widehat{t\xi}} \cup n] \right] : n \in \omega \ , \ F \in [\underline{c}]^{<\omega} \right\}$$

2. if $cf(\alpha) = \omega$ then let u_t be the ultrafilter generated by the filter base

$$\left\{ A_s - \left[\bigcup_{\xi \in F} [A_{\widehat{t\xi}} \cup n] \right] : n \in \omega \ , \ s \in \widehat{T}^+, s < t \ \ and F \in [\underline{c}]^{<\omega} \right\}$$

3. if $cf(\alpha) > \omega$ then fix an indexing of $X_t = \{x(t,\xi) : \xi < \kappa_t \leq \underline{c}\}$. Note that it is possible for κ_t to be 0. Define

$$u_t = \left\langle \left\{ A_s - B : s \in \widehat{T}^+ \ , \ s < t \ and \ \ B \in CO(X_t, t) \ \right\} \right\rangle$$

and for each $\xi < \kappa_t$ define

$$u_{t,\xi} = \langle \{B : B \in \mathcal{B}(X_t, t, \mathcal{A}_t) \cap CO(X_t, t) \ and \ \ x \in cl_{X_t} B\} \rangle$$

One can now check that the set of ultrafilters described above is in fact all of the ultrafilters of $\mathcal{B}$. The space is very similar to the branch space. Let us now investigate the new version of §2 Def. 8. First define, for $t \in \widehat{T}$ the set $t^\uparrow$ to be the set of all ultrafilters of the form u_s or $u_{s,\xi}$ such that $t \leq s$ and $s \in \widehat{T}$. Each of the assertions below follow easily from the description of the relevent ultrafilter bases.

Observation 4.5 *For each* $t \in \widehat{T}$

1. $t^\uparrow$ *is closed in* $S(\mathcal{B})$;

2. u_t *is the only point in* $t^\uparrow$ *which may be in the closure of* $S(\mathcal{B}) - [\omega \cup t^\uparrow]$ *(and it* *<u>is</u>* *exactly when* t *is a limit node);*

3. *if* $cf(t) = \omega$ *then* t *has a countable neighbourhood base in the subspace* $\{u_t\} \cup S(\mathcal{B}) - [\omega \cup t^{\uparrow}]$;

4. *if* $cf(t) > \omega$ *then* $S(\mathcal{B}) - [\omega \cup t^{\uparrow}]$ *is* $\aleph_0$*-bounded;*

5. *if* $cf(t) \leq \omega$ *then* u_t *is in the closure of* $Y \subset t^{\uparrow} - \{u_t\}$ *iff there are infinitely many immediate successors* r *of* t *such that* Y *contains of point from* $r^{\uparrow}$ *(in which case this subset of* Y *converges to* u_t*);*

6. *if* $cf(t) > \omega$ *then* $\mathcal{A}_t \cup X_t$ *is homeomorphic to the subspace of* $S(\mathcal{B})$ *consisting of all* u_s *such that* s *is an immediate successor of* t *together with (of course) the set of all* $u_{t,\xi}$ *– but not* u_t*;*

7. *if* $cf(t) > \omega$ *and* $Y \subset [t^{\uparrow} - X_t]$ *and if* $\xi < \kappa_t$ *then* $u_{t,\xi}$ *is in the closure of* Y *iff there is a set* $Z \subset \mathcal{A}_t$ *such that* $x(t,\xi) \in cl_{X_t} Z$ *and such that* $Y \cap s^{\uparrow} \neq \emptyset$ *for each* $s \in Z$ *(remember that we have identified the set of immediate successors of* t *with* $\mathcal{A}_t$*)* .

Definition 4.6 *Define* $\omega \cup X$ *to be the subspace of* $S(\mathcal{B})$ *where*

$$X = \{u_t : t \in \hat{T} \text{ and } cf(t) \leq \omega \} .$$

Theorem 4.7 $[\underline{b} = \underline{c}]$ *There is a space,* $\omega \cup X$*, which is a countably compact space of countable tightness containing* ω *as a completely divergent subsequence.*

PROOF. Of course we let X be the space constructed above (under the assumption that $\underline{b} = \underline{c}$). By 15 3),5) , it is clear that X has countable tightness at all points which are not of the form $u_{t,\xi}$. It should be clear from 15 6),7) , that X has countable tightness at a point of the form $u_{t,\xi}$ so long as $\omega \cup \mathcal{A}_t \cup X_t$ has countable tightness. Now these subspaces do have countable tightness by §1 6 and 2 . As remarked above ω is completely divergent in $S(\mathcal{B}_{\hat{T}})$, hence it is again since $\mathcal{B}_{\hat{T}} \subset \mathcal{B}$. For each $t \in \hat{T}^+$, it is easily checked that $u_t \in X$ is a limit point of A_t . Therefore it remains only to check that X is countably compact. Since the spaces X_t were all sequentially compact it is even the case that X is sequentially compact but we shall prove that it is countably compact directly from 15 . Suppose $Y \subset X$ is countably infinite and let z be a limit point of Y in the full Stone space. Since we may as well assume that $z \notin X$, choose a node $t \in \hat{T}$ with uncountable cofinality and assume that $u_t = z$. By 15-4 we may as well assume that $Y \subset t^{\uparrow}$. By 15-6 and the fact that $\mathcal{A}_t \cup X_t$ is countably compact we may as well assume that $Y \cap X_t = \emptyset$. Let Z be the set of immediate successors, s , of t such that $Y \cap s^\dagger \neq \emptyset$. Now Z must be infinite since otherwise u_t would not be a limit point of Z . But now Z must have a limit point in X_t by 15-7 and the fact that $\mathcal{A}_t \cup X_t$ is countably compact. ⊣

Remark 4.8 The reader may be interested in checking the fact that the only place we used the assumption $\underline{b} = \underline{c}$ was in Lemma 2 . That is, Theorem 17 holds in any model in which for every filter $\mathcal{F} \subset \mathcal{P}(\omega)$ with $|\mathcal{F}| < \underline{h}$, there is an $\mathcal{F}$-sequential ω-*-compact $\mathcal{F}$-space of countable tightness. The assumptions of local compactness and 0-dimensionality in §1 were purely for notational convenience.

References

[1] J. Pelant B. Balcar and P. Simon. The space of ultrafilters on N covered by nowhere dense sets. *Fundamenta Mathematica*, pages 11–24, 1980.

[2] Z. Balogh. On compact Hausdorff spaces of countable tightness. *Proc. Amer. Math. Soc.*, 1989.

[3] W. Fleissner and R. Levy. Stone-čech remainders which make continuous images normal. preprint.

[4] J. Pelant P. Nyikos and P. Simon. Sequential compactness and trees. preprint.

[5] P. Simon. Divergent sequences in bicompacta. *Sov. Math. Dokl.*, pages 1573–1577, 1978.

The Space of Minimal Prime Ideals of C(βN - N) is Probably Not Basically Disconnected

ALAN DOW[1] Department of Mathematics, York University, North York, Canada

1 Introduction

In 1961, Henriksen and Jerison [7] asked if $m(C(\beta N - N))$ is basically disconnected; where, for a ring R, $m(R)$ is the set of minimal prime ideals endowed with the usual hull kernel topology. In the case of $X = \beta N - N$ $(= N^*)$, or in fact X any compact F-space, $m(C(X))$ can be regarded as just X itself endowed with the topology generated by the (old) closures of cozero sets; this is due to Henriksen and Kopperman, (see [3] for more information). The obvious reformulation of the question for N^* is then very natural; recall that a space is *basically disconnected* if the closure of every cozero set is clopen. It was shown in [3] that there is a compact F-space X for which $m(C(X))$ is not basically disconnected and that it is consistent with the usual axioms of set-theory that $m(C(N^*))$ (henceforth mN^*) is not basically disconnected.

The results in this paper show that the consistency of mN^* being basically disconnected would imply the consistency of there being a measurable cardinal and the consistency of the failure of the Singular Cardinal hypothesis. However the choice of heading is not intended as a philosophical statement concerning the existence of

[1]Research supported by NSERC of Canada

measurable cardinals but rather a timid conjecture. The main idea employed in this paper is the same as in [3] and we merely tinker with the proof in order to weaken the set-theoretic hypotheses (from MA). The rationale behind the heading is largely due to the fact that very little analysis of the topology of mN^* has been done and the cozero set (with non-clopen closure) which we construct is the simplest type not ruled out by the results in [7]. Furthermore the cardinal assumptions we make are needed in order to construct P-sets; we do not know if there is another approach to the solution which does not involve P-sets.

There is some hope that our results may have (or lead to) some consequences concerning another interesting question about N^*. Is there a *ccc* (or even separable) P-set in N^* ([11]) ? The P-set we construct very likely admits an open mapping onto the unit interval.

2 A Set-theoretic Translation

In this section we formulate a simple set-theoretic principle and show that it is sufficient to guarantee that mN^* is not basically disconnected. This principle is derived by simply extracting exactly what is needed for the proof in [3] to go through.

Let Q denote the set of rationals in the unit interval I. Call $\mathcal{F} \subset \mathcal{P}(Q) \setminus \{\emptyset\}$ a *P-filter* if $\mathcal{F}$ is a countably complete mod finite filter which contains the cofinite filter (i.e. $\forall \{F_n : n \in \omega\} \subset \mathcal{F}\ \exists F \in \mathcal{F} \forall n \in \omega(|F - F_n| < \omega)$).
A non-empty set $K \subset \beta N - N$ is a *P-set* if $\mathcal{F} = \{X \in \mathcal{P}(N) : K \subset X^*\}$ forms a P-filter (on N). (For $X \subset N$, X^* denotes the remainder of X in βN, i.e. $X^* = \text{cl}_{\beta N} X \setminus N$.)

Let **Mel** denote the assertion that there are disjoint countable dense subsets A, B of $I - Q$ and a P-filter $\mathcal{F}$ on Q such that, for each $X \in \mathcal{F}^+$, both $\overline{X} \cap A$ and $\overline{X} \cap B$ are non-empty, where $\overline{X} = \text{cl}_I X$. ($\mathcal{F}^+ = \{X \subset Q : X \cap F \neq \emptyset\ \forall F \in \mathcal{F}\}$)

Theorem 2.1 Mel *implies mN^* is not basically disconnected.*

PROOF. Fix A, B and $\mathcal{F}$ as in **Mel**. Let Q_d denote the space Q endowed with the **discrete** topology and let Q_d^* denote $\beta Q_d \setminus Q_d$. Clearly $Q_d^* \approx N^*$. We shall use Q_d^* to refer to this topology and mQ_d^* to refer to the $m(C(Q_d^*))$ topology. For each $x \in I$, let

$$\begin{aligned} Z_x &= \{p \in Q_d^* \mid \forall S \in p\ (x \in \text{cl}_I S)\} \\ &= \bigcap \{[Q \cap (x - 1/n, x + 1/n)]^* : n \in \omega - \{0\}\} . \end{aligned}$$

Since $[Q \cap (x - 1/n, x + 1/n)]^*$ is clopen in Q_d^*, each Z_x is a zero-set in Q_d^*. It will suffice to show that

$$C_A = \bigcup_{a \in A} int_{Q_d^*} Z_x \quad \text{and} \quad C_B = \bigcup_{b \in B} int_{Q_d^*} Z_b$$

do not have disjoint closure in the mQ_d^* topology.

Define $K = \{p \in Q_d^* : \mathcal{F} \subset p\} = \bigcap\{F^* : F \in \mathcal{F}\}$.

Claim: K is contained in $\mathrm{cl}_{mQ_d^*} C_A \cap \mathrm{cl}_{mQ_d^*} C_B$.

Let $p \in K$. Let C be a cozero set of Q_d^* such that $p \in cl_{Q_d^*}C$ – the set of such closures forms a mQ_d^*-neighbourhood base for p. Choose $\{X_n \in \mathcal{P}(Q) : n \in \omega\}$ so that $C = \bigcup_n X_n^*$. Since $p \in \mathrm{cl}_{Q_d^*} C$ and $\mathcal{F} \subset p$, it follows that

$$\forall F \in \mathcal{F} \exists n (F \cap X_n \neq^* \emptyset) .$$

Furthermore $\mathcal{F}$ is countably complete mod finite, hence there is an $n \in \omega$ so that $X_n \in \mathcal{F}^+$. But now, $\mathrm{cl}_I X_n \cap A \neq \emptyset$ and $\mathrm{cl}_I X_n \cap B \neq \emptyset$. We show that $X_n^* \cap C_A \neq \emptyset$; the proof that $X_n \cap C_B \neq \emptyset$ is the same. Choose $a \in A \cap \mathrm{cl}_I X_n$. By definition of Z_a, $X_n^* \cap Z_a \neq \emptyset$. But now, Z_a is regular closed [6], hence

$$\emptyset \neq X_n^* \cap \mathrm{int}_{Q_d^*} Z_a \subset X_n^* \cap C_A . \square$$

3 When does Mel hold?

In this section we show that **Mel** holds in any model in which $cf([\mathfrak{d}]^\omega, \subset) = cf([\mathfrak{d}]^\omega) = \mathfrak{d}$. That is, $[\kappa]^\omega$ is the set of countable subsets of κ and, of course, $cf([\kappa]^\omega, \subset)$ is the minimum of the set $\{|\mathcal{X}| : \mathcal{X} \subset [\kappa]^\omega$ is such that $\forall Y \in [\kappa]^\omega \exists X \in \mathcal{X}\ Y \subset X\}$. $\mathfrak{d}$, the dominating number, is the minimum cardinality of a *dominating* family $\mathcal{D} \subset {}^\omega\omega$, i.e. $\forall f \in {}^\omega\omega \exists g \in \mathcal{D} \forall n (f(n) < g(n))$.

Let us briefly remark on the consistency strength of the assumption $cf([\mathfrak{d}]^\omega, \subset) = \mathfrak{d}$; hence of **Mel**. It is straighforward to show (and has been done in several articles) that $cf([\aleph_n]^\omega, \subset) = \aleph_n$ for all $n \in \omega$. Clearly $cf([\kappa]^\omega, \subset) \leq \mathfrak{c}$ for all $\kappa \leq \mathfrak{c}$. It follows easily from, what is known as, the Covering Lemma that $cf([\mathfrak{d}]^\omega, \subset) = \mathfrak{d}$. The failure of the Covering Lemma implies the existence of inner models with measurable cardinals [1].

Let us recall two well-known facts.

Proposition 3.1 *If $X \subset I$ is a countable dense set, then $I - X$ is homeomorphic to the Baire space ${}^\omega\omega$.*

PROOF. Brouwer [9] and implicitly Fréchet [5] prove that for any countable dense subset, A, of $(0,1)$ there is an autohomeomorphism of $(0,1)$ taking A to Q. See Problem 4.3.H in [4].

Proposition 3.2 *There is a family $\mathcal{K}$ of compact subsets of ${}^{\omega}\omega$ such that $|\mathcal{K}| = \underline{d}$ and, for any compact $C \subset {}^{\omega}\omega$ there is a $K \in \mathcal{K}$ with $C \subset K$.*

PROOF. See [2, 8.2].

Corollary 3.3 *If X is a countable dense subset of I there is a family $\mathcal{K}_X$ of compact nowhere dense subsets of $I - X$ such that $|\mathcal{K}_X| = \underline{d}$ and for each compact $J \subset I - X$ there is a $K \in \mathcal{K}_X$ with $J \subset K$.*

We are now ready for **Mel.**

Theorem 3.4 $cf([\underline{d}]^{\omega}, \subset) = \underline{d} \implies$ **Mel.**

PROOF. Let A, B and Q be disjoint countable dense subsets of I where Q is the rational points of I. Fix $\mathcal{K}_A = \{K_\alpha : \alpha < \underline{d}\}$ and $\mathcal{K}_B = \{J_\alpha : \alpha < \underline{d}\}$ as in Corollary 4. Choose, by assumption, $\mathfrak{I} \subset [\underline{d}]^{\omega}$ a cofinal set with cardinality $\underline{d}$. Fix an indexing $\{I_\alpha : \alpha < \underline{d}\}$ for $\mathfrak{I}$ and assume without loss of generality that $I_n = \emptyset$ for $n \in \omega$ and $I_\alpha \subset \alpha$ for all $\alpha < \underline{d}$.

Our plan is to define a base $\{F_\alpha : \alpha < \underline{d}\} \subset \mathcal{P}(Q)$ for $\mathcal{F}$ by induction on $\alpha < \underline{d}$. In order to ensure that $\mathcal{F}$ is a P-filter we will construct F_α so that $|F_\alpha - F_\beta| < \omega$ for each $\beta \in I_\alpha$. In order to ensure that $\text{cl}_I X \cap A \neq \emptyset$ and $\text{cl}_I X \cap B \neq \emptyset$ for each $X \in \mathcal{F}^+$, we will construct F_α to be a subset of $Q - (K_\alpha \cup J_\alpha)$. However the difficulty here is to be sure that we did not already choose some F_β to be a subset of $K_\alpha \cup J_\alpha$. This is accomplished by ensuring that each $F \in \mathcal{F}$ will be a dense subset of Q (note that $K_\alpha \cup J_\alpha$ is nowhere dense).

The induction is based on an old idea of Ketanon's [8] . For $n \in \omega$, let $F_n = Q - (K_n \cup J_n)$. Now we show that $\{F_\alpha : \alpha < \underline{d}\}$ can be inductively constructed to satisfy the following inductive hypotheses:

1. $F \in \mathcal{F}_\alpha \implies \text{cl}_I F = I$ where $\mathcal{F}_\alpha$ is the filter base generated by $\{F_\beta : \beta < \alpha\}$;

2. $F_\alpha \subset Q - (K_\alpha \cup J_\alpha)$; and

3. $|F_\alpha - F_\beta| < \omega$ for each $\beta \in I_\alpha$.

Let $\alpha < \underline{d}$ and suppose that $\{F_\beta : \beta < \alpha\}$ have been chosen so as to satisfy the inductive hypotheses. Choose $\{G_n : n \in \omega\} \subset \mathcal{F}_\alpha$ a descending sequence so that $\forall \beta \in I_\alpha \exists n \in \omega \; |G_n - F_\beta| < \omega$. We may of course assume that $G_n \cap n = \emptyset$; hence $\bigcap_n G_n = \emptyset$. Now define $H_n = G_n - G_{n+1}$ for $n \in \omega$. Suppose we pick an $f \in {}^{\omega}\omega$ and define $X_f = \bigcup_n H_n \cap [0, f(n)]$. Then we'd have that $X_f - G_n$ is finite for each $n \in \omega$; hence

$|X_f - F_\beta| < \omega$ for each $\beta \in I_\alpha$. If any H_n is finite, we can just define $f(n) = \max H_n$ so we may as well assume that each H_n is infinite. Let $\{h(n,m) : m \in \omega\}$ be an indexing of H_n. Also let $\{B_k : k \in \omega\}$ be a listing of all non-empty open rational intervals in I.

For each $F \in \mathcal{F}_\alpha$ and $k \in \omega$, define a function $g_{F,k} \in {}^\omega\omega$ as follows. For each $n \in \omega$, choose $m = g_{F,k}(n)$ so that for some $\tilde{n} \geq n$, $h(\tilde{n}, m) \in F \cap B_k - (J_\alpha \cup K_\alpha)$; since $F \cap G_n$ is dense, B_k is open and $(J_\alpha \cup K_\alpha)$ is nowhere dense, there is such an m. Now the set $\{g_{F,k} : F \in \mathcal{F}_\alpha, k \in \omega\}$ is not dominating so we may choose an *increasing* $f \in {}^\omega\omega$ so that for each $F \in \mathcal{F}_\alpha$ and $k \in \omega$ the set $\{n : g_{F,k}(n) < f(n)\}$ is infinite.

We claim that with X_f defined as above, $F_\alpha = X_f - (J_\alpha \cup K_\alpha)$ does the job. It remains only to show that $F \cap F_\alpha$ is dense for each $F \in \mathcal{F}_\alpha$. To see this, let $k \in \omega$ and choose $n \in \omega$ so that $g_{F,k}(n) < f(n)$. Let $\tilde{n} \geq n$ be such that $h(\tilde{n}, g_{F,k}(n)) \in F \cap B_k - (J_\alpha \cup K_\alpha)$. Since f is increasing, we have that $g_{F,k}(n) < f(\tilde{n})$ and so $h(\tilde{n}, g_{F,k}(n)) \in X_f$ as required.

□

4 Questions

1. Can the closure in mQ^*_d of $\bigcup_{q \in Q} Z_q$ ever be open?
   It follows from **MEL** that, e.g. $\bigcup_{q \in Q} Z_q$ does not have open closure in the mQ-topology. The consistency strength of **MEL** appears so weak and the definition of $\bigcup_{q \in Q} Z_q$ is so canonical one would have to conjecture NO.

   Let $\mathcal{N}_Q$ denote the ideal $\{A \subset Q : \mathrm{cl}_I A \text{ is nowhere dense}\}$. Let us now switch from discussing P-filters to P-ideals: i.e. an ideal $\mathcal{I}$ is a P-ideal if the dual filter $\mathcal{I}^* = \{A \subset Q : Q - A \in \mathcal{I}\}$ is a P-filter. For an ideal $\mathcal{I}$ let $K_\mathcal{I}$ be the canoncal subset of Q^*_d associated with $\mathcal{I}$: i.e. $K_\mathcal{I} = \bigcap_{A \in \mathcal{I}} [Q - A]^*$. Also let $f_\mathcal{I}$ be the canonical map restricted to $K_\mathcal{I}$ induced by the equation $f_\mathcal{I}[A^*] = \mathrm{cl}_I A$.

2. Can $\mathcal{N}_Q$ be extended to a P-ideal?
   Let us show that **MEL** follows from a YES answer. If the answer is YES then we may choose a P-ideal $\mathcal{I}_B$ as above so that $B \cap Q \in \mathcal{I}_B$ for each non-empty open rational interval B. Let $\mathcal{I}$ be the intersection of these countably many P-ideals. Since $\mathcal{N}_Q \subset \mathcal{I}$, it follows that if $X \in \mathcal{I}^+$, then $\mathrm{cl}_I X$ has interior hence would meet any dense set. We need only show that $\mathcal{I}$ is a P-ideal. Suppose that $\mathcal{A} \subset \mathcal{I}$ is countable. For each B as above choose $X_B \in \mathcal{I}_B$ so that $X_B - A$ is finite for each $A \in \mathcal{A}$. But now it is clear that we may choose a set $X \subset Q$ so that for each $A \in \mathcal{A}$ we have $X - A$ is finite and for each B as above $X_B - X$ is finite. It follows that $X \in \mathcal{I}$. Let us observe that in this case the map $f_\mathcal{I}$ is pseudo-open.

3. Does it follow from $cof([\underline{d}]^\omega, \subset) = \underline{d}$ that there is a (separable) *ccc* P-set?
   If $\underline{d} = \underline{c}$ then one can easily show that there is an $\mathcal{I}$ as in 4.1 so that in addition

$f_{\mathcal{I}}$ is an irreducible map from $K_{\mathcal{I}}$ onto I. Does such an K_I remain (separable) *ccc* under ${}^{\omega}\omega$-bounding forcing? (A forcing is ${}^{\omega}\omega$-bounding if every *new* member of ${}^{\omega}\omega$ is dominated by a function from the ground model. See [10]).

4. Can **MEL** ever be wrong?
   MEL follows from the seemingly weaker principle:
   MEL*: there are disjoint countable dense sets A, B of $I-Q$ and a P-ideal $\mathcal{I} \supset \mathcal{I}_{AB}$ where $\mathcal{I}_{AB}$ is the ideal generated by $\mathcal{I}_A \cup \mathcal{I}_B$ and $\mathcal{I}_A = \{X \subset Q : \mathrm{cl}_I X \cap A = \emptyset\}$.

References

[1] A.J. Dodd and R.B. Jensen. The covering lemma for $L[U]$. *Ann. Math. Logic*, 22:127–135, 1982.

[2] E.K.van Douwen. *The integers and topology*, pages 111–168. North-Holland, 1984.

[3] A. Dow, M. Henriksen, R. Kopperman, and J. Vermeer. The space of minimal prime ideals of $C(X)$ need not be basically disconnected. *Proc. AMS*, 104:317–320, 1988.

[4] R. Engelking. *General Topology*. PWN - Polish Scientific Publishers, 1977.

[5] M. Fréchet. Les dimensions d'un ensemble abstrait. *Math. Ann.*, pages 145–168, 1910.

[6] L. Gillman and M. Jerison. *Rings of Continuous Functions*. Van Nostrand, 1960.

[7] M. Henriksen and M. Jerison. The space of minimal prime ideals of a commutative ring. *Trans. AMS*, 115:110–130, 1965.

[8] J. Ketonen. On the existence of P-points in the Stone-Čech compactification of the integers. *Fund. Math.*, 92:91–94, 1976.

[9] L.E.J.Brouwer. Über den natürlichen Dimensionsbegriff. *Journ. für die reine und angew. Math.*, pages 146–152, 1913.

[10] S.Shelah. *Proper Forcing*. Springer Lecture Notes, 1982.

[11] Jan van Mill. An introduction to βN. In K. Kunen and J.E. Vaughan, editors, *Handbook of Set-Theoretic Topology*. North-Holland, 1984.

Convex and Pseudoprime Ideals in C(X)

LEONARD GILLMAN Department of Mathematics, The University of Texas, Austin, Texas

1 INTRODUCTION

This paper presents the fundamental results about pseudoprime ideals put forth in Gillman-Kohls (1960), along with some new ones I obtained in 1988. For convenience, I include proofs; those of Theorems 3.1 and 6.1 in particular streamline the original ones by bypassing some of the equivalences considered therein. Proofs I omit from background discussions can be found in Gillman-Jerison (1960).

The paper has benefitted from some enlightening comments by Mel Henriksen.

In what follows, X is a completely regular space and $C(X)$ is the ring of all continuous functions from X to $\mathbf{R}$. The zero-set of f is $Z(f) \equiv \{x \in X: f(x) = 0\}$. An ideal P in a commutative ring is *prime* if $ab \in P$ implies $a \in P$ or $b \in P$.

The present discussion stems from the following proof (Gillman-Henriksen, 1954). The general theorem was stated for arbitrary M^p and O^p, but all we need here is the following special case for which the picture is easy to draw: $X = \mathbf{R}$ and $p = 0$, so that M (= M^0) is the ideal of all continuous functions on $\mathbf{R}$ to $\mathbf{R}$ that vanish at 0, and O (= O^0) is the ideal of those functions whose zero-set is a neighborhood of 0.

THEOREM 1.1. *If P is a prime ideal in $C(\mathbf{R})$ and $P \subset M$, then $P \supset O$.*

Proof: Given $f \in O$, pick g so that $g(0) = 1$, while $g = 0$ outside Int $Z(f)$. Then $fg = \mathbf{0}$ and hence belongs to the ideal P. Since $g \notin P$ and P is prime, $f \in P$. Thus, $O \subset P$. ♦

The point is that the hypothesis that P be prime is not used full force—the reason we know that $fg \in P$ is that $fg = \mathbf{0}$, which is a very special way of belonging to P.

To emphasize the point, let us say that an ideal in a commutative ring is *pseudoprime* if $ab = 0$ implies $a \in P$ or $b \in P$. Thus the theorem holds with *pseudoprime* in the hypothesis (and in the proof) in place of *prime*; since every prime ideal is (obviously) pseudoprime, the new theorem is formally stronger than the original.

We are thus led to investigate pseudoprime ideals and, in particular, their relation with prime ideals. We are especially interested in results that hold in $C(X)$, and in fact it will turn out that pseudoprime ideals play a role in describing its structure, particularly its order structure. (See for example Theorem 4.2 and Corollary 6.2.)

2 ELEMENTARY QUESTIONS ABOUT PSEUDOPRIME IDEALS

When we look at the ring of integers, we find that the concept of pseudoprime is a total washout: *every* ideal in $\mathbf{Z}$ is pseudoprime—for if $ab = 0$, then either a or b is 0 and hence belongs to the ideal. Salvaging what we can, we notice two things. First, a pseudoprime ideal need not be prime (since $\mathbf{Z}$ contains nonprime ideals). Second, to say that $\mathbf{Z}$ has no zero-divisors is to say that the zero ideal in $\mathbf{Z}$ is (pseudo)prime; the above argument thus shows that any ideal containing a (pseudo)prime ideal is pseudoprime. Of course, this raises another question: does every pseudoprime ideal contain a prime ideal?

Every student knows (?) by induction that if P is prime, then for all n, if a product of n factors belongs to P, one of the factors must belong to P. Let us define an ideal P to be *universally pseudoprime* if for all n, whenever a product of n factors equals 0, one of the factors must belong to P. The observation just made implies that every ideal containing a prime ideal is universally pseudoprime. Thus we will be considering the following four propositions about an ideal P in a commutative ring:

(Pr)	P **is prime**: if $ab \in P$ then $a \in P$ or $b \in P$
($\supset$ Pr)	P **contains a prime** ideal
(UPs)	P **is universally pseudoprime**: $\forall\, n$, if $a_1 a_2 \cdots a_n = 0$, then $a_1 \in P$ or $\cdots$ or $a_n \in P$
(Ps)	P **is pseudoprime**: if $ab = 0$ then $a \in P$ or $b \in P$

At the moment, we have:

For commutative rings, **(Pr)** $\rightarrowtail$ **($\supset$ Pr)** $\rightarrow$ **(UPs)** $\rightarrow$ **(Ps)**,

where the symbol $\rightarrowtail$ signifies that the implication cannot be reversed.

The following lemma will be useful.

LEMMA 2.1. *Let P be a pseudoprime ideal in $C(X)$.*

(a) *For any f, either $f - |f| \in P$ or $f + |f| \in P$.*

(b) *For any f, $f \in P$ if and only if $|f| \in P$.*

Proof: To get (a), note that the product of the two functions is $\mathbf{0}$; to get (b), apply (a). ♦

3 z-IDEALS

The next theorem shows that for the class of *z-ideals* in $C(X)$, the four propositions are equivalent (Gillman-Jerison, 2.9; cf. Kohls, 1958a, 2.20). We recall that an ideal I in $C(X)$ is a *z-ideal* if $\mathbf{Z}(h) \supset \mathbf{Z}(k)$ and $k \in I$ implies $h \in I$. The z-ideals form an important class, as they are in one-one correspondence with the z-filters on X. All maximal ideals and all "neighborhood" ideals O^p are z-ideals. A simple example of a non-z-ideal is the principal ideal generated by the identity function $\mathbf{i}(x) = x$ in $C(\mathbf{R})$. There are examples of prime ideals that are not z-ideals, but they are much more complicated (Gillman-Jerison). The ideal O of Theorem 1.1 is a z-ideal that is not prime.

THEOREM 3.1. *Every pseudoprime z-ideal is prime.*

Proof: Let I be pseudoprime and let $gh \in I$. By Lemma 2.1(a), we have, say,

$$l \equiv (g^2 - h^2) - |g^2 - h^2| \in I.$$

Then

$$k \equiv l^2 + (gh)^2 \in I.$$

For $x \in Z(k)$,

$$l(x) = 0 \quad \text{and} \quad g(x)h(x) = 0;$$

and the first of these implies that

$$g^2(x) \geq h^2(x).$$

It follows that $h(x) = 0$. Thus $\mathbf{Z}(k) \subset \mathbf{Z}(h)$. Since I is a z-ideal, $h \in I$. Thus, I is prime. ♦

In summary:

THEOREM 3.2. *For z-ideals in C(X),* **(Pr)** $\longleftrightarrow$ **($\supset$ Pr)** $\longleftrightarrow$ **(UPs)** $\longleftrightarrow$ **(Ps)**.

4 CONVEX IDEALS

Recall that an ideal I in $C(X)$ is

convex if $0 \leq f \leq g$ and $g \in I$ implies $f \in I$;
absolutely convex if $0 \leq |f| \leq |g|$ and $g \in I$ implies $f \in I$.

Convexity is the necessary and sufficient condition that $C(X)/I$ be partially ordered; absolute convexity, that it be a lattice, with the canonical homomorphism from $C(X)$ onto $C(X)/I$ a lattice homomorphism. Prime ideals and z-ideals are absolutely convex (the latter is obvious), and an absolutely convex ideal is convex (but none of these implications can be reversed). A convex ideal I is absolutely convex if and only if $f \in I$ implies $|f| \in I$. It follows from Lemma 2.1(b) that convex *pseudoprime* ideals are absolutely convex.

Every z-ideal is an intersection of prime ideals (Gillman-Henriksen, 1956, 1.4; Gillman-Jerison, 2.8). Theorem 4.2 to follow states, correspondingly, that absolutely convex ideals are intersections of pseudoprime ideals. We denote by $AC(I)$ the smallest absolutely convex ideal containing an ideal I.

LEMMA 4.1. $AC(I) = \{f \in C(X): |f| \leq \sum_i |g_i|$ *for some* $g_1, \cdots, g_n \in I\}$.

The proof is straightforward.

THEOREM 4.2. *Every absolutely convex ideal I in C(X) is an intersection of pseudoprime ideals (in fact, of absolutely convex pseudoprime ideals).*

Proof: We show that if I is absolutely convex and $f \notin I$, then there is an absolutely convex pseudoprime ideal containing I but not f. Let P be maximal in the class of absolutely convex ideals containing I but not f; we prove that P is pseudoprime. Suppose on the contrary that there exist $g \notin P$ and $h \notin P$ such that $gh = \mathbf{0}$. Then $f \in AC(P, g)$ and $f \in AC(P, h)$. By the lemma, there exist $p_i, q_j \in P$ and $s_i, t_j \in C(X)$ such that

$$|f| \leq \sum_i |p_i + s_i g| \quad \text{and} \quad |f| \leq \sum_j |q_j + t_j h|.$$

Since for each x, either $g(x) = 0$ or $h(x) = 0$,

$$\text{either} \quad |f(x)| \leq \sum_i |p_i(x)| \quad \text{or} \quad |f(x)| \leq \sum_j |q_j(x)|.$$

Therefore

$$|f| \leq \sum_i |p_i| + \sum_j |q_j|.$$

Since P is absolutely convex, this implies $f \in P$, contrary to assumption. ♦

We saw in the preceding section that if a pseudoprime ideal is a z-ideal, then it is prime. With the weaker condition of (absolute) convexity, it need not be prime, as we will see in Section 7. In the other direction, we now show that a pseudoprime ideal need not be convex at all.

THEOREM 4.3. *If some point of X is the limit of a sequence of distinct points, then C(X) contains a pseudoprime ideal that is not convex.*

Proof: For simplicity of notation, we work in $C(\mathbf{R})$. Let

$$S = \{1, 1/2, \ldots, 1/n, \ldots\}.$$

Express S as a union of disjoint infinite sets S_k $(k \in \mathbf{N})$. Let $\mathcal{F}$ denote the filter of all subsets of S that include all but a finite number of points of S_k for all but finitely many k. Let $\mathcal{U}$ be any ultrafilter containing $\mathcal{F}$. Then

For each $U \in \mathcal{U}$, $U \cap S_k$ is infinite for infinitely many k.

Now define

$$P = \{f \in C(X): \mathbf{Z}(f) \cap S \in \mathcal{U}\}.$$

Since $\mathcal{U}$ is an ultrafilter, P is a prime ideal in $C(X)$ (cf. Gillman-Jerison, 14I.*3*).

Let $\mathbf{i}$ be the identity function on $\mathbf{R}$: $\mathbf{i}(x) = x$. We will show that the pseudoprime ideal $(P, \mathbf{i})$ is not convex. Define $f \in C(\mathbf{R})$ by

$$f(x) = x/k \text{ for } x \in S_k,$$

$f(0) = 0$, and then by extension from the compact set $S \cup \{0\}$. Then $0 \le f(x) \le x$ on S and we may assume that $0 \le f \le |\mathbf{i}|$ in $C(\mathbf{R})$. Consequently, if $(P, \mathbf{i})$ were convex, and hence absolutely convex, we should have $f \in (P, \mathbf{i})$. We now prove this is impossible.

If $f \in (P, \mathbf{i})$, then there exists $h \in C(X)$ such that $f - h\mathbf{i} \in P$, i.e., the set

$$U \equiv \{x \in S: f(x) = xh(x)\}$$

is a member of $\mathscr{U}$. Consider any $k \in \mathbf{N}$. For $x \in U \cap S_k$, we have

$$x/k = xh(x);$$

then $h(x) = 1/k$. Pick $k \neq l$ such that $U \cap S_k$ and $U \cap S_l$ are both infinite; then h assumes each value $1/k$ and $1/l$ infinitely often, so cannot be continuous at 0. ♦

COROLLARY 4.4. *A pseudoprime ideal in C(X) need not be prime.*

5 UNIVERSALLY PSEUDOPRIME IDEALS

Does every pseudoprime ideal in a commutative ring contain a prime ideal? More modestly, is every pseudoprime ideal universally pseudoprime? I always enjoy the picture of the impatient professor explaining matters to the careful student who was unable to produce a proof. To illustrate the induction, let's assume the case $n = 2$ (P pseudoprime) and establish the case $n = 3$. Suppose $abc = 0$. Then $a(bc) = 0$, so either $a \in P$ or—uh—wait a minute—uh—or bc——and the professor's voice trails off, followed by a desperation attempt with $(ab)c$.

Returning to the original question—with the formally stronger hypothesis that the ideal is universally pseudoprime, the answer is yes:

THEOREM 5.1. *Every universally pseudoprime ideal I (in an arbitrary commutative ring) contains a prime ideal.*

Proof: Consider the set

$$S = \{s_1 \cdot \cdot \cdot s_n: s_k \notin I\}.$$

It is closed under multiplication. Since I is universally pseudoprime, $0 \notin S$. In these circumstances, there is a prime ideal P disjoint from S. (Take a maximal ideal in the class of ideals disjoint from S and verify that it is prime.) Since $S \supset \mathscr{C}I$, $I \supset P$. ♦

The answer to the two questions as stated is no:

EXAMPLE 5.2. *A pseudoprime ideal that is not universally pseudoprime (and hence does not contain a prime ideal).* In the ring $\mathbf{Z}_8$, the principal ideal $(4) = \{0, 4\}$ is pseudoprime. But it is not universally pseudoprime, since $2 \times 2 \times 2 = 0$ but $2 \notin (4)$.

Of course, it is also obvious directly that (4) does not contain a prime ideal (the only ideals it contains being itself and (0)).

Our array of implications is now complete:

THEOREM 5.3. *For commutative rings,* **(Pr)** $\rightarrowtail$ **(⊃ Pr)** $\longleftrightarrow$ **(UPs)** $\rightarrowtail$ **(Ps)**.

While writing this up, I was musing about the fact that the difficulty in the induction step from $n = 2$ to $n = 3$ persists at every higher level as well. And, indeed, for each $n \geq 2$, the ideal (4) in the ring $\mathbf{Z}_{2^{n+1}}$ is "pseudoprime at the level n" but not at $n + 1$—that is, if the product of $\leq n$ factors is 0 then one of the factors belongs to the ideal (4), whereas $2^{n+1} = 0$ although $2 \notin (4)$. In this sense, there are infinitely many steps between pseudoprime and universally pseudoprime.

6 PSEUDOPRIME IDEALS IN *C(X)*

We shall prove that every pseudoprime ideal in *C(X) does* contain a prime ideal. Along the way, we obtain some additional results of interest in themselves. Recall that the radical of an ideal is the intersection of the prime ideals containing it.

THEOREM 6.1. *For any ideal I in C(X), the following are equivalent.*

(a) *I contains a prime ideal.*
(b) *I is pseudoprime.*
(c) *For every convex ideal $J \supset I$, C/J is a totally ordered ring.*
(d) *The radical of I is prime.*

Proof: (a) *implies* (b). Trivial.

(b) *implies* (c). Let J satisfy the hypotheses of (c). Since J is convex, C/J is partially ordered. And since $J \supset I$, J is pseudoprime, so every f satisfies either $f \equiv |f|$ or $f \equiv -|f|$ (mod J); therefore C/J is totally ordered.

(c) *implies* (d). Let J denote the radical of I. Then $J \supset I$. Also, J is convex, as it is an intersection of prime, hence convex, ideals. By hypothesis, then, C/J is totally ordered. Now, the prime ideals in C/J, being the images of the prime ideals in C that contain J, are convex. Consequently, they are symmetric intervals in C/J, and so form a chain. Hence the prime ideals in C that contain J—which are the same as those that contain I—form a chain. Therefore J, their intersection, is prime.

(d) *implies* (a). Since the radical, J, is prime, it contains a prime z-ideal Q (Kohls, 1958b, 1.1; Gillman-Jerison, 14.7); we show that in fact $I \supset Q$. Let $f \in Q$. We may assume f is bounded. Put $h = \sum_n 2^{-n} |f|^{1/n}$. Then $h \in C(X)$, and $Z(h) = Z(f)$. Since Q is a z-ideal, the latter implies that $h \in Q$. So $h \in J$. Then (as is well known) $h^n \in I$ for some $n > 1$. Now, $2^{-n} |f|^{1/n} \leq h$; so $2^{-n^2} |f| \leq h^n$. This implies that f is a multiple of h^n (Gillman-Jerison, 1D.3); hence $f \in I$. Thus, $Q \subset I$. ♦

After a recent talk of mine, Dennis Kletzing of Stetson University, Deland, Florida asked what one can say about residue rings of pseudoprime ideals. I forgot at the time that my own paper (Gillman-Kohls) contains an answer, at least for convex ideals (i.e., for ordered residue rings).

COROLLARY 6.2. *A convex ideal I in C(X) is pseudoprime if and only if C/I is totally ordered.*

Proof: If C/I is totally ordered, then C/J is totally ordered for every convex ideal $J \supset I$. The result now follows from the equivalence of (b) and (c).

A commutative ring with unity is called a *valuation ring* if of any two elements, one divides the other. If $C(X)/P$ is a valuation ring, where P is prime, then of any two positive elements, the larger divides the smaller, as follows from the fact that every element > 1 is a unit; see also (Kohls, 1958a, p. 524). It follows readily that C/P is a valuation ring if and only if every ideal in C/P is convex. And this is true if and only if in $C(X)$, every ideal containing P is convex. So we have:

COROLLARY 6.3. *The following are equivalent in C(X).*

(a) *Every pseudoprime ideal is convex.*
(b) *For every prime ideal P, C/P is a valuation ring.*

A pseudoprime ideal in $C(X)$ need not be prime (Corollary 4.4). Thus we have the following complete table of implications in $C(X)$:

COROLLARY 6.4. *For rings C(X),* **(Pr)** $\rightarrowtail$ **($\supset$ Pr)** $\longleftrightarrow$ **(UPs)** $\longleftrightarrow$ **(Ps)**.

At another talk, Curtis Herink of Mercer University, Macon, Georgia asked me about ideals I that are intermediate between prime and pseudoprime, as follows: for $I \supset J, fg \in J$ implies $f \in I$ or $g \in I$. In such a case, call *I prime with respect to J* . When $J = I$, this says I is prime; when $J = (0)$, pseudoprime.

Evidently, if $I \supset Q \supset J$, Q a prime, then I is prime with respect to J.

The following corollary is a partial answer to Herink's query.

COROLLARY 6.5.
(a) *If I is prime with respect to some ideal $J \subset I$, then it is pseudoprime.*
(b) *If I is pseudoprime, then it is prime with respect to every* ***z-ideal*** *$J \subset I$.*

Proof: (a). Obvious. (b). The proof of Theorem 6.1 actually shows more generally that every pseudoprime ideal *containing a z-ideal H* contains a prime ideal *containing H*—this is because the theorem of Kohls quoted in the proof that (d) implies (a) shows in fact that if $J \supset H$, then $Q \supset H$. The result then follows from the remark preceding the corollary. ♦

In (b), note that some extra condition on J is necessary—for instance, if $J = I$, not prime, then I is *not* prime with respect to J.

7 *P*-SPACES AND *F*-SPACES

Henriksen and I introduced the classes of P-spaces and F-spaces (Gillman-Henriksen, 1954, 1956; for an outline, see Gillman (1989) in this volume). We established a number of characterizations of these spaces, and Kohls (1957, 6.3) and Gillman-Jerison (1960, 14.25 and 14.29) added others. Table 7.1 shows several of them, along with a later one involving pseudoprime ideals.

Table 7.1. *P*-spaces and *F*-spaces

P-spaces		***F-spaces***
Every **P**rime ideal is maximal	**1**	The prime ideals contained in a given maximal ideal form a chain
Every finitely generated ideal is generated by an idempotent	**2**	Every **F**initely generated ideal is principal
Every ideal O^p is maximal	**3**	Every ideal O^p is prime
Cozero-sets are C-embedded	**4**	Cozero-sets are C^*-embedded
Cozero-sets are (open and) closed	**5**	Disjoint cozero-sets are completely separated
$\forall f, \exists k, f = kf^2 \ (= k\|f\|^2)$	**6**	$\forall f, \exists k, f = k\|f\|$—whence $\|f\| = kf$
Every ideal is a z-ideal	**7**	Every ideal is convex
Every ideal is an intersection of prime ideals	**8**	Every ideal is an intersection of pseudoprime ideals

Each entry on the left characterizes P-spaces, and on the right, F-spaces. The entries are paired, the left one obviously implying the right one. Thus, all P-spaces are F-spaces. The converse is not true: $\beta\mathbf{N} \setminus \mathbf{N}$ is a counterexample. We remark that Euclidean spaces are a far cry from these spaces: any metric F-space must be discrete.

Combining **6** and **7** in the right-hand column of the table, we see that X is an F-space if and only if every ideal is absolutely convex.

We now look at several propositions involving pseudoprime ideals.

THEOREM 7.2 (Gillman-Kohls). *X is an F-space if and only if every ideal in C(X) is an intersection of pseudoprime ideals.*

Proof: *Necessity.* In an F-space, all ideals are absolutely convex, hence, by Theorem 4.2, are intersections of pseudoprime ideals.

An alternative proof first establishes the general formula

$$I = \bigcap\nolimits_p (I, O^p),$$

then observes that in an F-space, all O^p are prime, so that all (I, O^p) are pseudoprime.

Sufficiency. For every f, the principal ideal (f) is an intersection of pseudoprime ideals P_α. Hence for each $\alpha, f \in P_\alpha$; and since P_α is pseudoprime, $|f| \in P_\alpha$. It follows that $|f| \in (f)$. Therefore X is an F-space. ♦

The next result was sitting right there in Gillman-Kohls, but we never raised the question. I thought of it just in time to mention it at the 1988 Hewitt Conference.

COROLLARY 7.3. *If X is an F-space but not a P-space, then C(X) contains an absolutely convex pseudoprime ideal that is not prime.*

Proof: Since X is an F-space, every ideal in $C(X)$ is an intersection of pseudoprime ideals. Hence if all pseudoprime ideals were prime, every ideal would be an intersection of prime ideals, and X would be a P-space. Since this is not the case, $C(X)$ contains a pseudoprime ideal that is not prime; as X is an F-space, this ideal is absolutely convex. ♦

The following characterization of P-spaces was obtained at the Henriksen Conference one night at 1:30 a.m. as I was tossing around in bed, thinking about my talk for the following afternoon. Incidentally, it yields another immediate proof of Corollary 7.3.

THEOREM 7.4. *X is a P-space if and only if every pseudoprime ideal in C(X) is prime.*

Proof: *Necessity*. In any case, all pseudoprime *z-ideals* are prime (Theorem 3.1). In a P-space, all ideals are z-ideals, and so all pseudoprime ideals are prime.

Alternatively, every pseudoprime ideal Q contains a prime ideal (Theorem 6.1). Hence, if X is a P-space, this prime ideal is maximal, whence Q is maximal and therefore prime.

Sufficiency. If X is not a P-space, there exists a prime ideal P contained properly in a maximal ideal M. Pick $f \in M \setminus P$. Then $f^2 \in M \setminus P$. The ideal (P, f^2) is pseudoprime.

If (P, f^2) is prime, then $f \in (P, f^2)$, and there exists k such that $f - kf^2 \in P$. Then

$$f \times (1 - kf) \in P, \quad \text{whence} \quad (1 - kf) \in P \subset M, \quad \text{so that} \quad 1 \in M,$$

which is not possible. Therefore (P, f^2) is not prime. ♦

Theorem 4.3 establishes the existence of a nonconvex pseudoprime ideal whenever some point is a limit of a sequence of distinct points. At one point in the course of writing up this paper, I thought I could establish the far-reaching generalization that X is an F-space if (and only if) every pseudoprime ideal in $C(X)$ is convex. That would have made an elegant counterpart to Theorem 7.4. But I couldn't get past one "tiny" hurdle, and I asked Henriksen for help. He located a counterexample in Cherlin-Dickmann (1986, Example 10): a non-F-space X in which every C/P, P prime, is a valuation ring—hence, by Corollary 6.3 above, in which every pseudoprime ideal is convex.

REFERENCES

1. Cherlin, G. L. and Dickmann, M. A. (1986). *Real closed rings*. I. *Residue rings of rings of continuous functions*, Fund. Math. 126, 147–183.
2. Gillman, L. (1989). *The Gillman-Henriksen papers,* this volume.
3. Gillman, L. and Henriksen, M. (1954). *Concerning rings of continuous functions,* Trans. Amer. Math. Soc. 77, 340–362.
4. Gillman, L. and Henriksen, M. (1956). *Rings of continuous functions in which every finitely generated ideal is principal,* Trans. Amer. Math. Soc. 82, 366–391.
5. Gillman, L. and Jerison, M. (1960 [1976]). *Rings of continuous functions,* Van Nostrand [Springer].
6. Gillman, L. and Kohls, C. (1960). *Convex and pseudoprime ideals in rings of continuous functions,* Math. Zeitschr. 72, 399-409.
7. Kohls, C. W. (1957). *Ideals in rings of continuous functions,* Fund. Math. 45, 28–50.
8. Kohls, C. W. (1958a). *Prime ideals in rings of continuous functions,* Illinois J. Math. 2, 505–536.
9. Kohls, C. W. (1958b). *Prime ideals in rings of continuous functions,* II, Duke Math. J. 25, 447–458.

The Gillman-Henriksen Papers

LEONARD GILLMAN Department of Mathematics, The University of Texas, Austin, Texas

Mel reads voraciously and remembers what he read. It should be no surprise, then, that he is the one who read the papers of Hewitt and Kaplansky that suggested problems and got us going.

During the middle fifties, spurred by Ed Hewitt's seminal paper (Hewitt, 1948), we published four joint papers in rings of continuous functions, plus a related one in pure algebra. Most of the principal results were incorporated into Gillman-Jerison (1960).

The two best known of the papers are Gillman-Henriksen (1954) and (1956b), on P-spaces and F-spaces, which sparked a number of topological and set-theoretic investigations—though, interestingly enough, our own motivation was algebraic. (For a survey of these and later results, see Gillman (1989) in this volume.)

The results in Erdös-Gillman-Henriksen (1955) on isomorphisms have been of interest to logicians, because of their relation to nonstandard analysis.

Gillman-Henriksen (1954), *P*-spaces

Kaplansky (1947) had remarked that when X is discrete, all Prime ideals in $C(X)$ are maximal—in our terminology, every discrete space is a P-space. Mel suggested to me that we investigate whether the converse is true, and, if not, that we find alternative properties to characterize P-spaces. We showed that X is a P-space if and only if: all zero-sets are open; all G_δ-sets are open; $C(X)$ is regular (i.e., for every f there exists k such that $f = kf^2$; every ideal is an intersection of maximal ideals. We also constructed P-spaces with no isolated points at all.

Our paper starts off with some introductory results about $C(X)$ in general: Every prime ideal contained in M^p contains O^p; every prime ideal is contained in a unique maximal ideal.

The paper also contains several results about linearly ordered spaces: All linearly ordered spaces are countably paracompact. A linearly ordered space is paracompact if and only if every gap is a "Q-gap", and is realcompact if and only if every gap is a nonmeasurable Q-gap. *Corollary*: Every realcompact linearly ordered space is paracompact. Finally, if X is a linearly ordered space for which every Q-gap is nonmeasurable, then υX can be constructed in a particularly simple way (though in general it will not be a linearly ordered space).

Gillman-Henriksen-Jerison (1954), *The Gelfand-Kolmogoroff theorem*

This paper had its origin in a 1953 seminar organized by Mel, where our late colleague Merrill Shanks uncovered an error in a paper he was reporting on. The question concerned the relation between the maximal ideals in $C(X)$ and those in $C^*(X)$. Mel, Jerry, and I got to work on the problem and rediscovered the theorem of Gelfand and Kolmogoroff (1939): at the time, we knew their paper only by title. Then we read it, to find they had scooped us by 14 years. But they had announced their theorem without proof. Our paper begins by supplying a proof. Then it introduces the extension f^* and uses it in clarifying the order structure of the fields $C(X)/M^p$ and thence the relationship between M^p and M^{*p} (Gillman-Jerison, 7.6, 7.9(a)). Next come two constructions of υX, followed by further remarks on the maximal ideals in $C(X)$ and $C^*(X)$. Finally, the paper establishes Hewitt's conjecture that every "m-closed" ideal in $C(X)$ is an intersection of maximal ideals, and shows that the corresponding result holds in $C^*(X)$ if and only if X is pseudocompact. (The Gelfand-Kolmogoroff theorem states that $M^p = \{f \in C(X): p \in \mathrm{Cl}_{\beta X} Z(f)\}$. In Gillman-Jerison, this theorem is essentially the definition of the topology of βX.)

Erdös-Gillman-Henriksen (1955), *Real-closed fields*

A *hyper-real* field is a field $C(X)/M$, where M is a hyper-real maximal ideal. It was known that all hyper-real fields are real-closed (Henriksen-Isbell, 1953; Isbell, 1954). The symbol η_α refers to a particular type of order structure defined by Hausdorff. *Results*: For $\alpha > 0$, all real-closed η_α-fields of cardinal $\aleph_\alpha$ are isomorphic. All hyper-real fields are η_1. *Corollary*: Under the continuum hypothesis, all hyper-real fields of cardinal $\mathfrak{c}$ are isomorphic. For every infinite discrete space X, there exist hyper-real fields $C(X)/M$ of cardinal $\mathfrak{c}$. For X the discrete space of cardinal $\mathfrak{c}$ and M any maximal ideal in $C(X)$ for which every zero-set has cardinal $\mathfrak{c}$, the field $C(X)/M$ is of cardinal $> \mathfrak{c}$. For X the discrete space of any cardinal $\mathfrak{m} \geq \mathfrak{c}$, there exists a residue class field of $C(X)$ of cardinal $> \mathfrak{m}$.

Gillman-Jerison vastly simplifies the proof of this last result. An immense amount of research has since been published on such set-theoretic questions.

Gillman-Henriksen (1956a), *Elementary divisor rings*

Another paper of Kaplansky's on commutative rings with identity (Kaplansky, 1949) led to our study of F-spaces. Kaplansky was interested in the condition **D**: every matrix over the ring can be reduced to **D**iagonal form ("elementary divisor" rings). A necessary condition is **F**: all **F**initely generated ideals are principal; but it was not known whether **F** was sufficient. An intermediate condition is **T**: every matrix over the ring can be reduced to **T**riangular form ("Hermite" rings). Mel suggested we work on Kaplansky's problem: does **F** imply **T** and does **T** imply **D**? In our paper, we obtain necessary and sufficient conditions for **D** and for **T**, workable conditions that we were able to apply in the succeeding paper to obtain a solution.

Gillman-Henriksen (1956b), *F-spaces*

General results: O^p coincides with the intersection of all the prime ideals contained in M^p. To each function in $C(X)$ there is associated a bounded function belonging to the same ideals.

An F-space is a space X for which $C(X)$ satisfies the condition **F** of the preceding paper; T-space and D-space are defined correspondingly. A number of necessary and sufficient conditions are given that X be an F-space: several that connect f and $|f|$, and the criteria that βX be an F-space, that all ideals O^p be prime, and that all cozero-sets be C^*-embedded. The theorem is established that when X is locally compact and σ–compact, $\beta X \setminus X$ is an F-space.

Next, the paper derives a topological characterization of T-spaces and constructs an F-space that does not satisfy the condition; then derives a necessary topological condition for a D-space and constructs a T-space that does not satisfy this condition. Kaplansky's problem is thus solved: **F** does not imply **T**, and **T** does not imply **D**. However, as far as I know, the problems remain open for integral domains.

I conclude with a comment about the proof that $\beta X \setminus X$ is an F-space for X locally compact and σ–compact. Mel and I first worked out a straightforward proof for the case $X = \mathbf{R}$, using the criterion that Pos f and Neg f are completely separated, then extended the ideas to the general case. When Jerry and I got to writing up the result for our book, we brought in the Gelfand-Kolmogoroff theorem; using the same criterion for an F-space, we obtained a more sophisticated proof, half as long as the original—although still the better part of a page. When we showed it proudly to Mel, he said, yes, that was surely the way the proof should go. Then in the summer of 1963, Stelios Negrepontis, at the time a graduate student at Rochester, came up with the following gem, based on the criterion that cozero-sets are C^*-embedded (Negrepontis, 1967):

> Any cozero-set A in $\beta X \setminus X$ is an F_σ and therefore, since $\beta X \setminus X$ is compact, is σ-compact. Therefore $X \cup A$ is σ-compact and hence normal. Since X is locally compact it is open in $X \cup A$, whence A is closed. So A is C^*-embedded in $X \cup A$, hence in βX, hence in $\beta X \setminus X$.

REFERENCES

1. Erdös, P., Gillman, L., and Henriksen, M. (1955). *An isomorphism theorem for real-closed fields,* Ann. of Math. 61, 542–554.
2. Gelfand, I., and Kolmogoroff, A. N. (1939). *On rings of continuous functions on topological spaces,* C. R. (Doklady) Acad. Sci. URSS 22, 11–15.
3. Gillman, L. (1989). *Convex and pseudoprime ideals in C(X),* this volume.
4. Gillman, L. and Henriksen, M. (1954). *Concerning rings of continuous functions,* Trans. Amer. Math. Soc. 77, 340–362.
5. Gillman, L. and Henriksen, M. (1956a). *Some remarks about elementary divisor rings,* Trans. Amer. Math. Soc. 82, 362–365.
6. Gillman, L. and Henriksen, M. (1956b). *Rings of continuous functions in which every finitely generated ideal is principal,* Trans. Amer. Math. Soc. 82, 366–391.

7. Gillman, L., Henriksen, M., and Jerison, M. (1954). *On a theorem of Gelfand and Kolmogoroff concerning maximal ideals in rings of continuous functions,* Proc. Amer. Math. Soc 5, 447–455.
8. Gillman, L. and Jerison, M. (1960 [1976]). *Rings of continuous functions,* Van Nostrand [Springer].
9. Henriksen, M. and Isbell, J. (1953). *On the continuity of the real roots of an algebraic equation,* Proc. Amer. Math. Soc. 4, 431–434.
10. Hewitt, E. (1948). *Rings of real-valued continuous functions,* I, Trans. Amer. Math. Soc. 64, 54-99.
11. Isbell, J. (1954). *More on the continuity of the real roots of an algebraic equation,* Proc. Amer. Math. Soc. 5, 439.
12. Kaplansky, I. (1947). *Topological rings,* Amer. J. Math. 69, 153–183.
13. Kaplansky, I. (1949). *Elementary divisors and modules,* Trans. Amer. Math. Soc. 66, 464–491.
14. Negrepontis, S. (1967). *Absolute Baire sets,* Proc. Amer. Math. Soc 18, 691–694.

Properties of Families of Special Ranks

ELISE M. GRABNER

ANDRZEJ SZYMANSKI

Slippery Rock University, Slippery Rock, Pennsylvania

INTRODUCTION

P. Fletcher and W. F. Lindgren [FL] as well as essentially R. McCoy [Mc] have shown the following:

If R is a point-finite open family in a Baire space X, then the points where R is locally finite constitute an open and dense subset of X.

As a matter of fact they have shown that this property characterizes Baire spaces. We will refer in the sequel to this characterization as the FLM theorem. Let us recall that a family R is: *point-finite* if for every x, the set $R_x = \{A \in R : x \in A\}$ is finite and *locally finite* at a point x, if there exists a neighborhood U of a x such that the set $R_U = \{A \in R : A \cap U \neq \emptyset\}$ is finite.

Our paper is a variation on the FLM theorem. As in most cases involving the Baire Category theorem, we use the FLM theorem as a method rather than a fact. Motivated by its structural feature, we distinguish the class of PF spaces and we show that each space of first category is modelled on a PF space via a continuous skeletal map (see Theorem 2). Since every compact Hausdorff space is Baire and every ideal on a cardinal generates a compact Hausdorff space (namely the Stone space of the quotient Boolean algebra) we can adopt and interpret the FLM theorem in the theory of ideals. This enables us to show that semiregular ideals coincide with ideals satisfying Fodor's property (Theorem 4). We are also able to estimate cardinalities of point-finite and point-finite mod I families through the saturatedness of ideals (Theorem 5).

The last section has another character. We estimate cardinalities of special families consisting of sets from a field of sets. This has been motivated by previous considerations concerning (σ-)fields of measurable sets or Baire sets. Let us add that the book [CN] by W. Comfort and S. Negrepontis is a valuable source of many other related results.

1. A CHARACTERIZATION OF FIRST CATEGORY SPACES

A partially ordered set $(P, \leq)$ is said to be a *PF set* if for each $p \in P$ the set $(\leftarrow, p) = \{g \in P : g \leq p\}$ is finite.

Let $[X]^{<\omega} = \{z : z \subset X \text{ and } |z| < \omega\}$. We always treat $[X]^{<\omega}$ as a partially ordered set with inclusion. Any subset of $[X]^{<\omega}$ is an example of a PF set. It turns out that such examples exhaust, up to isomorphism, all possible instances of PF sets. For if $(P, \leq)$ is a partially ordered set, then P is isomorphic to a subset of $(P(P), \subseteq)$ via the identification of p with $\{q \in P : q \leq p\}$. If now $(P, \leq)$ is a PF set, this shows that P is isomorphic to some subset of $([P]^{<\omega}, \subseteq)$.

Let $\mathcal{U}$ be a point-finite cover of a set Y. The set $P_{\mathcal{U}} \subset [\mathcal{U}]^{<\omega}$ consisting of all $U_y = \{U \in \mathcal{U} : y \in U\}, y \in Y$, is a PF set; let us call it a *PF set associated with a (point-finite) cover* $\mathcal{U}$.

There exists a function $h : Y \to P_{\mathcal{U}}$ given by $h(y) = U_y$; let us call it a *natural function associated with (point-finite) cover* $\mathcal{U}$.

Let S be a family of subsets of a set X such that $X \in S$. An *S-filter* on X is any $F \subset S$ satisfying the following conditions:

a) If $A \in F$ and $A \subset B \in S$, then $B \in F$.

b) If $A, B \in F$, then $A \cap B \in F$.

c) $\emptyset \notin F$ and $X \in F$.

If $S = P(X)$, then any S-filter is called simply a *filter*. A filter which is maximal (with respect to inclusion) is called an ultrafilter. If S is a family of all open (resp. closed, measurable, etc.) subsets, then an S filter is called an *open* (resp. *closed measurable*, etc.) *filter*.

Let ϕ be a function on $[X]^{<\omega}$ such that $\phi(s)$ is a nonempty S-filter on X, for every $s \in [X]^{<\omega}$. With any such function ϕ we define a topology τ_ϕ on $[X]^{<\omega}$ in the following way:

U is open in τ_ϕ if and only if for each $s \in U$,
$\{x \in X : s \cup \{x\} \in U\} \in \phi(s)$.

Let us compare the topology τ_ϕ with some other more familiar topologies and give some of its properties.

1. One can define in a similar way a topology on the set Seq(X) of all finite sequences in X. Many authors have considered such topologies and these ideas go back to P. Alexsandroff and P. Urysohn [AU].

2. Let X be a topological space. The Pixley–Roy space over X is the set $[S]^{<\omega}$ endowed with the topology generated by the sets of the form $[s, W] = \{t \in [X]^{<\omega} : s \subset t \subset W\}$, where $s \in [X]^{<\omega}$ and W is open in X (see Pixley and Roy [PR] and van Douwen [vD] for a survey of basic properties of Pixley–Roy spaces).

Let ϕ be a function on $[X]^{<\omega}$ such that $\phi(s)$ is the open filter of all open sets containing s. Sets of the form $[s, W]$ are open in the topology τ_ϕ, for if W is an open set in X and $t \in [s, W]$, then $s \subset t \subset W$. Thus $\{x \in X : t \cup \{x\} \in [s, W]\} = W$ is an open neighborhood of s, that is, for all $t \in [s, W]$, $\{x \in X : t \cup \{x\} \in [s, W]\} \in \phi(s)$. Hence the topology τ_ϕ is finer than that of the Pixley–Roy topology. Note that if X is a T_1 space, then the topology τ_ϕ is always strictly finer than the Pixley–Roy topology.

3. There are subsets of $[X]^{<\omega}$ that are open in every topology τ_ϕ on $[X]^{<\omega}$. Namely, if $s \in [X]^{<\omega}$, then $[s, X]$ is an example of such a subset. This is an obvious consequence of the fact that $X \in F$ for arbitrary S filter F on X. Thus every space $([X]^{<\omega}, \tau_\phi)$ is hereditarily metacompact.

4. Let P be a point-finite open cover of a space X. Let $Y \subset [P]^{<\omega}$ be a PF-set associated with the cover P and let $h : X \to Y$ be the natural map. If the topology τ_ϕ on $[P]^{<\omega}$ is generated by the function ϕ such that $\phi(s) = \{P\}$ for every $s \in [P]^{<\omega}$, then the function h is continuous because $h^{-1}([P_x, P]) = \cap P_x$, for every $x \in X$.

Theorem 1 A topological space X is Baire if and only if every continuous function from an open subset of X onto a PF space is constant on some open nonempty subset of X.

Proof: Let $f : W \to Y$ be a continuous function from an open subset W of a Baire space X onto a PF space Y. Suppose to the contrary that $f(U)$ contains at least two points for every open nonempty subset U of W. Since Y is a PF space we can treat it as a subspace of the space $([Z]^{<\omega}, \tau_\phi)$. Let us set $L_n = \{y \in Y : |\{z \in Z : z \subset y\}| \leq n\}$ for every $n \in N$. Clearly, the sets L_n, $n \in N$, cover Y. We shall show inductively that $f^{-1}(L_n)$ is nowhere dense in W for every $n \in N$.

The sets of the form $[z, Z] = \{t \in [Z]^{<\omega} : z \subset t\}$ are open in every topology τ_ϕ (see 3). Hence the sets L_n are closed, for every $n \in N$. Since $L_1 \cap [z, Z] = \{z\}$ for every $z \in L_1$, L_1 is a closed discrete subset of Y. Thus $f^{-1}(y)$ is a nowhere dense subset of W for every $y \in L_1$ and by the Banach Category Theorem (see [K]), $f^{-1}(L_1)$ is a nowhere dense subset of W.

Suppose that $f^{-1}(L_n)$ is a nowhere dense subset of W but $f^{-1}(L_{n+1})$ is not for some $n > 1$. Since the sets $f^{-1}(L_n)$ and $f^{-1}(L_{n+1})$ are closed in W there exists nonempty open subset U of W such that $U \subset f^{-1}(L_{n+1}) - f^{-1}(L_n) = f^{-1}(L_{n+1} - L_n)$. Since $L_{n+1} - L_n$ is a discrete space and $f : U \to L_{n+1} - L_n$ is continuous, f is constant on some open subset of U, which is impossible. Hence $f^{-1}(L_{n+1})$ is also nowhere dense. As a consequence, W is an open set of first category in a Baire space: a contradiction.

Suppose now that a topological space X is not Baire. By the FLM theorem, there exists a point-finite open cover P of X and an open nonempty subset W of X such that every open nonempty subset of W intersects infinitely many members of P. If we take a PF space Y associated with cover P and a natural function h associated with cover P, then h is a continuous function onto a PF space which is nonconstant on any open subset of W. This follows immediately from (4). □

We say that a continuous function $f : X \to Y$ is *semiskeletal* if it is constant on no open nonempty subset of X. Now theorem 1 can be reformulated in the following way.

Theorem 2 A topological space X is of the first category if and only if there exists a semiskeletal function from X onto a PF space.

In this light, every space of first category can be obtained from some canonical one (namely a PF space) by a semiskeletal function.

2. SEMI-REGULAR IDEALS

An S *ideal* on X is any $I \subset P(X)$ such that the family $\{X - A : A \in I\}$ is an S filter on X.

An *ideal on an infinite cardinal* κ is any family $I \subset P(\kappa)$ such that $\{\kappa - A : A \in I\}$ is a filter on κ and $[\kappa]^{<\kappa} \subset I$. For an S ideal I on X let $I^+ = S - I$ be the sets of "positive I-measure" and $I^* = \{X - A : A \in I\}$ be the sets of "I-measure one."

We say that I is λ *complete* if $\bigcup J \in I$ for every $J \in [I]^{<\lambda}$; ω_1-complete ideals are called *countably complete*.

A family $R \subset P(X)$ is said to be *point-finite* mod$I$, if for every $S \in [R]^\omega$ there exists $F \in [S]^{<\omega}$ such that $\bigcap F \in I$. If I is an ideal on cardinal κ, and if $P(\kappa)/I$ denotes the Boolean algebra of subsets of κ mod$I$ and $*$ is the Stone isomorphism between $P(\kappa)/I$ and the field of all clopen subsets of the Stone space $\mathrm{St}(P(\kappa)/I)$ of $P(\kappa)/I$, then

Proposition 1 A family $R \subset P(\kappa)$ is point-finite mod$I$ if and only if the family $[R]^* = \{[A]^* : A \in R\}$ is point-finite in the space $\mathrm{St}(P(\kappa)/I)$.

We have the following version of the FLM theorem for ideals.

Theorem 3 Let I be an ω_1-complete ideal on κ and R be a point-finite (mod$I$) family of subsets of a cardinal κ. Then for every set A of positive I measure there exists a set B of positive I measure such that $B \subset A$ and the set $\{C \in R : C \cap B \in I^+\}$ is finite.

Proof: In the case of R being point-finite mod$I$, the conclusion follows from the FLM theorem for $\mathrm{St}(P(\kappa)/I)$. So assume that R is point finite. Let $A_n = \{\alpha \in \kappa : |R_\alpha| \le n\}$ and $m \in \omega$ be the minimal number such that $A_m \in I^+$. There exists $Z \in [R]^{<m+1}$ (not excluding $Z = \emptyset$) such that $A_m \cap \bigcap Z \in I^+$ but $A_m \cap \bigcap Z' \in I$ for every $Z' \not\subset R$ with $Z \subset Z'$ (otherwise, there would be an $\alpha \in A_m$, such that $|R_\alpha| > m$). It suffices to let $B = A_m \cap \bigcap Z$. □

An ideal I on κ is λ*-saturated* if every almost disjoint family $R \subset I^+$ has cardinality less than λ. Let $\mathrm{sat}(I) = \inf\{\lambda : I \text{ is } \lambda\text{-saturated}\}$.

Proposition 2 Let I be an ideal on κ. If $R \subset I^+$ is point-infinite (mod I), then $[R] < \text{sat}(I)$.

Proof: Let S be the maximal almost disjoint family consisting of sets satisfying the conclusion of Theorem 3. By virtue of Theorem 3, for every $A \in I^+$ there exists $B \in S$ such that $A \cap B \in I^+$. Let $\{B_\alpha : \alpha < \lambda\}$ be an enumeration of the family S. Then $\lambda < \text{sat}(I)$. Define a function $f : R \to \lambda$ by setting $f(A) = \min\{\alpha : A \cap B_\alpha \in I^+\}$, for $A \in R$. Then F is a finite-to-one function and thus $|R| \leq \lambda < \text{sat}(I)$. □

Following A. Taylor [T] we say that an ideal I on κ satisfies *Fodor's property* if every collection indexed by κ of positive I-measure sets has a disjoint refinement (that is, if for every family $\{A_\alpha : \alpha < \kappa\} \subset I^+$ there exists a pairwise disjoint family $\{R_\alpha : \alpha < \kappa\} \subset I^+$ such that $R_\alpha \not\subset A_\alpha$ for every $\alpha < \kappa\}$; if every such collection has a point-finite (point-finite mod I) refinement, then I is called a *regular (semiregular) ideal*. Fodor's property and regularity of ideals have been intensely studied for a long time. The reader can consult three basic papers in this topic [BV], [BHM], and [T], for information on this topic.

It is known that Fodor's property and regularity are equivalent in the case of κ-complete ideals on κ (see [T; Theorem 5.2]). We shall show that semiregularity can be added to that list.

Theorem 4 Let I be a κ-complete ideal on κ. Then I satisfies Fodor's property if and only if I is regular if and only if I is semiregular.

Proof: The proof we are going to present uses only the conclusion of Theorem 3 and therefore it works for both regular and semiregular ideals. We shall present details for semiregular ideals, only. In this case only the fact that κ-complete semiregular ideals satisfy Fodor's property requires a proof. So let $A = \{A_\alpha : \alpha < \kappa\} \subset I^+$ be given. Without loss of generality we may assume that A is point-finite mod I.

Let $C_0 \subset A_0$ satisfies the conclusion of Theorem 3. Suppose that $\beta < \kappa$ and that C_α, $\alpha < \beta$, have been already defined. To define C_β consider two cases:

1. if $A_\beta - \cup\{C_\alpha : \alpha < \beta\} \in I$, then $C_\beta = \varnothing$;
2. if $A_\beta - \cup\{C_\alpha : \alpha < \beta\} \in I^+$, then C_β is a subset of $A_\beta - \cup\{C_\alpha : \alpha < \beta\}$ satisfying the conclusion of Theorem 3. Note that by κ-completeness of I it follows that for every $\alpha < \kappa$ there exists a $\xi \leq \alpha$ such that $C_\xi \cap A_\alpha \in I^+$.

Let R_β be a disjoint refinement of the finite family $\{C_\beta \cap A_\xi : \xi < \kappa \text{ and } C_\beta \cap A_\xi \in I^+\}$, for every $\beta < \kappa$. Such a refinement exists since the semiregularity of I guarantees that $\{C \cap D : D \in I^*\}$ is not an ultrafilter on κ, for any $C \in I^+$. We set $B_\alpha = \cup\{R : R \subset A_\alpha \text{ and } R \in R_\beta \text{ for some } \beta < \kappa\}$. Clearly, $B_\alpha \subset A_\alpha$ for every $\alpha < \kappa$ and $\{B_\alpha : \alpha < \kappa\}$ is a disjoint family. By the note above, $B_\alpha \in I^+$ for every $\alpha < \kappa$. □

3. FIELDS

In this section we attempt to give a sort of generalization of the FLM theorem to fields of sets.

A field of sets is a family S of subsets of a nonempty set X such that:

a) $X \in S$.

b) If $A \in S$, then $X - A \in S$.

c) If $A, B \in S$, then $A \cup B \in S$.

A field S of subsets of X is said to be *κ-complete* if $\cup F \in S$ for every $F \in [S]^{<\kappa}$.

Theorem 5 Let S be a field of subsets of a set X and let I be an S-ideal such that:

(i) Both S and I are κ-complete for some uncountable regular cardinal κ,

(ii) Each family consisting of pairwise disjoint subsets of I^+ is of cardinality $< \kappa$,

(iii) No element of I^+ is a subset of a sum of κ members of I.

Then every point-κ family $R \subset I^+$ is of cardinality $< \kappa$.

Proof: Without loss of generality we may assume that $|R| \leq \kappa$. Let P be a maximal pairwise-disjoint family such that $P \subset I^+$ and if $B \in P$ then $|\{R \in R : R \cap B \in I^+\}| < \kappa$. We shall show that $X - \cup P \in I$. Suppose it is not true. Since $|P| < \kappa$ and S is a κ-complete field, $A = X - \cup P \in I^+$. By the maximality of P we get that

($*$) if $B \in I^+$ and $B \subset A$, then $|\{R \in R : R \cap B \in I^+\}| = \kappa$. Let $R = \{R_\alpha : \alpha < \kappa\}$ and let $E_\beta = A - \cup\{R_\alpha : \beta \leq \alpha < \kappa\}$ for every $\beta < \kappa$. It is obvious that E_β intersects at most the sets among R_α's where $\alpha < \beta$. We will come up with a contradiction if we show that every set E_β is a subset of a set that belongs to the ideal I. So let us fix $\beta < \kappa$. By virtue of the conditions (i) and (ii) there exists $\nu < \kappa$ such that $R_\xi - \{R_\alpha : \beta \leq \alpha \leq \nu\} \in I$ for every $\xi > \nu$. Then the set $B = A - \cup\{R_\alpha : \beta \leq \alpha \leq \nu\}$ is in S. Since $B \cap R_\xi \in I$ for every $\xi > \nu$, $B \in I$. Thus $E_\beta \subset B$ and the fact that $X - \cup P \in I$ has been proven.

Let $R_B = \{R \in R : R \cap B \in I^+\}$ for every $B \in P$. Then $|R_B| < \kappa$ and $\cup\{R_B : B \in P\} = R$ (because $X - \cup P \in I$). Since $|P| < \kappa$ and κ is regular, $|R| < \kappa$. □

The above theorem was known for the σ-field of measurable sets and the ideal of null sets (with respect to a σ-additive measure) or for the σ-field of Baire sets and the ideal of first category sets (in a topological space) (see for instance [CN], [CSW], [J], [Ta]).

Theorem 6 Let S be a field of subsets of a set X and let I be an S-ideal such that:

(i) Both S and I are κ^+-complete for some cardinal κ.

(ii) Each family consisting of pairwise disjoint subsets of I^+ is of cardinality $\leq \kappa$.

(iii) No element of I^+ is a subset of the sum of κ members of I.

If R is a point-κ family and $R \subset I^+$, then for every $A \in I^+$ there exists $B \in I^+$ such that $B \subset A$ and $|\{R \in R : R \cap B \in I^+\}| < \kappa$.

Proof: Suppose to the contrary that some $A \in I^+$ does not satisfy the conclusion of our theorem.

Claim If $B \in I^+$ and $B \subset A$, then $|\{R \in R : R \cap B \in I^+\}| > \kappa$.

Proofof the claim: Suppose otherwise and let $B \in I^+$ be such that $B \subset A$ and that the family $R' = \{R \in R : R \cap B \in I^+\}$ is of cardinality $\leq \kappa$; as a matter of fact it is then of cardinality exactly κ. Let $\{R_\alpha : \alpha < \kappa\}$ be an enumeration of the family R'. We set $E_\alpha ; B - \cup\{R_\xi : \alpha \leq \xi < \kappa\}$ for every $\alpha < \kappa$. By (i), $E_\alpha \in S$ for every $\alpha < \kappa$. Since E_α intersects only $\leq |\alpha|$ sets among R_ξ and $|\alpha| < \kappa$, $E_\alpha \in I$ for every $\alpha < \kappa$. But then $B = \cup\{E_\alpha : \alpha < \kappa\}$, which is impossible; the claim has been proved.

For every $\alpha < \kappa$ we are going to define R_α and f_α so that the following hold.

1. R_α is a pairwise disjoint family and $R_\alpha \subset I^+$ and $A - R_\alpha \in I$
2. f_α is a 1-1 function and $\mathrm{dom} f_\alpha = \cup\{R_\xi : \xi \leq \alpha\}$ and $\mathrm{rng} f_\alpha \subset R$;
3. $C \subset f_\alpha(C)$ for every $C \in \mathrm{dom} f_\alpha$;
4. If $\alpha < \beta < \kappa$, then $f_\alpha \subset f_\beta$.

Suppose that R_α and f_α have already been defined for every $\alpha < \beta$, where $\beta < \kappa$. Let us consider the family $R' = R - \cup\{f_\alpha(R_\alpha) : \alpha < \beta\}$. Since $|R_\alpha| \leq \kappa$ and $\beta < \kappa$, $|\cup \{f(R_\alpha) : \alpha < \beta\}| \leq \kappa \cdot |\beta| \leq \kappa$. By virtue of the claim we have:

(*) if $B \in I^+$ and $B \subset A$, then $|\{R \in R' : R \cap B \in I^+\}| > \kappa$.

Let $\{R_\xi : \xi < \lambda\}$ be a faithful indexation of the family R'. Let $\xi = \inf\{\xi : R_\xi \cap A \in I^+\}$. Suppose that we have defined $\xi(\alpha) < \lambda$ for every $\alpha < \nu$, $\nu < \kappa^+$, in such a way that $\xi(0) < \xi(1) < \cdots < \xi(\alpha) < \cdots$ and $C_\delta = R_{\xi(\delta)} \cap (A - \cup\{R_{\xi(\alpha)} : \alpha < \delta\} \in I^+$ for every $\delta < \nu$. Let us consider the set $A^\sim = A - \cup\{R_{\xi(\alpha)} : \alpha < \nu\}$. Certainly, $A^\sim \in S$. If $A^\sim \in I$, then we stop the construction.

If $A^\sim \in I^+$, then by virtue of (*), there exists $\xi(\alpha) = \inf\{\xi : R_\xi \cap A^\sim \in I^+\}$ and the induction continues. By virtue of condition (ii) the construction has to stop before reaching κ^+, that is, there exists a $\eta < \kappa^+$ such that $A - \cup\{R_{\xi(\alpha)} : \alpha, \eta\} \in I$. We set $R_\beta = \{C_\alpha : \alpha < \eta\}$ and $f_\beta = \cup\{f_\alpha : \alpha < \beta\} \cup \{(C_\alpha, R_{\xi(\alpha)}) : \alpha < \delta\}$. Then R_α and f_α satisfy conditions (1)–(4) for every $\alpha \leq \beta$; the induction is finished.

By virtue of condition (1) and κ^+-completeness of I, there exists a point p belonging to the set $A \cap \cap\{\cup R_\alpha : \alpha < \kappa\}$. For every $\alpha < \kappa$ let C_α be the unique element of R_α containing p. By virtue of the condition (3), $p \in \cap\{f_\alpha(C_\alpha) : \alpha < \kappa\}$. Hence by virtue of condition (2), $|\{R \in R : p \in R\}| \geq \kappa$; a contradiction. □

ACKNOWLEDGEMENT

We would like to thank the referee for many helpful and useful suggestions.

REFERENCES

[AU] P. Alexsandroff and P. Urysohn, *Mémoire sur les éspaces topologiquesa compacts*, Verh. Nederl. Akad. Wetensch. Sec. I, 14, 1929, 1–96.

[BV] B. Balcar and P. Vojtás, *Refining systems on Boolean algebras*, Set theory and hierarchy theory V (Proc. 3rd Conf. Bierutowice), Lecture Notes in Math., 619, 1977, 45–58.

[BHM] J. Baumgartner, A. Hajnal, and A. Mátè, *Weak saturation properties of ideals*, Colloq. Math. Soc. Janós Bolyai 10, Infinite and finite sets, Keszthely, 1973, 137–158.

[CN] W. Comfort and S. Negrepontis, *Chain conditions in topology*, Cambridge University Press, 1982.

[CSW] J. Cichon, A. Szymanski, and B. Weglorz, *On intersection of sets of positive Lebesgue measure*, Coll. Math. 52, 1987, 173–174.

[vD] E. K. van Douwen, *The Pixley–Roy topology on spaces of subsets*, Set-Theoretic Topology, Academic Press, 1977, 111–134.

[FL] P. Fletcher and W. F. Lindgren, *A note on spaces of second category*, Arch. Math, 24, 1973, 186–187.

[J] H. J. K. Junnila, *On countability of point-finite families of sets*, Canadian J. Math., 31, 1979, 673–679.

[Mc] R. McCoy, *A Baire space extension*, Proc. Amer. Math. Soc., 33, 1972, 199–202.

[PR] C. Pixley and P. Roy, *Uncompletable Moore spaces*, Proc. 1969 Auburn University Conf. (Auburn, 1969).

[T] A. Taylor, *Regularity properties of ideals and ultrafilters*, Annals of Mathematical Logic, 16, 1979, 33–55.

[Ta] F. Tall, *The density topology*, Pacific J. Math. 62, 1976, 275–284.

The Extremally Disconnected Group

Ananda V. Gubbi

Introduction: In 1967 A.V. Arhangelskii [A] asked whether a nondiscrete infinite extremally disconnected(ED) topological group could be constructed within the axioms of ZFC. In their 1976 survey paper Ponomarev & Shapiro asked a more concrete version of the same: Is there a 'naive' example of a **countable** nondiscrete ED group? This is listed as Problem 6 in their survey paper [PS]. Comfort 1984 [C], and Comfort & van Mill 1987 [CM], have listed this as an open problem. Thus the problem has been unsolved for 22 years. To date, Sirota [S1], Louveau [L] and Gubbi & Szymanski [GS] have examples, but always assuming some axiom outside of ZFC. In this paper the author has constructed a countably infinite nondiscrete ED space with an addition operation. This operation is a homeomorphism in each of the variables. The additive inverse operation is also an autohomeomorphism. The author confines himself to the discussion of Problem 6 cited above and considers spaces satisfying the T_1 axiom of separation. It may be noted that such a topological group is already Tychonov, i.e., completely regular and Hausdorff [B].

The Projective cover of the Cantor Space:

Let 2^c be the Cantor space obtained by taking the Cartesian product of c number of copies of the discrete space $2 = \{0,1\}$ and by giving it the Tychonov product topology, where c is the cardinality of the continuum. It is known that 2^c is separable, of weight and π–weight c [E]. Let $P2^c$ be the projective cover of 2^c(see [G1]). Our aim in this section is to first obtain a homogeneous countably infinite ED subspace E of $P2^c$ and then to explore the possibilities of defining an algebraic operation on E that would be compatible with the topology of E.

Let S be a countable dense subset of 2^c and let D be the set of all finite sums of elements of S. This is the subgroup generated by S. The additive inverse of any element in D is itself. Since D consists of finite sums only, D is also countably infinite. Let $D = \{ d_1$ $d_2 \ldots\ldots d_n \ldots\}$ with d_1 the zero element. Consider the commutative diagram:

$$\begin{array}{ccc} P2^c & \overset{\overline{f}_k}{- - - - - >} & P2^c \\ \downarrow & & \downarrow \\ 2^c & \xrightarrow[f_k]{\hspace{3em}} & 2^c \end{array}$$

Here f_k is the autohomeomorphism such that $f_k(d_1) = d_k$ $\forall\ k \geq 2$, $k \in \mathbf{N}$ and f_1 is the identity autohomeomorphism of 2^c. By using the basic properties of projectivity, the properties of irreducibility, surjectivity and the commutativity of the diagram above, it follows that $\overline{f}_k$ is a perfect irreducible and onto map. [A continuous closed map such that preimages of singletons are compact sets, is a perfect map. A map is irreducible if there does not exist a proper closed subset of the domain whose image coincides with that of the domain. p.i.o = perfect, irreducible and onto.] Using a result of Gleason [G1], viz., a perfect irreducible surjection from a compact Hausdorff space onto an ED compact Hausdorff space is a homeomorphism, it follows that $\overline{f}_k$ is also an autohomeomorphism. Let now $A_k = p^{-1}(d_k)$ $\forall$ k $\in$ **N**, in $P2^c$. We now claim that the map $\overline{f}_k(A_1) = A_k$ $\forall$ k = 1,2,3,..... In order to see this, let a $\in A_1$ and fix k. We first note that p(a) = d_1 and $f_k \circ p(a) = f_k(d_1) = d_k$ by choice of f_k. By commutativity, $p \circ \overline{f}_k(a) = d_k ==> \overline{f}_k(a) \in p^{-1}(d_k) = A_k ==> f_k(A_1) \subset A_k$. On the other hand, if x $\in A_k$, then $\overline{f}_k$ being a homeomorphism, $\exists$ an a such that $\overline{f}_k(a) = x$. $==> p \circ \overline{f}_k(a) = p(x) = d_k$. However, $f_k \circ p(a) = d_k ==> f_k(p(a)) = d_k ==> p(a) = d_1$ as f_k is a homeomorphism and $f_k(d_1) = d_k$ by choice of f_k $==> a \in p^{-1}(d_1) = A_1$. Thus $f_k(A_1) = A_k$. Let A be a choice set so that $A \cap A_k = \{a_k\}$, for every k $\in$ **N**. This means A represents all the fibres $p^{-1}\{s_k\}$, containing precisely one element from each fibre. It is therefore obvious that p(A) = S and $p(a_k)$ s_k $\forall$ k $\in$ **N**. From our discussion on the existence of homeomorphisms above, we now have:

Lemma 1: The set $A = \bigcup_{k \in N} \{a_k\}$ is homogeneous in $P2^c$.

Remark 2: A being the preimage of a dense subset in 2^c, under an irreducible map, it follows that A is dense.

Remark 3: The map a |—> –a in 2^c being the identity homeomorphism, by our discussion above, lifts to an autohomeomorphism, which we designate as $\overline{f}_1$.

Remark 4: We have thus succeeded in obtaining a countable, perfect ED homogeneous subspace A in $P2^c$. We now merely need to note that the operation +: D x D ——> D in 2^c, when restricted to **either** of the two variables is a homeomorphism of D onto D [B] and this lifts to an autohomeomorphism of A as proved in Lemma 1.

Remark 5: In the above discussion we have actually proved that every autohomeomorphism of 2^c lifts to an autohomeomorphism of $P2^c$. One can also see that if a compact Hausorff space X is rigid so is its projective cover PX. Rigid Stone spaces exist within ZFC in abundance [DGS].

Remark 6: A above, has exponent 2.

Remark 7: If B is any clopen neighborhood of the zero element a_1 in A, the preimage of a_1 under + contains the diagonal elements of A x A. In fact in A x A the diagonal is precisely the preimage of a_1. Unless we assume the existence of measurable cardinals, which is outside of ZFC, B x B is **not** ED.([G], [M], [R], [S2]).

Remark 8: Suppose there exists a nondiscrete countably infinite ED topological group G satisfying the T_1 separation axiom. Using a result of Malykhin [M], G has a clopen(closed and open) subgroup H with a + a = 0 for every a $\in$ H. This makes H abelian. Being open, H is ED. It is also easy to see that H is infinite. H can be considered as a vector space over the 2 element field $Z_2 = \{ 0, 1 \}$. It is well–known that H is algebraically isomorphic to the additive group $\oplus_\omega Z_2$. This leads to the following charactereization of Arhangelskii's problem: An ED topological group satisfying the conditions of Problem 6 of Ponomarev & Shapiro exists iff one can define a nondiscrete ED topology on $\oplus_\omega Z_2$ with the desired properties.

One can also complete H into a Boolean ring. Then the Stone space X of H is a totally disconnected and metrizable compact Hausdorff space [H]. These are two possibilities for X: (1) X is countably infinite. In this case X is a scattered metrizable compact space. Among the infinitely many possible examples, one may mention $\alpha\mathbf{N}$ the one–point compactification of **N**. (2) A second possibility is the Cantor space $\mathbf{2}^{\aleph_0}$.

Acknowledgements: The author thanks, Professor Les Reid for the many valuable discussions he has had, concerning the algebraic aspects of this problem. He thanks the referree of this paper for the many suggestions.

:B I B L I O G R A P H Y:

[A] Arhangelskii, A.V., Every extremally disconnected bicompactum of weight c is inhomogeneous. Soviet Math Dokl, 8, 897–900, 1967.

[B] Bourbaki, N., Elements of Mathematics–General Topology, Part 1, Addison–Wesley Pub Co Reading MA 1966.

[C] Comfort, W.W., Topological Groups, in: K. Kunen and J.E. Vaughan, eds, Handbook of Set–theoretic Topology, North Holland, Amsterdam, 1984.

[CM] Comfort, W.W. & van Mill, J. , A homogeneous extremally disconnected countably compact space, Topology and its Applications 65–73d, 1987.

[DGS] Dow, A., Gubbi, A.V. & Szymanski, A., Rigid Stone spaces within ZFC. Proc AMS, vol 102, No 3, 1988.

[E] Efimov, B., Extremal disconnectedness and dyadicity. Gen Top and its Aplcns to Modern An and Alg–Proc. 2nd Prague Top. Symp 1966.

[G1] Gleason, A.M., Projective topological spaces, Ill J Math. 2, 482–489, 1958.

[G2] Gubbi, A.V., Products of extremally disconnected spaces. Preprint.

[GS] Gubbi A.V. & Szymanski, A. Some Stone spaces and their applications. Preprint.

[H] Halmos, P.R., Lectures on Boolean Algebras. Springer–Verlag, New York, 1974. (Published earlier in 1966 by van Nostrand.)

[L] Louveau, A., Sur un article de S. Sirota, Bul Sci Math (France) Ser 2 , 96, 3–7, 1972.

[M] Malykhin, V.I., Extremally disconnected and nearly extrermally disconnected groups, Soviet Math Dokl. 16–25, 1975.

[PS] Ponomarev, B.I. & Shapiro, L.B., Absolutes of topological spaces and their continuous maps. Russian Math Surveys 31:5 (1976), 138–154.

[R] Rajagopalan, M., personal communication.

[S1] Sirota, S.M., The products of topological groups and extremal disconnectedness. Math USSR Sbornik, 8, 169 – 180, 1969.

[S2] Szymanski, A., Products and measurable cardinals. Proc 13th Winter School on Abstract Analysis, Supp ai Rendi del Circolo Mat di Palermo Serie II, no 11, 1985.

Ideals in General Topology

T. R. Hamlett and Dragan Janković

Department of Mathematics
East Central University
Ada, Oklahoma, USA, 74820

§1. THE SETTING

An *ideal* on topological space (X,τ) is a nonempty collection of subsets of X satisfying the following two properties:

I. If $A \in I$ and $B \subseteq A$, then $B \in I$ (heredity).

II. If $A \in I$ and $B \in I$, then $A \cup B \in I$ (additivity).

A σ-*ideal* on a space (X,τ) is an ideal which satisfies:

III. If $\{A_i: i = 1, 2, 3, \ldots\} \subseteq I$ then $\bigcup\{A_i: i = 1, 2, 3, \ldots\} \in I$ (countable additivity).

If $X \notin I$ then I is called a *proper ideal*. The collection of complements of a proper ideal is a filter, hence proper ideals are sometimes called dual filters.

We will denote by (X,τ,I) a topological space (X,τ) and an ideal I on X.

Given a space (X,τ,I) a set operator $(\ )^*:P(X) \to P(X)$, called the *local function of* I *with respect to* τ in (Vaidyanathaswamy, 1945), is defined as follows: for $A \subseteq X$, $(A)^*(I,\tau) = \{x \in X : U \cap A \notin I \text{ for every } U \in N(x)\}$, where $N(x) = \{U \in \tau : x \in U\}$. A Kuratowski closure operator Cl* for a topology $\tau^*(I)$ finer than τ is defined as follows: $Cl^*(A) = A \cup (A)^*(I,\tau)$. When there is no ambiguity, we will simply write $A^*(I)$ or A^* for $(A)^*(I,\tau)$ and τ^* for $\tau^*(I)$. A basis $\beta(I,\tau)$ for $\tau^*(I)$ can be described as follows: $\beta(I,\tau) = \{U-I: U \in \tau, I \in I\}$ (Vaidyanathaswamy, 1947). Examples are provided in (Vaidyanathaswamy, 1945) and (Janković, a) showing that β is not, in general, a topology.

In what follows we will denote by Int(A) and Cl(A) the interior and closure of $A \subseteq (X,\tau)$ respectively.

We will denote by Int*(A) and Cl*(A) the interior and closure of $A \subseteq (X,\tau,I)$ with respect to τ^* respectively. We abbreviate the phrase "if and only if" by "iff", we abbreviate "neighborhood" by "nbd", and $N(x)$ denotes $\{U \in \tau : x \in U\}$ where $x \in (X,\tau)$.

§2. HISTORICAL BACKGROUND AND THE BANACH LOCALIZATION PROPERTY

One historical line of investigation which led to a study of ideals was motivated by the desire to generalize the concepts of "limit point", "closure point", and "condensation point" (see Martin, 1961). This is accomplished quite nicely in the following way. Given a topological space (X,τ), denote by I_f the ideal of finite subsets of X, denote by I_c the ideal of countable subsets of X, and by $\langle A \rangle = \{B \subseteq X: B \subseteq A\}$ the principal ideal of all subsets of A where $A \subseteq X$. Then $x \in X$ is: a limit point of $A \subseteq X$ iff $x \in A^*(I_f)$, when (X,τ) is a T_1 space; a closure point of A iff $x \in A^*(\langle\phi\rangle)$; and a condensation point of A iff $x \in A^*(I_c)$. An interesting application of this idea is in the setting of the usual real line, denoted by R, with the σ-ideal of subsets of Lebesgue measure zero, or null sets, denoted by N. A point x is called a point of "non-null approach" (Martin, 1961) by a set E if every nbd of x intersects E in a set of positive outer Lebesgue measure. Equivalently, x is a point of non-null approach by E iff $x \in E^*(\mathcal{N})$. Let τ denote the usual topology on the reals; i.e., our space is $(\mathcal{R},\tau,\mathcal{N})$. Scheinberg (1971) showed that the Borel sets of $\tau^*(\mathcal{N})$ are precisely the Lebesgue measurable sets.

Another historical line of investigation which naturally involves the concept of an ideal is motivated by the observation that many interesting theorems concern description of global properties of sets or functions from their local properties. If P is a property of sets in a topological space (X,τ), Kuratowski (1948, 1966) defines a subset A of X to have the property P at a point $x \in X$ if there exists a nbd U of x such that $U \cap A$ has the property P. Semadeni (1963) states: "In spaces satisfying the second axiom of countability inference from local belonging to a class to global belonging is often trivial and efforts are made to extend this to more general cases." Semadeni cites several examples to substantiate the above claim including (Montgomery, 1935), (Kuratowski, 1935), (Michael, 1954) and (Mrowka, 1957). Also mentioned in (Semadeni, 1963) are "Bourbaki's integration theory in locally compact spaces the requirement of which is that locally negligible sets be of measure zero, and Brelot's (1958) results on locally polar sets in the general potential theory...".

Semadeni (1963) and others have recognized a crucial property of ideals which was first proved for the ideal of meager sets by Banach (1930).

Definition. Given a space $(X,\tau,\mathcal{I})$, $\mathcal{I}$ is said to have the *Banach localization property with respect to* τ, denoted $\mathcal{I} \sim \tau$, if the following holds for every $A \subseteq X$: if for every $a \in A$ there exists a $U \in \mathcal{N}(a)$ such that $U \cap A \in \mathcal{I}$, then $A \in \mathcal{I}$.

The above property of ideals has been called "compatible" (Njastad, 1966), (Janković, a), (Hamlett, a), "super-compact"

(Vaidyanathaswamy, 1945), "adherence" (Vaidyanathaswamy, 1960), and "Strong Banach's localization property" (Semadeni, 1963). The fact that the ideal of meager sets, denoted by I_m, has this property was first proven for metric spaces by Banach (1930), extended to general topological spaces by Kuratowski (1948), and is known as the *Banach Category Theorem* (Oxtoby, 1980, p. 62, p. 96-97). It is known that the ideal of nowhere dense sets, denoted I_n, and the ideal of scattered sets, denoted I_s, also satisfy this property in any topological space (Vaidyanathaswamy, 1945), and in T_1 spaces respectively. It is also known that in a hereditarily Lindelöf space, any σ-ideal satisfies this property (Jankovic´, a, Theorem 4.2). These investigators (Jankovic´, c) have devised a technique for extending any given ideal I to a larger ideal $\tilde{I}$, $I \subseteq \tilde{I}$, which has this property, and extending $\tilde{I}$ to a σ-ideal $(\tilde{I})\sigma$, $I \subseteq \tilde{I} \subseteq (\tilde{I})\sigma$, which also has this property. Consider, for example, the ideal of finite subsets of a space (X,τ), denoted by I_f. In general I_f does not have the Banach localization property (specifically, $I_f \sim \tau$ iff (X,τ) is hereditarily compact), denoted $I_f \not\sim \tau$. The extension is defined as $\tilde{I}_f = \{A \subseteq X : A^*(I_f) \varepsilon \tilde{I}_n\}$. Note that $A^*(I_f)$ is the derived set of A if (X,τ) is T_1. $(\tilde{I}_f)\sigma$ is simply all countable unions of members of $\tilde{I}_f$.

To illustrate the importance of this property we offer the following two Theorems.

Theorem (Generalized Cantor-Bendixson) (Freud, 1958) Let (X,τ,I) be a space such that $I \sim \tau$ and $\{x\} \varepsilon I$ for every $x \varepsilon X$. If a set $A \subseteq X$ is closed with respect to τ^*, then A is the union of a set which is perfect with respect to τ and a set in I.

The original Cantor-Bendixson Theorem is a special case of the above theorem for the usual real line, which is second countable and hence hereditarily Lindelof, and the ideal I_c of countable subsets.

Theorem (Jankovic´, a, Theorem 7.9) Let $f:(X,\tau,I) \to (Y,\sigma)$ be a function with $I \sim \tau$ and Y regular. The following are equivalent:

(a) $f:(X,\tau^*) \to (Y,\sigma)$ is continuous,

(b) $f|X^*:(X^*, \tau|X^*) \to (Y,\sigma)$ is continuous, and

(c) $f^{-1}(V)\in\beta(I,\tau)$ for every $V\in\sigma$.

We should also observe that if $I\sim\tau$ then $\beta = \tau^*$ and all the open sets in τ^* have the "simple form" U-I where $U\in\tau$, $I\in I$.

In the latter theorem, X^* is closed and $X-X^*\in I$. In the case then that $I \cap \tau = \{\phi\}$ (in this case, I is said to be "τ-boundary"), we have that the formally weaker continuity of $f:(X,\tau^*) \to (Y,\sigma)$ is equivalent to continuity. These hypotheses will be satisfied in several important cases such as: the usual real line and the ideal N of null sets, any space and the ideal of nowhere dense sets, any Baire space and the ideal of meager sets, and any hereditarily Lindelöf space in which countable sets are not open and the ideal of countable subsets.

§3. THE IDEAL OF CODENSE SETS AND SUBMAXIMAL SPACES

For any nonempty collection A of subsets of a space (X,τ) the smallest ideal on X containing A will be denoted $I(A)$ and $I(A) = \{A:A \subseteq \bigcup_{i=1}^{n} A_i$ for some finite collection $\{A_i:i = 1, 2, \ldots, n\}\subseteq A\}$. Let $C = \{A \subseteq X:\text{Int}(A) = \phi\}$ be the collection of codense sets and let $B = \{\text{Bd}(A):A \subseteq X\}$ be the collection of subsets of the space that are boundaries. The collection B is finitely additive (Scott, 1982) and the collection C is hereditary; consequently, $I(B) = \{A \subseteq B:B\in B\}$ and $A\in I(C)$ iff A is a finite union of codense sets.

The following theorem contains the interesting and surprising fact that $I(B) = I(C)$. Recall that a space is *resolvable* (Hewitt, 1943) if it is the disjoint union of two dense (or codense) subsets. A subset of a space is resolvable if it is resolvable as a subspace (ϕ is vacuously resolvable).

Theorem (Janković, b, Theorem 2.2). The following are equivalent for a subset of a space (X,τ): (a) $A\in I(B)$, (b) $\text{Cl}(A)\in B$, (c) $\text{Int}(\text{Cl}(A))$ is resolvable, (d) $\text{Int}(A)$ is resolvable, and (e) $A\in I(C)$.

Denote the ideal $\mathcal{I}(\mathcal{B}) = \mathcal{I}(\mathcal{C})$ by $\mathcal{R}$. It has been shown by these investigators (Jankovic´, b) that $\mathcal{R}\sim\tau$ and that $\mathcal{R}_\sigma\sim\tau$, where $\mathcal{R}_\sigma$ denotes the countable extension of $\mathcal{R}$. It is clear from the previous theorem that a space $X\in\mathcal{R}$ iff X is resolvable, and several characterizations of resolvable spaces are known. An open question is the following:

Question: Under what conditions is the space $X\in\mathcal{R}_\sigma$?
Recall that spaces having the property that their dense subsets are open are called *submaximal* by Bourbaki (1966), and submaximal dense-in-themselves spaces are called *maximally irresolvable* by Hewitt (1943).

These investigators have characterized submaximal spaces in the following theorem. Denote by $\mathcal{I}_{cd}$ the ideal of closed and discrete subsets of a space.

Theorem (Jankovic´, b, Theorem 4.1) The following are equivalent for a space (X,τ): (a) (X,τ) is submaximal, (b) $\tau = \tau^*(\mathcal{R})$, (c) $\mathcal{C} \subseteq \mathcal{I}_{cd}$, (d) $\mathcal{B} \subseteq \mathcal{I}_{cd}$, and (e) $\mathcal{R} \subseteq \mathcal{I}_{cd}$.

§4. $\mathcal{I}$-COMPACT SPACES

Consider the following definition.

Definition (Newcomb, 1967), (Rančin, 1972) A subset A of a space $(X,\tau,\mathcal{I})$ is said to be *$\mathcal{I}$-compact* if for every open cover $\{U_\alpha:\alpha\in\Delta\}$ of A, there exists a finite subcollection $\{U_{\alpha_i}:i = 1, 2, \ldots, n\}$ such that $A-\bigcup\{U_{\alpha_i}\}\in\mathcal{I}$. $(X,\tau,\mathcal{I})$ is said to be *$\mathcal{I}$-compact* if X is $\mathcal{I}$-compact as a subset.

A space is obviously compact iff it is $\langle\phi\rangle$-compact. The following theorem (Hamlett, a, Theorem 1.4) shows that H-closedness is also a special case. Recall that a space is said to be quasi H-closed, abbreviated QHC, if every open cover of the space has a finite subcollection whose closures cover the space; and that a space is H-closed iff it is QHC and Hausdorff.

Theorem Let $(X,\tau,\mathcal{I})$ be a space with $\mathcal{I}_n$ the ideal of nowhere dense subsets of X.

(a) If (X,τ,I) is I-compact and $\tau \cap I = \{\phi\}$, then (X,τ) is QHC.

(b) If $I_n \subseteq I$ and (X,τ) is QHC, then (X,τ) is I-compact.

The following result is an immediate corollary (Hamlett, a, Corollary 1.5).

Corollary Let (X,τ) be a space.

(a) (X,τ) is I_n-compact iff (X,τ) is QHC.

(b) If (X,τ) is Hausdorff, then (X,τ) is I_n-compact iff (X,τ) is H-closed.

Lambrinos (1973) defines a subset A of a topological space (X,τ) to be *bounded* if every open cover of X contains a finite subcover of A. Clearly a subset A of a space X is bounded iff it is $\langle X-A \rangle$-compact.

We see then that the study of I-compact spaces has compactness, H-closedness, and boundedness as special cases, unified in one theory.

The standard convergence type characterizations of compactness generalize quite nicely. We denote by $P(X)$ the power set of all subsets of X. Given a space (X,τ,I) and a collection $A \subseteq P(X)$, A is said to have the *finite intersection property modulo* I, denoted I-FIP, if for every finite subfamily $\{A_i\}$ of A we have $\bigcap A_i \notin I$.

Theorem (Hamlett, a, Theorem 1.13) Let (X,τ,I) be a space. The following are equivalent.

(a) (X,τ) is I-compact.

(b) For every family $\{F_\alpha\}$ of closed sets such that $\bigcap F_\alpha \neq \phi$, there exists a finite subfamily $\{F_{\alpha i}\}$ such that $\bigcap F_{\alpha i} \in I$.

(c) (Newcomb, 1967) For every family $\{F_\alpha\}$ of closed sets with I-FIP, $\bigcap F_\alpha \neq \phi$.

(d) (Newcomb, 1967) Every filterbase B with $B \cap I = \phi$ has an accumulation point.

(e) (Newcomb, 1967) Every filter F with $F \cap I = \phi$ has an accumulation point.

(f) Every maximal filterbase M with $M \cap I = \phi$ converges.

(g) (Newcomb, 1967) Every ultrafilter $\mathcal{U}$ with $\mathcal{U} \cap \mathcal{I} = \phi$ converges.

(h) For every filterbase $\mathcal{B}$ with $\mathcal{B} \cap \mathcal{I} = \phi$, $\bigcap\{B^*: B \in \mathcal{B}\} \neq \phi$.

It is interesting to compare these characterizations with the known characterizations of H-closed spaces such as: (Berri, 1970) A Hausdorff space is H-closed iff every open filterbase has an accumulation point. Condition (d) of the previous theorem gives that this is equivalent to: Every filterbase $\mathcal{B}$ with $\mathcal{B} \cap \mathcal{I}_n = \phi$ has an accumulation point.

Newcomb (1967) shows that if $f:X \to Y$ is a function and $\mathcal{I}$ is an ideal on X, then $f(\mathcal{I}) = \{f(I): I \in \mathcal{I}\}$ is an ideal on Y. Newcomb (1967) also shows that if $f:(X,\tau,\mathcal{I}) \to (Y,\sigma)$ is a continuous surjection, and (X,τ) is $\mathcal{I}$-compact, then (Y,σ) is $f(\mathcal{I})$-compact. It follows (Hamlett, a, Theorem 1.14) that if a product of spaces is $\mathcal{I}$-compact, then each factor is $P(\mathcal{I})$-compact where P is the appropriate projection map (which is a continuous surjection). A generalized Tychonoff Theorem can then be stated as follows.

Theorem (Hamlett, a, Theorem 1.15) Let $\{(X_\alpha,\tau_\alpha,\mathcal{I}_\alpha): \alpha \in \Delta\}$ be a collection of spaces with (X_α,τ_α) $\mathcal{I}_\alpha$-compact for each $\alpha \in \Delta$, and let P_α denote the projection in coordinate α. Let X denote $(\Pi X_\alpha, \Pi\tau_\alpha)$ and let $\mathcal{I}$ be an ideal on X. The following are equivalent.

(a) $\mathcal{I}(\mathcal{A}) \subseteq \mathcal{I}$ where $\mathcal{A} = \{P_\alpha^{-1}(I_\alpha): I_\alpha \in \mathcal{I}_\alpha, \alpha \in \Delta\}$.

(b) (Rančin, 1972) $P_\alpha^{-1}(\mathcal{I}) \subseteq \mathcal{I}$ for every $\alpha \in \Delta$.

(c) (Newcomb, 1967) $M \notin \mathcal{I}$ implies $P_\alpha(M) \notin \mathcal{I}_\alpha$ for every $\alpha \in \Delta$.

(d) There exists a $\beta \in \Delta$ such that $P_\beta(I) \in \mathcal{I}_\beta$ implies $I \in \mathcal{I}$.

Furthermore, if $\mathcal{I}$ satisfies one of the above equivalent conditions then X is $\mathcal{I}$-compact.

If each $\mathcal{I}_\alpha$ in the above theorem is $\langle\phi\rangle$, we obtain the classical Tychonoff Theorem. Also by letting $\mathcal{I}_\alpha = \mathcal{I}_n(\tau_\alpha)$ for each α, we obtain the well known result that H-closedness is productive as a corollary.

A function $f:(X,\tau,I) \to (Y,\sigma)$ is said to be *pointwise I-continuous*, denoted *PIC*, (Kaniewski, 1986) if $f:(X,\tau^*) \to (Y,\sigma)$ is continuous. The following theorem generalizes a well known and important result on mappings of compact spaces.

Theorem (Hamlett, a, Cor. 2.8 and Theorem 2.9) Let $f:(X,\tau,I) \to (Y,\sigma)$ be *PIC* with (X,τ) I-compact and (Y,σ) Hausdorff. Then

(a) $f:(X,\tau^*(I)) \to (Y,\sigma^*(f(I)))$ is closed.

Furthermore, if f is a bijection, then

(b) $f:(X,\tau^*) \to (Y,\sigma^*)$ is a homeomorphism.

It is well known that compact Hausdorff spaces are maximal compact and minimal Hausdorff. These two results generalize as follows.

Theorem (Newcomb, 1967) If (X,τ,I) is I-compact and Hausdorff, then τ^* is the largest I-compact topology containing τ.

Theorem (Hamlett, a, Theorem 2.12) Let (X,τ,I) be I-compact Hausdorff and let σ be a topology for X such that $\sigma \subseteq \tau^*$. If (X,σ) is Hausdorff, then $\tau^* = \sigma^*$.

Newcomb (1967) has shown that if (X,τ,I) is I-compact and $I \cap \tau = \{\phi\}$, then (X,τ) is pseudocompact.

We conclude our sampling of known results on I-compactness with the following decomposition theorem (Hamlett, a, Corollary 2.6) which generalizes a result of Rančin (1972).

Theorem Let (X,τ) be an I_m-compact space. Then X^* (which is τ-closed) is QHC as a subspace of (X,τ), and $X-X^*$ is meager.

REFERENCES

1. S. Banach, *Théorème sur les ensembles de première catégorie*, Fund. Math. 16 (1930), 395-398.

2. M. P. Berri, J. R. Porter, and R. M. Stephenson, Jr., *A survey of minimal topological spaces*, Proc. Kanpur Topological Conference 1968, General Topology and its Relations to Modern Analysis and Algebra III, Academic Press, New York, 1970.

3. N. Bourbaki, *General Topology*, Addison-Wesley, Mass., 1966.

4. M. Brelot, *Axiomatique des fonctions harmoniques et subharmoniques dans un espace localement compact*, Séminaire de Théorie du Potentiel, Vol. 2 (1958), No. 1, 40.

5. G. Freud, *Ein beitrag zu dem satze von Cantor and Bendixson*, Acta Math. Hung., 9 (1958), 333-336.

6. a. T. R. Hamlett and D. Jankovic´, *Compactness with respect to an ideal*, submitted.

7. b. T. R. Hamlett and D. Jankovic´, *Ideals in topological spaces and the set operator* ψ, submitted.

8. E. Hewitt, *A problem in set theoretic topology*, Duke Math. J., 10 (1943), 309-333.

9. a. D. Jankovic´, and T. R. Hamlett, *New topologies from Old via Ideals*, to appear in the Amer. Math. Monthly.

10. b. D. Jankovic´, and T. R. Hamlett, *The ideal generated by codense sets and the Banach localization property*, submitted.

11. c. D. Jankovic´, and T. R. Hamlett, *Extensions of ideals*, submitted.

12. J. Kaniewski and Z. Piotrowski, *Concerning continuity apart from a meager set*, Proc. Amer. Math. Soc., 98 (1986), 324-328.

13. K. Kuratowski, *Quelques problèmes concernant les espaces métriques non-séparables*, Fund. Math. 25 (1935), 534-545.

14. K. Kuratowski, *Topologie I*, 2- ème éd., Warszawa, 1948.

15. K. Kuratowski, *Topology*, Vol. 1, Academic Press, New York, 1966.

16. P. Lambrinos, *A topological notion of boundedness*, Manuscripta Math., 10 (1973), 289-296.

17. N.F.G. Martin, *Generalized condensation points*, Duke Math. J., 21 (1961), 163-172.

18. E. Michael, *Local properties of topological spaces*, Duke Math. J., 21 (1954), 163-172.

19. D. Montgomery, *Non-separable metric spaces*, Fund. Math., 25 (1935), 527-533.

20. S. Mrówka, *On local topological properties*, Bull. Acad. Pol. Sci., Cl. III, 5 (1957), 951-956.

21. R. L. Newcomb, *Topologies which are compact modulo an ideal*, Ph.D. dissertation, Univ. of Cal. at Santa Barbara, 1967.

22. O. Njastad, *Remarks on topologies defined by local properties*, Avh. Norske Vid.-Akad. Oslo I (N.S.), 8 (1966), 1-16.

23. J. C. Oxtoby, *Measure and Category, second edition*, Springer-Verlag, 1980.

24. D. V. Rancin, *Compactness modulo an ideal*, Soviet Math. Dokl., Vol. 13 (1972), No. 1.

25. Z. Semadeni, *Functions with sets of points of discontinuity belonging to a fixed ideal*, Fund. Math., LII (1963), 25-39.

26. S. Scheinberg, *Topologies which generate a complete measure algebra*, Advances in Math., 7 (1971), 231-239.

27. B. M. Scott and Z. Robinson, *The boundary topology of a space*, Amer. Math. Monthly, 89 (1982), 307-309.

28. R. Vaidyanathaswamy, *The localization theory in set-topology*, Proc. Indian Acad. Sci., 20 (1945), 51-61.

29. R. Vaidyanathaswamy, *Treatise on Set Topology*, Part I, Indian Math. Soc., Madras, 1947.

30. R. Vaidyanathaswamy, *Set Topology*, Chelsea Publishing Co., 1960.

Countably Compact Groups with Non-Countably-Compact Products

Klaas Pieter HART Department of Mathematics and Informatics, Delft University of Technology, Delft, the Netherlands.

Jan VAN MILL Department of Mathematics and Computer Science, Vrije Universiteit, Amsterdam, the Netherlands.

0 INTRODUCTION

The question whether the product of countably compact (topological) groups is countably compact is an old and a natural one.

To put it into perspective consider the following three topological properties: compactness, countable compactness and pseudocompactness.

Of course the product of compact spaces is again compact. The other two properties do not behave so well with respect to products: in (1953) Novák constructed two countably compact spaces X and Y such that $X \times Y$ is not even pseudocompact.

For topological groups the situation is a bit better: in (1966) Comfort and Ross proved that any product of pseudocompact groups is pseudocompact, thus showing that a group structure can have considerable effect on the product behaviour of topological properties.

This left open the question whether there are countably compact groups G and H with $G \times H$ not countably compact. In (1980) van Douwen constructed such groups assuming **M**(artin's) **A**(xiom).

As pointed out by van Douwen, his construction can not work without some form of **MA**. To be precise, the combination of $2^\omega = 2^{\omega_1}$ and an ultrafilter on ω of character ω_1 makes his construction impossible.

We present here a new construction which needs a much weaker form of **MA**, but which is still impossible under the conditions mentioned above.

This note is organized as follows:

In section 1 we sketch the constructions by Novák and van Douwen, and we discuss their similarities. In section 2 we describe our construction and point out the differences with those in section 1. In section 3 we elaborate on the obstruction mentioned above, that prevents us from doing our construction in **ZFC**.

Undefined terms can be found in (Engelking, 1977) and (Comfort, 1984). Other notions will be defined when needed. We let $\mathfrak{c} = 2^{\omega}$, the cardinality of $\mathbb{R}$.

1 NOVÁK AND VAN DOUWEN

The key idea in these constructions is to start with a countably compact space Z without converging sequences. It is easy to see that then for every countably infinite set $F \subseteq Z$ we have $|\overline{F}| \geq \mathfrak{c}$.

One then constructs two countably compact subspaces X and Y such that $D = X \cap Y$ is countably infinite. Then $\Delta D = \{ \langle d, d \rangle : d \in D \}$ is a closed subset of $X \times Y$: it is the intersection of ΔZ and $X \times Y$. Certainly ΔD is not countably compact, being countably infinite; it follows that $X \times Y$ is not countably compact.

1.0 Novák

In (1953) Novák starts with the compact space $Z = \beta\omega$. In this case we even have $|\overline{F}| = 2^{\mathfrak{c}}$, whenever $F \subseteq Z$ is countably infinite. It is then not hard to construct $X, Y \subseteq Z$ with X and Y countably compact and with $X \cap Y = \omega$. In this case $\Delta\omega$ is even a clopen discrete subspace of $X \times Y$, ensuring that $X \times Y$ is not even pseudocompact.

1.1 van Douwen

In (1980) van Douwen constructs, assuming **MA**, a countably compact subgroup S of the group ${}^{\mathfrak{c}}2$, without converging sequences. A relatively straightforward modification of Novák's arguments then gives us in **ZFC** countably compact subgroups G and H of S with $G \cap H$ countably infinite.

Note that since we are working inside ${}^{\mathfrak{c}}2$ these groups are topological groups automatically.

1.2 Squares

Novák's construction immediately gives us a countably compact space with a non-pseudocompact square: $X \oplus Y$.

For topological groups things are not so easy: try to make a topological group out of $G \oplus H$. In (1980) van Douwen did announce the construction of a single countably compact group K with K^2 not countably compact, but this example was never published.

2 A NEW CONSTRUCTION

In this section we outline the construction from (Hart and van Mill, 1988) of two countably compact subgroups H_1 and H_2 of ${}^{\mathfrak{c}}2$ such that $H_1 \times H_2$ is not countably compact. We will also indicate how to modify this construction so as to yield a countably compact group H with H^2 not countably compact.

2.0 A picture of H_1 and H_2

Actually we construct three subgroups D, G_1 and G_2 of ${}^{\mathfrak{c}}2$ satisfying:

1) D is countably infinite,
2) G_1 and G_2 are ω-bounded,
3) if $d \in D, g_1 \in G_1$ and $g_2 \in G_2$, and if $d + g_1 + g_2 = \underline{0}$ then $d = g_1 = g_2 = \underline{0}$, and
4) if $E \subseteq D$ is infinite then it has accumulation points in G_1 and G_2.

We let $H_1 = D + G_1$ and $H_2 = D + G_2$. Then $H_1 \cap H_2 = D$ by 3), and H_1 and H_2 are countably compact by 2) and 4).

Note: a space is called ω-bounded if every countable subset of it has compact closure.

2.1 A sketch of the construction

To begin we take a collection $\mathcal{A}$ of size $\mathfrak{c}$ of subsets of $\mathfrak{c}$ satisfying:

(*) if $\mathcal{A}', \mathcal{A}'' \subseteq \mathcal{A}$ are countable and disjoint then $|\bigcap \mathcal{A}' \setminus \bigcup \mathcal{A}''| = \mathfrak{c}$.

Such a family is called ω-independent.

Split $\mathcal{A}$ into two disjoint parts $\mathcal{A}_1$ and $\mathcal{A}_2$ both of size $\mathfrak{c}$.

We let for $i = 1,2$ G_i be the smallest ω-bounded subgroup of ${}^{\mathfrak{c}}2$ containing $\mathcal{A}_1$ (here we identify a set $X \subseteq \mathfrak{c}$ with its characteristic function $\chi_X : \mathfrak{c} \to 2$). Then 3) already holds for G_1 and G_2.

The main problem then is to construct D. The points of D are constructed in an induction of length $\mathfrak{c}$, one coordinate at a time.

We let $I = \{ x \in {}^{\omega}2 : |x^{\leftarrow}(1)| < \omega \}$ and we will have $D = \{ d_x : x \in I \}$ with $d_x \restriction \omega = x$ and $d_{x+y} = d_x + d_y$, for all x and y in I.

At every stage $\alpha \geq \omega$ we must find $d_x(\alpha)$ for all $x \in I$. Note that the map $x \mapsto d_x(\alpha)$ is to be a homomorphism from I to 2. At the same time we begin to make sure that 4) will hold for one more infinite $E \subseteq D$. This gives us each time fewer than $\mathfrak{c}$ conditions to satisfy. Every such condition determines a dense open set in the space Hom of homomorphisms from I to 2. Now Hom is homeomorphic to the Cantor set, so we may apply the weakest form of **MA**:

"The real line can not be covered by fewer than $\mathfrak{c}$ nowhere dense sets,"

to find a homomorphism satisfying all conditions.

Some minor modifications have to be made to achieve 3) (and in all honesty also 4)), but this is the main idea.

2.2 Towards one group

The idea here is to consider the subgroups $H_1' = H_1 \times \{\underline{0}\}$ and $H_2' = \{\underline{0}\} \times H_2$ of ${}^{\mathfrak{c}}2 \times {}^{\mathfrak{c}}2$. Now $H_1' + H_2'$ being homeomorhic to $H_1 \times H_2$ is not a countably compact group, but we can throw in an extra ω-bounded subgroup G_3 of ${}^{\mathfrak{c}}2 \times {}^{\mathfrak{c}}2$ with $G_3 \cap (H_1' + H_2') = \{\underline{0}\}$ and such that $H = G_3 + H_1' + H_2'$ is countably compact.

Then $H_1 \times H_2$ is a closed subgroup of H^2 so that H^2 is not countably compact.

Again we have sketched the main idea of the construction; the details of 2.1 and 2.2 appear in (Hart and van Mill, 1988).

2.3 Differences

Here we want to point out the differences between van Douwen's example and ours.

To begin the strategies of the respective construction are different: van Douwen begins with a countable group D and in an induction of length $\mathfrak{c}$ extends this group to two countably compact groups whose intersection is D. We on the other hand have our two groups almost ready, except for the countable intersection.

Also the resulting groups are quite different: van Douwen's groups have no non-trivial converging sequences, whereas our groups are full of them. Every countably infinite subset of either G_1 or G_2 is contained in a compact subgroup of G_1 or G_2 respectively, and compact groups contain many non-trivial converging sequences.

We see that, although countably compact spaces with a non-countably compact product must contain sequences without converging subsequences, they still may contain lots of non-trivial converging sequences.

3 WHY MA?

Both in the construction by van Douwen and in ours some form of **MA** seems to be necessary:

First define a space to be initially ω_1-compact if every open cover of it of size at most ω_1 has a finite subcover.

If one assumes $2^\omega = 2^{\omega_1}$ in addition to the form of **MA** needed for the construction then in both cases one can construct the groups in such a way that they become initially ω_1-compact, without any essential modifications. Thus we get two initially ω_1-compact groups whose product is not even countably compact.

On the other hand it is consistent with all cardinal arithmetic that every product of initially ω_1-compact spaces is countably compact. We sketch the argument (see [vD]):

Assume u is an ultrafilter on ω of character ω_1. By Kunen (1980; VIII.A10) the existence of such an ultrafilter is consistent with any consistent cardinal arithmetic.

Let X be initially ω_1-compact and let $\langle x_n : n \in \omega \rangle$ be a sequence in X; since u has character ω_1 and since X is initially ω_1-compact

$$\bigcap_{U \in u} \overline{\{ x_n : n \in U \}} \neq \emptyset.$$

Actually since we tacitly have been assuming that all our spaces are Hausdorff there is exactly one point x in this intersection. We write $x = u\text{-}\lim x_n$.

Now if $\{ X_i : i \in I \}$ is any family of initially ω_1-compact spaces and $\langle x_n : n \in \omega \rangle$ is a sequence in $\prod_{i \in I} X_i$ then $u\text{-}\lim x_n$ exists and is an accumulation point of the set $\{ x_n : n \in \omega \}$.

REFERENCES

1. Comfort, W. W. (1984). "Topological Groups," *Handbook of Set-Theoretic Topology*, (K. Kunen and J. E. Vaughan eds.), North-Holland, Amsterdam, pp.1143–1263.
2. ______________ and Ross, K. A. (1966). "Pseudocompactness and uniform continuity in topological groups," *Pacific J. Math.*, **16**, 483–496.
3. van Douwen, E. K. (1980). "The product of two countably compact topological groups," *Trans. Amer. Math. Soc.*, **262**, 417–427.
4. Engelking, E. (1977). *General Topology*, PWN, Warszawa.
5. Hart, K. P. and van Mill, J. (1988), "A countably compact topological group H such that H^2 is not countably compact," to appear.
6. Kunen, K. (1980). *Set Theory, an introduction to independence proofs*, North-Holland, Amsterdam.
7. Novák, J. (1953). "On the Cartesian product of two compact spaces," *Fund. Math.*, **40**, 106–112.

Bitopological Spaces of Ideals

Melvin Henriksen, Harvey Mudd College

Ralph Kopperman, City College of New York

1 - Introduction and Notation

In [HK] we generalized the idea of a structure space on a set of prime ideals of a commutative ring in the following way: If R is an arbitrary set and X is any collection of subsets of R, then for $r \in R$, we define the *hull* of r, $h(X,r) = \{I \in X : r \in I\}$, and the *hull-complement of r*, $h^c(X,r) = X-h(X,r)$. The topology generated by the family of all hulls of elements of R is called the *dual hull-kernel* topology, that generated by the family of all the hull-completments is the *hull-kernel* topology, and their join is the *patch* topology. We denote these topologies respectively, as S, H, and P.

In [HK], 5.2, it is shown that the set of hulls is closed under finite unions, if and only if there is a binary operation * on R such that (R,*) is an abelian semigroup and X is a set of prime ideals of (R,*). It is further shown there that the set of hulls of elements is closed under finite unions, if and only if there is a subset R' of R and a binary operation * on R' such that (R',*) is an idempotent abelian semigroup with identity element and X is a set of prime ideals of (R',*).

Recall that a *bitopological space* $\underline{X} = (X,T^0,T^1)$ is simply a set with two topologies, and a map $f:X\text{->}X'$ is *pairwise continuous* from $\underline{X}$ to $\underline{X'}$ if f is continuous from (X,T^0) to

(X',T'^0) and from (X,T^1) to (X',T'^1). The idea of pairwise homeomorphism is similarly defined. Bitopological spaces were introduced in [Ke] and further discussed in [La] and elsewhere (see [Sa] for a detailed development and references, and [Ne] for a recent application of these spaces). Our notation on these spaces is consistent with that in the early papers.

Below we characterize (X,S,H) as a bitopological space and we use results from [HK] to show that each (X,S,H) is pairwise homeomorphic to a set of prime ideals of an abelian idempotent semigroup with identity.

2 - Pairwise 0-dimensional Bitopological Spaces and 0-Bases

1 Definition: A bitopological space $\underline{X} = (X,T^0,T^1)$ is:

Pairwise T_0 if whenever x and y are distinct elements of X, there is a $P \in T^0 \cup T^1$ such that $\{x,y\} \cap P$ is a singleton (that is, one of x,y fails to be in the closure of the other in at least one of the T^i).

Pairwise 0-dimensional if pairwise T_0 and for $x \in P \in T^i$, i in $\{0,1\}$, there is a $Q \in T^i$ such that $x \in Q \subset P$ and $X-Q \in T^{1-i}$.

Various notions of pairwise 0-dimensionality were collected and considered by Reilly [Re]. Notice that if $\underline{X}$ is pairwise 0-dimensional, then (X,T^i) is a T_0-space for each $i \in \{0,1\}$. With no loss of generality, let i = 0 and suppose x and y are distinct points of X. If there is a $P \in T^0$, such that $x \in P$ and $y \notin P$ (renaming x,y if necessary), then we are done, otherwise there is such a $P \in T^1$. Find $Q \in T^1$ such that $x \in Q \subset P$ and $X-Q \in T^0$; then $y \in X-Q$, $x \notin X-Q$, completing the proof.

We now look at some facts that will be useful in our work:

We now look at some facts that will be useful in our work:

2 Definition: A *0-base* for $\underline{X}$ is a base B for T^0 such that {X-B : B ∈ B} is a base for T^1, and such that if x,y ∈ X there is a B ∈ B containing exactly one of them.

Given bitopological spaces together with *0-bases* ($\underline{X}$,B), ($\underline{X'}$,B'), a *0-base map* (or simply *0-map*) is an f:X->X' such that if B ∈ B' then $f^{-1}[B]$ ∈ B.

3 Lemma: A bitopological space is pairwise 0-dimensional if and only if it has a 0-base. Further, each 0-map is pairwise continuous.

Proof: First let X be pairwise 0-dimensional; we show that C = {B ∈ T^0 : X-B ∈ T^1} is a 0-base:

If x ∈ P ∈ T^0, our assumption gives us a B ∈ C such that x ∈ B ⊂ P, so C is a base for T^0. Similarly, if x ∈ P ∈ T^1, there is a Q such that x ∈ Q ⊂ P, Q ∈ T^1 and X-Q ∈ T^0; but then B = X-Q ∈ C, x ∈ X-B ⊂ P, so {X-B : B ∈ C} is a base for T^1. Finally, if x ≠ y find P ∈ T^i such that {x,y}∩P contains exactly one element, which, by renaming if necessary, we let be x. But then we can find some B ∈ C such that x ∈ B ⊂ P, so {x,y}∩B = {x}, completing the proof that C is a 0-base. We leave proof of the converse to the reader.

Pairwise continuity comes from the fact that if the inverse image of each basic open set is open, then a map is continuous.

As will be shown below, the assumption that $\underline{X}$ is pairwise 0-dimensional is not strong enough to imply that either (X,T^0) or (X,T^1) is 0-dimensional. But we will show that the join of T^0 and T^1 must be 0-dimensional:

4 Definition: Given a bitopological space X =

(X,T^0,T^1), its *supremum topology* is $T^S = T^0 \vee T^1$ (the join, or supremum of the topologies T^0 and T^1 in the lattice of topologies on X).

5 Proposition: If $\underline{X}$ is a pairwise 0-dimensional bitopological space, then (X,T^S) is a 0-dimensional Hausdorff space.

Proof: By lemma 3 we can choose a 0-base B for $\underline{X}$. In T^S, the elements of B are clopen, and thus the base $\{B \cap (X-B') : B,B' \in B\}$ for T^S is one of clopen sets which separate points by definition of 0-base.

6 Lemma: If R is any set and $X \subset 2^R$, let B be the smallest set closed under finite unions and intersections which contains $\{h(X,r) : r \in R\}$. B is a 0-base for (X,S,H), which is therefore a pairwise 0-dimensional bitopological space.

Proof: By definition, B is a base for S and $\{X-B : B \in B\}$ is one for H. Further, if I and J are distinct elements of X, then there is some $r \in R$ such that $r \in I-J$ or $r \in J-I$; in the first case, $h(r,X) \in S$, $\{I,J\} \cap h(r,X) = \{I\}$, in the second, $h(r,X) \in S$, $\{I,J\} \cap h(r,X) = \{J\}$. Thus B is a 0-base for (X,S,H), and the other assertion follows from lemma 3.

7 Examples: (a) If (X,T^0) is any T_0-space, and B' is any subbase for T^0, let B be the closure of B' under finite unions and intersections. If T^1 is the topology generated by the complements of elements of B, then by definition, B is a 0-base for (X,T^0,T^1). Thus any T_0-topology is a topology of a pairwise 0-dimensional bitopological space.

(b) If X is any compact Hausdorff space, then X is homeomorphic to the space of maximal ideals of C(X) under the hull-kernel topology. This is a more familiar way of expressing many non

0-dimensional topologies as one of the topologies in a pairwise 0-dimensional bitopological space.

8 Definition: A *concrete pairwise 0-dimensional space* (or simply *concrete space*) is one of the form (X,S,H) for some R and $X \subset 2^R$. Since S,H are determined by R,X, we denote the concrete space by (R,X).

Since the class of 0-maps is closed under composition, the class of bitopological spaces with 0-bases, together with these maps forms a category which we call *OB*. Similarly, the class of all concrete pairwise 0-dimensional bitopological spaces forms the class of objects of a category *CO*, in which a map from (R,X) to (R',X') is an φ:R->R' such that for each J $\in$ X', $\varphi^{-1}[J] \in X$.

We now define Con:*OB*->*CO* by letting Con($\underline{X}$,B) = (B,Y), where Y = {h(B,I) : I $\in$ X}, and for f:($\underline{X}$,B)->($\underline{X}$',B'), Con(f):Con($\underline{X}$',B')->Con($\underline{X}$,B) via Con(f)(B') = f^{-1}[B'] whenever B' $\in$ B'.

The next theorem uses these notations to associate concrete pairwise 0-dimensional bitopological spaces with arbitrary 0-dimensional bitopological spaces in a functorial manner:

9 Theorem: The map Con:*OB*->*CO* is a contravariant functor. Moreover, for any ($\underline{X}$,B) in *OB*, $\underline{X}$ is pairwise homeomorphic to (Y,S,H).

Proof: This result is verified by showing the following facts:

If we define s:($\underline{X}$,B)->(B,Y) by setting s(I) = h(B,I) for I $\in$ X, then s:(X,T^0,T^1)->(Y,S,H) is a pairwise homeomorphism.

If f:(X,B)->(X',B') and g:(X',B')->(X",B") are 0-maps, then Con(g∘f) = Con(f)∘Con(g).

If J ϵ X and f:(X,B)->(X',B') is a 0-map, then $(\mathrm{Con}(f))^{-1}[J] = s(f(J))$.

All the proofs are straightforward.

3 - Applications

Here we combine the results of the last section with some in [HK] to say more about concrete representations of pairwise 0-dimensional bitopological spaces. First let *SB* (Semigroup Base) be the full subcategory of *OB* whose objects are those pairs (X,B) in *OB* for which $\emptyset$ ϵ **B** and **B** is closed under finite unions (and whose maps are all those in *OB* between two objects in *SB*). Notice that if f is such a map then Con(f) is a semigroup homomorphism preserving the identity. Also let *CS* (concrete Semigroups) be the subcategory of *CO* consisting of those objects (R,X) such that R is an abelian idempotent semigroup with identity element and X is a set of prime ideals of R, and those maps which are semigroup homomorphisms preserving the identity. We show:

10 Theorem: The restriction Pid (prime ideal) of Con to *SB* has range in *CS*. Each pairwise 0-dimensional bitopological space is pairwise homeomorphic to a space of prime ideals of an abelian, idempotent semigroup with identity.

Proof: For the second assertion, notice that for each pairwise 0-dimensional bitopological space *X* there is a 0-base B which is closed under finite unions and contains $\emptyset$ (eg., B might be the collection of all sets open in T^0 whose complement is open in T^1). Then $* = \cup$ is an idempotent, abelian semigroup operation on **B** with identity $\emptyset$. Also, since Con(X,B) = (B,Y), the following statements are equivalent for B,C ϵ **B**:

B*C ϵ s(I),

$I \in B \cup C$,

$I \in B$ or $I \in C$,

$B \in s(I)$ or $C \in s(I)$.

The equivalence of the first and last shows that each element of Y is a prime ideal.

For R a commutative ring with identity element, let Spec(R) denote the set of proper prime ideals of R. In [H1], M. Hochster characterized the topological spaces (Spec(R),H) and called them *spectral spaces*. He showed also that this cannot be done functorially, in that the functor Spec from the catgory of commutative rings with identity element to the category of topological spaces cannot be inverted. In [H2], Hochster characterizes the spaces (Minspec(R),H), where Minspec(R) denotes the set of minimal prime ideals of R. In [NR], S. Niefield and K. Rosenthal study some categories of sheaves of rings in which a functor related to Spec can be inverted. Using different terminology, Prodanov [Pr] considers pairwise 0-dimensional spaces in which both topologies are quasicompact, and shows that a variety of situations give rise to such spaces.

Our correspondence between pairwise 0-dimensional spaces and spaces of prime ideals of an idempotent abelian semigroup with identity is functorial by theorem 9, but our results are not easily comparable to Hochster's because:

our spaces are bitopological,

our ideals are on certain semigroups, rather than rings, and

we do not obtain <u>all</u> prime ideals of this semigroup.

For our last application we need a brief review of some definitions and results which are discussed and thoroughly motivated in section 3 of [HK]:

11 Definition: Suppose R is a nonempty set, m,n,k are nonnegative integers, and $f \subset R^{m+n+1}$ is a relation.

(a) The quadruple (m,n,k,f) is called a *guided relation.* A *guided function* (m,n,k,f) is a guided relation such that $f:R^{m+n} \to R$ is a function.

(b) A subset I of R is said to be *closed under* (m,n,k,f) if $f[I^m \times (R-I)^n] \subset Co^k(I)$ (where $Co^k(I) = R-I$ if k is odd, I if k is even).

(c) If D is a set of guided relations and X is a collection of subsets of R, then Fe(X,D) denotes the collection of all elements of X which are closed under each element of D.

(d) If $X \subset 2^R$ then a subset $Y \subset X$ is *relationally (functionally) closed in X* if Y = Fe(X,D) for some set D of guided relations (functions). X is *relationally (functionally) closed* if X is relationally closed in 2^R.

12 Theorem: Suppose X is a concrete space. Then:

(a) X is relationally closed if and only if (X,P) is compact.

(b) If additionally, $\cap X \neq \emptyset \neq R-\cup X$, then X is functionally closed if and only if (X,P) is compact.

At the end of [HK], section 3, an example is given of an X,R such that $X \subset 2^R$, (X,P) is compact, but X is not functionally closed. Our problem here is different, since we are allowed to choose R:

13 Corollary: Let $\underline{X}$ be a pairwise 0-dimensional bitopological space. Then:

(a) T^S is compact if and only if whenever $\underline{X}$ is pairwise homeomorphic to a concrete space (R,Y), Y is relationally closed.

(b) T^S is compact if and only if $\underline{X}$ is pairwise homeomorphic to a concrete space (R,Y) with Y functionally closed.

Proof: (a) is immediate from 12 (a). For (b), let B be a 0-base for the pairwise 0-dimensional bitopological space *X*. Then $C = B \cup \{\emptyset, X\}$ is also a 0-base for *X*. If $I \in X$, then $I \notin \emptyset$, so $\emptyset \notin h(C,I) = s(I)$; similarly $X \in s(I)$. Let $Y = s[X] = \{s(I) : I \in X\}$. Then $X \in \cap Y$, $\emptyset \in R - \cup Y$, so by 7 (b), since $Con(X,C) = (C,Y)$, Y is functionally closed, and *X* is pairwise homeomorphic to (Y,S,H) by 5.

References

[HK] M. Henriksen and R. Kopperman, "A General Theory of Structure Spaces with Applications to Spaces of Prime Ideals", to appear in Alg. Universalis.

[H1] M. Hochster, "Prime Ideal Structure in Commutative Rings", Trans. of the AMS 142 (1969), 43-60.

[H2] M. Hochster, "The Minimal Prime Spectrum of a Ring", Can. J. Math. 23 (1972), 749-758.

[Ke] J. C. Kelly, "Bitopological Spaces", Proc. London Math. Soc. 13 (1963), 303-315.

[La] E. P. Lane, "Bitopological Spaces and Quasi-uniform Spaces", Proc. London Math. Soc. 17 (1967), 341-356.

[Ne] C. W. Neville, "A Loomis Sikorski Theorem for Locales", *Papers on General Topology and Related Category Theory and Topological Algebra*, Annals of the N. Y. Acad. Sci., 99-108.

[NR] S. B. Niefield and K. I. Rosenthal, "Sheaves of Semiprime Ideals", to appear in Cah. de top. et geom. diff.

[Pr] I. Prodanov, "An Abstract Approach to the Algebraic Notion of Spectrum", AMS Translations, 1985.

[Re] I. Reilly, *Zero Dimensional Bitopological Spaces*, Indag. Math. 35 No. 2 (1973), 127-131.

[Sa] S. Salbany, *Bitopological Spaces, Compactifications and Completions.* (1970), *Math. Monographs of the Univ. of Cape Town, No. 1.*

Ever Finer Partitions

H. H. Hung

Department of Mathematics
Concordia University
Montréal, Québec, Canada H4B 1R6

THE ideas of weak bases, networks (Arhangel'skiĭ, 1966) and k-networks (Michael, 1966; O'Meara, 1970) as generalizations of bases are not new. Here we propose the idea of an *accumulation capacity retaining* family that generalizes that of the base. The virtue of an ACR family is that, unlike the base, it can, under fairly general circumstances, take the form of (a countable union of) *ever finer partitions*. We are to give three applications of our idea.

The beautiful theorem of Stone (1956), Morita and Hanai (1956) on the metrization of closed images of metrizable spaces, known as Lašnev spaces, had, in the eighteen years since its inception, received much attention and been much improved on, when Siwiec (1974) looked at all the improvements and saw the need for a 'last word' and gave it in Corollary 5. Our idea of an ACR family allows us to factor, in a manner most direct, metrizability into two components, one being a property common to all Lašnev spaces, the other Siwiec's property (given in his Lemma 4) pared down to its essentials, neither of which contains the all important notion of a Fréchet space (at least not in an obvious way), and arrive at an improvement on Siwiec without nearly as much effort.

Conditions for metrizability on compact spaces have also drawn much attention. Among them are the G_δ-diagonals of Šneĭder (1945) and Miščenko's (1962) point-countable *p*-bases (Theorems VII.5 and 6 of Nagata (1985)). In terms of our ever finer partitions, we have the following exceedingly simple characterization of metrizability among (countably) compact spaces, to which Šneĭder, and therefore Miščenko, are easy corollaries: Metrizability is the availability of ever finer *finite* partitions, closures of members of (infinite) nests on which have *singleton* intersections.

The concept of the rank of a base originated with Nagata (1963) and was used to characterize dimensionality: A metric space has $\dim \le n \Leftrightarrow$ it has a base of rank $\le n+1$. Here we show that the concept of rank can be applied to our generalization of the base to characterize metrizability and dimensionality at once, a result that brings to mind vividly Lebesgue's vision of *brickwork*, with *little mortar* (see e.g., Engelking (1978)).

NOTATIONS AND TERMINOLOGY.

1. For any family $\mathcal{Q}$, we write $\overline{\mathcal{Q}}$ for $\{Cl Q: Q \in \mathcal{Q}\}$.
2. Given a sequence $\langle P_i \rangle$ of *ever finer* partitions of a topological space X, i.e., partitions such that P_i refines P_j whenever $i \geq j$. Let $P \equiv \cup\{P_i: i \in \omega\}$. This P, a col-

lection of subsets, is to be called a *partitions collection*, unless we want to draw attention to the natural partial order on it, in which case it is called a *partitions tree*. i) For any $\xi\in X$ and any open neighbourhood U, we write $P(\xi,U)$ for the family of all elements P of P such that a) $\xi\in \mathcal{C}\ell P\subset U$, b) P is *maximal* in terms of set inclusion (so that members of $P(\xi,U)$ are pairwise disjoint).
ii) For any $i\in\omega$ and any $\xi\in X$, $P_i(\xi)\equiv\{P\in P_i: \xi\in \mathcal{C}\ell P\}$, $P_i(-\xi)\equiv\{P\in P_i: \xi\notin \mathcal{C}\ell P\}$. iii) A branch B of the partitions tree P is *free* if $\cap B=\emptyset$. If $|\cap\overline{B}|=1$, we say B is *pointed*.
3. A collection C of (not necessarily open) subsets on a topological space X is said to be *accumulation capacity retaining* (ACR) if for any $\xi\in X$ and any A in any open neighbourhood U of ξ so that $\xi\in \mathcal{C}\ell A$, there is $C\in C$ such that $\mathcal{C}\ell C\subset U$ and $\xi\in \mathcal{C}\ell(C\cap A)$. ACR partitions collections can always be constructed on Lašnev spaces (Hung, 1988). On regular spaces, (open) bases are of course ACR collections as are k-networks, when the spaces are also Fréchet.
4. Given ξ on a topological space X. A *wedging* at ξ is a partition W of some neighbourhood of ξ such that $\xi\in \mathcal{C}\ell W$, for every $W\in W$. We say X is *tight around* ξ if no infinite wedging at ξ is hereditarily closure preserving. Strongly Fréchet spaces (Siwiec, 1971) (and of course first countable spaces), spaces of pointwise-countable type (Arhangel'skiı, 1965), q-spaces, r-spaces, countably bi-quasi-k spaces (Michael, 1964, 1966, 1972), locally countably compact spaces and spaces with property described in 9.1 (b) of Michael (1972) are all tight around every point.

PRELIMINARY RESULT.

The following theorem takes to a particularly pleasing form Nagami's theorem (Theorem 6 of Nagami (1969), Corollary 2.5 of Hung (1977)) and certainly does its part identifying metrizability.

THEOREM 0. A necessary and sufficient condition for a T_0-space X to be metrizable is that there be a sequence $\langle \mathcal{P}_i \rangle$ of ever finer partitions of X such that (†) given any open neighbourhood U of $\xi \in X$, for some $n \in \omega$, $\mathcal{P}_n(\xi)$ is a *wedging* at ξ such that $\cup \overline{\mathcal{P}}_n(\xi) \subset U$.

Proof. To prove the *sufficiency* part, we invoke the main theorem of Hung (1977), noting that in the hypothesis the disjoint pair $\{\sim \cup \overline{\mathcal{P}}_n(\xi), Int \cup \mathcal{P}_n(\xi)\}$ *separates* $\sim U$ and $Int \cup \mathcal{P}_n(\xi)$. For any $x \in Int \cup \mathcal{P}_n(\xi)$, we have $x \notin C\ell P$ if $P \in \mathcal{P}_n(-\xi)$ and therefore $\mathcal{P}_n(x) \subset \mathcal{P}_n(\xi)$ and $\cup \mathcal{P}_n(x) \subset \cup \mathcal{P}_n(\xi)$. For any $y \notin \cup \overline{\mathcal{P}}_n(\xi)$, we have $y \notin C\ell Q$ if $Q \in \mathcal{P}_n(\xi)$ and therefore $\mathcal{P}_n(y) \subset \mathcal{P}_n(-\xi)$ and $\cup \mathcal{P}_n(y) \cap \cup \mathcal{P}_n(\xi) = \emptyset$. The theorem cited above is therefore applicable and X is metrizable.

That the condition is *necessary* can be seen if one notes that a σ-discrete open base $\cup\{\mathcal{B}_i: i \in \omega\}$ gives rise to a sequence $\langle \mathcal{P}_i \rangle$ of ever finer partitions when one lets $\mathcal{P}_j$, for all $j \in \omega$, be $\wedge_{i \le j} [\mathcal{B}_i \cup \{X \setminus \cup \mathcal{B}_i\}]$. □

EXAMPLE. On $\mathbb{R}$, let $\mathbb{Q}$ be enumerated. For every $p \in \mathbb{Q}$, we write L_p for $\{r: r \le p\}$ and R_p for $\{r: p < r\}$ and have a partition $\mathcal{P}_p \equiv \wedge \{\{L_q, R_q\}: q \text{ preceeds } p\}$. Clearly, $\langle \mathcal{P}_p \rangle_{p \in \mathbb{Q}}$ is a sequence of ever finer partitions that satisfies (†). For, for any ξ and any open interval (s,t) around ξ, there are $p, q \in \mathbb{Q}$ such that $s < p < \xi < q < t$. For any $\rho \in \mathbb{Q}$ preceeded by both p and q, $\mathcal{P}_\rho(\xi)$ is a wedging at ξ and we have $\cup \overline{\mathcal{P}}_\rho(\xi) \subset (s,t)$.

REMARKS. From Theorem 0 and the Example that follows, one can see how *closure preserving* is brought into metrizability, the role it plays in metrization and why it is sufficient for metrization in the case of Nagami's closed covers, but not in the case of Ceder's M_1- and M_2-spaces (1961). Note that Bing's Theorem is an easy corollary to our Theorem 0.

MAIN RESULTS.

THEOREM 1. T_3-spaces are metrizable if (and only if) they are *tight around every point* and on them can be constructed an ACR *partitions collection* $\mathcal{P}$.

Proof. First, we prove that given any open neighbourhood U of any ξ (on these spaces), $\cup\mathcal{P}(\xi,U)$ is a neighbourhood of ξ and $\mathcal{P}(\xi,U)$ is a wedging at ξ. Suppose otherwise. Suppose $A\equiv U\backslash\cup\mathcal{P}(\xi,U)\neq\emptyset$ and $\xi\in\mathcal{C}\ell A$. We have a $G\in\mathcal{P}$ such that $\mathcal{C}\ell G\subset U$ and $\xi\in\mathcal{C}\ell(G\cap A)$ and maximal in terms of its inclusion in U. Clearly, $G\in\mathcal{P}(\xi,U)$ although $G\backslash\cup\mathcal{P}(\xi,U)\neq\emptyset$, a contradiction. $\mathcal{P}(\xi,U)$ is therefore a wedging at ξ.

Next, we are to prove that $\mathcal{P}(\xi,U)$ is finite. Suppose otherwise. Suppose $\mathcal{P}(\xi,U)$ is infinite. There are, within a closed neighbourhood of ξ within $\cup\mathcal{P}(\xi,U)$, a set B and an x such that $x\in\mathcal{C}\ell B$ and $x\notin\mathcal{C}\ell(B\cap F)$ for any $F\in\mathcal{P}(\xi,U)$. The fact that there is $G\in\mathcal{P}$ such that $\mathcal{C}\ell G\subset U$ and $x\in\mathcal{C}\ell(G\cap B)$ then violates the maximality of the members of $\mathcal{P}(\xi,U)$ in terms of their inclusion in U and proves that $\mathcal{P}(\xi,U)$ is finite.

Let $\mathcal{P}(\xi,U)=\{P_1,P_2,\ldots.P_N\}$. Let $\mathcal{P}_\mu$ be the first partition to contain one of these and $\mathcal{P}_\nu$ the last. For n=1, 2,3,...N, let U_n be an open neighbourhood of ξ such that $P_n\not\subset U_n$ (which is always possible unless $P_n=\{\xi\}$, in which case, the exercise of the present paragraph is hardly necessary). Let $V\equiv\cap\{U_n\cap U: 1\leq n\leq N\}$. Clearly, $\mathcal{P}(\xi,V)$ is again a finite wedging at ξ *refining* $\mathcal{P}(\xi,U)$ (in the sense not only members of the former are contained in the members of the latter, but also every member of the latter contains some member of the former), none of its members appearing in $\mathcal{P}_\mu$. In other words, we can find an open neighbourhood W of ξ so that the finite wedging $\mathcal{P}(\xi,W)$ refines $\mathcal{P}(\xi,U)$, none of its members appearing in the first ν partitions. Clearly, $\mathcal{P}(\xi,W)$ refines $\mathcal{P}_\nu(\xi)$ and $\mathcal{P}_\nu(\xi)$ refines $\mathcal{P}(\xi,U)$, and $\mathcal{P}_\nu(\xi)$ is a (finite) wedging at ξ such that $\cup\overline{\mathcal{P}}_\nu(\xi)\subset U$. Theorem 0 applies and metrizability is proved.□

REMARK. While it is true that on every Lašnev space we can construct an ACR partitions collection (Hung, 1988), it is not clear whether a space with such a collection has to be Fréchet and therefore Lašnev. In any case, there is the following corollary which is effectively Siwiec's (1974) result.

COROLLARY 2. Lašnev spaces are metrizable if (and only if) they are *tight around every point*.

THEOREM 3. Countably compact Hausdorff spaces $(X,\mathcal{T})$ are metrizable if (and only if) there is a sequence $\langle \mathcal{P}_i \rangle$ of ever finer *finite* partitions on X such that every branch of the partitions tree is *pointed*.

Proof. That $\mathcal{T}(\mathcal{P})$, that which is generated by $\mathcal{P}$, is a strengthening of $\mathcal{T}$ can be seen as follows. Let U be an open neighbourhood of a given $x \in X$. Let $\mathcal{B}$ be such a branch that $x \in \cap\mathcal{B}$. There must be one member B of $\mathcal{B}$ such that $\mathcal{C}\ell B \subset U$ (otherwise there is $y \in \cap\{\mathcal{C}\ell B\backslash U\colon B \in \mathcal{B}\}$ distinct from x). That the identity map $\imath\colon (X,\mathcal{T}(\mathcal{P})) \to (X,\mathcal{T})$ can be extended to $\check{2}^{\omega}$, the compactification of $(X,\mathcal{T}(\mathcal{P}))$ by adjoining to it all *free* branches in $\mathcal{P}$, can be shown similarly, if one makes use of Lemma 2.5 of Comfort and Negrepontis (1974). $(X,\mathcal{T})$ as a continuous image of $\check{2}^{\omega}$ is compact, as is $\check{2}^{\omega}$. As $(X,\mathcal{T})$ is Hausdorff, it is metrizable, as is $\check{2}^{\omega}$. □

REMARKS. In fact, Hausdorff spaces are compact and metrizable if on them can be constructed sequences of ever finer finite partitions such that, given a branch $\mathcal{B}$ and a closed subset C that intersects the closure of every member of $\mathcal{B}$, $C\cap\cap\overline{\mathcal{B}}$ is always a singleton. One may also notice that we have proved that compact metric spaces are *condensations* of dense subsets of $\check{2}^{\omega}$ (cf. Schoenfeld (1974)).

COROLLARY 4 (Šneǐder, 1945; Chaber, 1976). Every countably compact Hausdorff space X is metrizable, if it has a G_δ-diagonal.

Proof. X is compact according to Chaber (1976). There is therefore a sequence of *finite* open covers $\langle H_i \rangle$ of X such that, for every $x \in X$, $\cap\{St(x, \overline{H}_i): i \in \omega\} = \{x\}$. One has a sequence of ever finer finite partitions of the description in Theorem 3, if, for each $n \in \omega$, one lets $P_n \equiv \{\{H, \sim H\}: H \in \cup\{H_i: i \leq n\}\}$, which refines H_n. □

THEOREM 5. For a T_0-space X to be metrizable and of (covering) dimension $\leq n$, a necessary and sufficient condition is that there be a sequence $\langle P_i \rangle$ of ever finer partitions of X such that (††) given any open neighbourhood U of any $\xi \in X$, for some $\nu \in \omega$, $P_\nu(\xi)$ is a wedging at ξ of cardinality $\leq n+1$ such that $\cup \overline{P}_\nu(\xi) \subset U$.

Proof. Note that, for any $i<j$ and any $\xi \in X$, $P_i(\xi)$ is a wedging, if $P_j(\xi)$ is one, and $|P_i(\xi)| \leq n+1$, if $|P_j(\xi)| \leq n+1$. When (††) applies, X is of course metrizable and in particular T_1. It follows then $P_i(\xi)$ is a wedging of cardinality $\leq n+1$, for *any* $i \in \omega$ and any $\xi \in X$.

To prove that our condition is *sufficient* for X to have dimension $\leq n$, it suffices to recognize that $\overline{P}_i$ is locally finite, by virtue of (††), and invoke Theorem 2 of Morita (1955) (Theorem III.9 of Nagata (1965)). To prove that our condition is *necessary*, one notes that one can, in the structure of the Theorem of Morita quoted above, well order the index set Ω, and, for all $i \in \omega$, well order (lexicographically) Ω^i and by implication F_i. For each $F \in F_i$, we can define F' to be $F \backslash \cup\{G \in F_i: G < F\}$ and look at $P_i \equiv \{F': F \in F_i\}$, noting that $\overline{P}_i$ refines F_i and therefore is also locally finite. Clearly, $\{P_i: i \in \omega\}$ is a sequence of ever finer partitions satisfying (††). □

REMARK. In the above, while the *rank* of P is of course 1, that of $\overline{P}$ is $\leq n+1$.

REFERENCES

1. Arhangel'skiĭ, A. V. (1965), *Bicompact sets and the topology of spaces*, Trudy Moskov Mat. Obšč., 13: 3 (Trans. Moscow Math Soc., 1).

2. Arhangel'skiĭ, A. V. (1966), *Mappings and spaces*, Russian Math. Surveys, 21: 115.

3. Ceder, J. G. (1961), *Some generalizations of metric spaces*, Pac. J. Math., 11: 105.

4. Chaber, J. (1976), *Conditions which imply compactness in countably compact spaces*, Bull. Acad. Polon. Sci. Sér. Math. Astronom. Phys., 24: 993.

5. Comfort, W. W. and Negrepontis, S. (1974), The Theory of Ultrafilters, Springer-Verlag, New York, p.26.

6. Engelking, R. (1978), Dimension Theory, North-Holland, Amsterdam, p. 74.

7. Hung, H. H. (1977), *A contribution to the theory of metrization*, Can. J. Math., 29: 1145.

8. Hung, H. H. (1988), *A characterization of Lašnev spaces*, Proc. Amer. Math. Soc., 103: 1278.

9. Michael, E. A. (1964), *A note on closed maps and compact sets*, Israel J. Math., 2: 173.

10. Michael, E. A. (1966), $\aleph_0$-spaces, J. Math. Mech., 15: 983.

11. Michael, E. A. (1972), *A quintuple quotient quest*, Gen. Top. and its Appl., 2: 91.

12. Miščenko, A. S. (1962), *Spaces with point-countable bases*, Soviet Math. Dokl., 3: 855.

13. Morita, K. (1955), *A condition for the metrizability of topological spaces and for n-dimensionality*, Sci. Rep. Tokyo Kyoiku Daigaku Sec. A, 5: 33.

14. Morita, K. and Hanai, S. (1956), *Closed mappings and metric spaces*, Proc. Japan Acad., 32: 10.

15. Nagami, K. (1969), *σ-spaces and product spaces*, Math. Ann., 181: 109.

16. Nagata, J. (1963), *Two theorems for the n-dimensionality of metric spaces*, Compositio Math., 15: 227.

17. Nagata, J. (1965), Modern Dimension Theory, Interscience, New York, p. 125.

18. Nagata, J. (1985), Modern General Topology, Second revised edition, North-Holland, Amsterdam, pp. 402-404.

19. O'Meara, P. A. (1970), *A metrization theorem*, Math. Nach., 45: 69.

20. Siwiec, F. (1971), *Sequence-covering and countably bi-quotient mappings*, Gen. Top. and its Appl., 1: 143.

21. Siwiec, F. (1974), *On the theorem of Morita and Hanai, and Stone*, Topo 72 General topology and its applications, Lecture Notes in Math., 378, Springer, Berlin, pp. 449-454.

22. Schoenfeld, A. (1974), *Continuous surjections from Cantor sets to compact metric spaces*, Proc. Amer. Math. Soc., 46: 141.

23. Šneĭder, V. E. (1945), *Continuous images of Souslin and Borel sets. Metrization theorems*, Dokl. Acad. Nauk SSSR, 50: 77.

24. Stone, A. H. (1956), *Metrizability of decomposition spaces*, Proc. Amer. Soc., 7: 690.

Polyhedral Analogs of Locally Finite Topological Spaces

T. Y. Kong *
Department of Mathematics
City College of New York, CUNY

E. Khalimsky
Department of Computer Science
College of Staten Island, CUNY

Abstract

One approach to digital topology is to consider topological spaces whose points are the pixels or voxels of an image. This leads us to consider locally finite T_0 spaces; i.e., T_0 spaces in which each point has finite closure and a finite neighborhood. Given any locally finite topological space X we construct a polyhedral set $|K(X)|$ such that X is a quotient space of $|K(X)|$ under an open quotient map q. The homotopy properties of X are closely related to the homotopy properties of $|K(X)|$. For this reason we call $(|K(X)|, q)$ the ***polyhedral analog*** of X. The polyhedral analog is a special case of our notion of a ***metric analog*** of a topological space, which is defined categorically.

1 Introduction

Digital topology can be defined as the application of topological ideas to image arrays. Only finitely many pixels or voxels can be represented in

*Permanent Address: Department of Computer Science, Queens College (CUNY), Flushing, Queens, NY 11367

computer memory. But we may sometimes prefer to regard the pixels or voxels in memory as part of a theoretically infinite image array.

The conventional approach to digital topology does not use any topological space, but is based on the graph-theoretic notions of adjacency and connectedness. Refs. [Rosenfeld 1970], [Rosenfeld 1979], [Rosenfeld 1981], [Morgenthaler and Rosenfeld 1981], [Reed and Rosenfeld 1982], [Reed 1984], [Kong and Roscoe 1985] and [Kong 1988] illustrate this approach.

However, it is also possible to define a topology on the set of the pixels or voxels that make up the image. It is natural to require that topology to be such that whenever a two-point set $\{x, y\}$ is connected the pixels or voxels that correspond to x and y are near each other. So for each point x there should only be finitely many points y for which $\{x, y\}$ is connected. One simple class of spaces with this property is the following:

DEFINITION 1.1 *A* locally finite *topological space is a space in which every point has finite closure and a finite neighborhood.*

We are particularly interested in locally finite T_0 spaces. One example of a locally finite T_0 space is the space obtained by equipping the integers $\mathbf{Z}$ with the topology whose basic open sets are the sets of form $\{2k-1, 2k, 2k+1\}$ and $\{2k+1\}$ ($k \in \mathbf{Z}$). We denote this space by $\mathbf{Z}$ also. Since $\mathbf{Z}$ is a locally finite T_0 space so is the product space $\mathbf{Z}^n$ for all positive integers n. One can identify an infinite n-dimensional image array with the space $\mathbf{Z}^n$ and a finite n-dimensional image array with a subspace of $\mathbf{Z}^n$. These spaces have been investigated in the context of digital topology by the second author and others in [Khalimsky et al. 1988], [Kopperman et al. 1988], [Khalimsky 1986], [Khalimsky 1987] and [Kovalevsky 1986].

In image processing there is an important operation known as thinning. Thinning has a very substantial literature; a few references are [Stefanelli and Rosenfeld 1971], [Deutsch 1972], [Rosenfeld 1975], [Tamura 1978], [Lobregt et al. 1980], [Tsao and Fu 1982], [Toriwaki et al. 1982], [Hilditch 1983] and [Ronse 1986]. The effect of a thinning algorithm is to reduce the black point set of the input image down to a "topologically equivalent skeleton". This operation is in many ways analogous to deformation retraction, but is quite tricky to define precisely. In Figure 1 both the 1's and the S's are black points, and the 0's are white

0	0	0	0	0	0	0	0	0	0	0	0	0
0	1	1	1	1	1	1	1	1	1	1	1	0
0	1	S	S	S	S	S	S	S	S	S	1	0
0	1	S	1	1	1	1	1	1	1	S	1	0
0	1	S	1	1	1	1	1	1	1	S	1	0
0	1	S	1	1	0	0	0	1	1	S	1	0
0	1	S	1	1	0	0	0	1	1	S	1	0
0	1	S	1	1	0	0	0	1	1	S	1	0
0	1	S	1	1	1	1	1	1	1	S	1	0
0	1	S	1	1	1	1	1	1	1	S	1	0
0	1	S	S	S	S	S	S	S	S	S	1	0
0	1	1	1	1	1	1	1	1	1	1	1	0
0	0	0	0	0	0	0	0	0	0	0	0	0

Figure 1: Possible Effect of a Thinning Algorithm

points. A thinning algorithm might thin the black point set down to the set of S's. But regarded as subspaces of the space $\mathbb{Z}^2$, the set of S's is not a deformation retract of (and not even homotopy equivalent to) the black point set. The reason is that there are fewer points on the inner perimeter of the black point set than there are S's, so no map from the black point set to the set of S's can induce an isomorphism of the fundamental group.

To study the topological effect of thinning algorithms it is natural to study the homotopy properties of locally finite spaces. As the example in Figure 1 shows, things are not always what they seem. The polyhedral analog we construct in this paper provides a method of translating questions about the homotopy of locally finite spaces into questions about the homotopy of polyhedra.

2 Metric Analogs of Topological Spaces

In the rest of this paper every space is a *based* topological space (i.e., a space in which one point is singled out for special treatment — the distinguished point is called the *base point* of the space). All maps between spaces (including quotient maps) are continuous functions that map the base point of the domain space to the base point of the codomain space. Every homotopy is a fixed base point homotopy — thus if a and b are respectively the base points of A and B then a homotopy $h : A \times [0,1] \to B$ is assumed to satisfy $h(a,t) = b$ for all $t \in [0,1]$. For brevity, these assumptions will not always be made explicit. Thus "a map $f : A \to X$" will mean "a continuous function $f : A \to X$ that maps the base point of A to the base point of X".

DEFINITION 2.1 *Let q be a map with domain Y. We say that a homotopy $h : A \times [0,1] \to Y$* is ignored by q *if qh is a constant homotopy (i.e., if $qh(x,t) = qh(x,0) = qh(x,1)$ for all x in A and t in $[0,1]$).*

DEFINITION 2.2 *Let X be any topological space. A* metric analog *of X is a pair (M,q), where M is a metric space and q is an open quotient map of M onto X, such that if A is any metric space then:*

1. *For every map $f : A \to X$ there is a map $F : A \to M$ such that $f = qF$.*
2. *If $F_0 : A \to M$ and $F_1 : A \to M$ are arbitrary maps such that $qF_0 = qF_1$ then there is a homotopy $H : A \times [0,1] \to M$ between F_0 and F_1 that is ignored by q.*

The first part of this definition stipulates that for any based metric space A and any base point preserving continuous map $f : A \to X$ there is a base point preserving continuous map $F : A \to M$ such that the following diagram commutes:

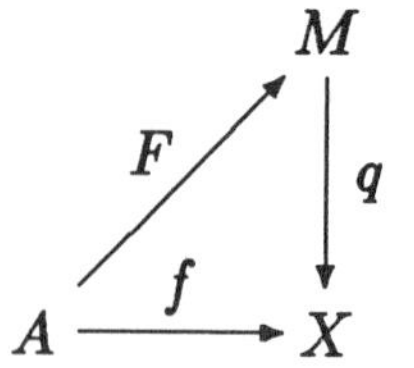

The second part of the definition stipulates that for any map $f : A \to X$ the map $F : A \to M$ which f factors through is unique up to a homotopy H that is ignored by q.

Note that if (M_1, q_1) and (M_2, q_2) are metric analogs of the same topological space X then M_1 and M_2 must have the same homotopy type. In fact, we can say more:

PROPOSITION 2.3 *If (M_1, q_1) and (M_2, q_2) are metric analogs of a topological space X then there are maps $F_1 : M_1 \to M_2$ and $F_2 : M_2 \to M_1$ such that F_1 is a homotopy equivalence, F_2 is a homotopy inverse to F_1, and the following diagram commutes:*

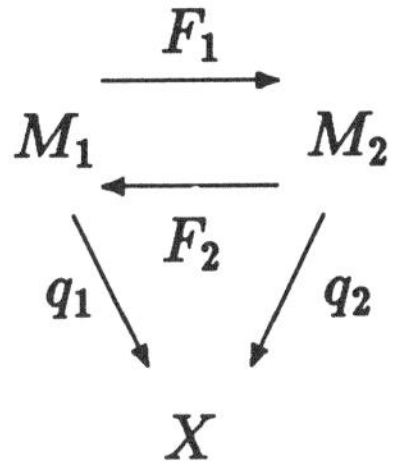

Moreover, F_1 and F_2 are unique up to homotopies that are ignored by q_2 and q_1, and there are homotopies of F_1F_2 and F_2F_1 to the identity maps on M_2 and M_1 that are ignored by q_2 and q_1 respectively.

Conversely, if (M_1, q_1) is a metric analog of X and such a commutative diagram of maps exists for a metric space M_2, where $q_2 : M_2 \to X$ is an open quotient map, and there are homotopies of F_1F_2 and F_2F_1 to the identity maps on M_2 and M_1 that are ignored by q_2 and q_1 respectively, then (M_2, q_2) is also a metric analog of X.

This result is easy to prove. It establishes the uniqueness, up to homotopy equivalences ignored by the quotient maps, of metric analogs.

We now address the question of existence. For $i = 0$ or 1 let G_i be the Sierpinski space on $\{0, 1\}$ with topology $\{\emptyset, \{1\}, \{0, 1\}\}$, and with i as its base point; also, let I_i be the closed unit interval $[0, 1]$ with the usual topology and with i as its base point. Then (I_i, q) is a metric analog of G_i, where q is the open quotient map defined by $q(0) = 0$ and $q(x) = 1$ for all x in $(0, 1]$. Since $[0, 1]^\omega$ is metrizable, $(I_0^\omega \times I_1^\omega, q^\omega \times q^\omega)$ is a metric analog of $G_0^\omega \times G_1^\omega$. It follows easily from [Engelking 1977, Theorem 2.3.26] that there is a base point preserving embedding of any second countable based space in $G_0^\omega \times G_1^\omega$. Hence we deduce the following theorem:

THEOREM 2.4 *Every second countable T_0 space has a metric analog.*

Another existence result is:

THEOREM 2.5 *Every locally finite T_0 space has a metric analog.*

We now outline a proof of this theorem. Suppose X is a locally finite space with base point x_0. We can associate each point in $X - \mathrm{cl}\,\{x_0\}$ with a distinct coordinate of $G_0^{\mathrm{card}(X)}$, and each point in $\mathrm{cl}\,\{x_0\}$ with a distinct coordinate of G_1^{ω}. This allows us to speak of the yth coordinate of a point in $G_0^{\mathrm{card}(X)} \times G_1^{\omega}$, for any point y in X. However, there will be coordinates of $G_0^{\mathrm{card}(X)} \times G_1^{\omega}$ that are not associated with any point in X.

There is an open quotient map $Q_X : I_0^{\mathrm{card}(X)} \times I_1^{\omega} \to G_0^{\mathrm{card}(X)} \times G_1^{\omega}$ given by $Q_X = q^{\mathrm{card}(X)} \times q^{\omega}$. There is also a base point preserving embedding of X in $G_0^{\mathrm{card}(X)} \times G_1^{\omega}$ which maps each point x in X to the point whose yth coordinate is 1 or 0 according as $y \in \mathrm{cl}\,\{x\}$ or $y \notin \mathrm{cl}\,\{x\}$, whose other coordinates of $G_0^{\mathrm{card}(X)}$ are all equal to 0 and whose other coordinates of G_1^{ω} are all equal to 1. Let T be the homeomorphic image of X in $G_0^{\mathrm{card}(X)} \times G_1^{\omega}$ under this embedding and let $W = Q_X^{-1}(T)$. Then one can show that W is metrizable (e.g. by the "Euclidean metric") and it follows from this that $(W, Q_X|W)$ is a metric analog of X.

We now construct a metric analog of a locally finite space that is more easily visualized than the one just described. Our construction is essentially due to P. S. Alexandroff [Alexandroff 1937], though he did not consider the notion of a metric analog.

Let X be a locally finite T_0 topological space. Let $K(X)$ be a (possibly infinite) simplicial complex with one vertex v_x for each point x in X, where the set of vertices $\{v_x | x \in V \subseteq X\}$ is the vertex set of a simplex of $K(X)$ if and only if every two-point subset of V is connected in X. As usual, let $|K(X)|$ be the polyhedron of $K(X)$. Then $|K(X)|$ is metrizable. If a is the base point of X then we take v_a as the base point of $|K(X)|$. From now on we shall identify each vertex v_x of $K(X)$ with the point x in X that it corresponds to.

For each x in X, let $K(x)$ be the set of all simplexes σ in $K(X)$ such that x is a vertex of σ and all other vertices of σ lie in $\mathrm{cl}\,\{x\}$. Also, let S_x be the union of $\{x\}$ with the interiors of the simplexes in $K(x)$. By the *interior* of a simplex σ we mean the set of points in σ that do not lie on any proper face[1] of σ. (Note that this is not the same as the *topological* interior of σ.) Then the sets $\{S_x | x \in X\}$ partition $|K(X)|$ and the map $q : |K(X)| \to X$ defined by $q^{-1}(\{x\}) = S_x$ is an open quotient map.

For a recent application of $|K(X)|$ in the case when X is a subspace of $\mathbf{Z}^3$ see [Kopperman et al. 1988].

We know $(W, Q_X|W)$ is a metric analog of X. One can deduce from this that $(|K(X)|, q)$ is also a metric analog of X, by applying the last part of Proposition 2.3 with $M_1 = W$ and $M_2 = |K(X)|$, where F_2 is the natural embedding of $|K(X)|$ in W and F_2F_1 is the natural deformation retraction of W to $F_2(|K(X)|)$. This gives us:

THEOREM 2.6 *Let X be a locally finite T_0 space. Then $(|K(X)|, q)$ is a metric analog of X.*

This theorem motivates the following definition:

DEFINITION 2.7 *Let X be a locally finite T_0 space. Then we call $(|K(X)|, q)$ the* polyhedral analog *of X.*

Observe that $|K(X)|$ is locally compact. It is a finite polyhedron when X is a finite T_0 space.

There are other ways of constructing geometrically satisfying metric analogs. The second author has verified that the construction in [Khalimsky 1988] actually yields an infinite family of such metric analogs of any locally finite T_0 space. Those metric analogs are in general different from the polyhedral analog we have just defined. Indeed, they need not be polyhedra, as they are built up of *open* simplexes of different dimensions. One should also be able to generalize the construction of [Khalimsky 1988] to build metric analogs from arbitrary open polytopes, not just open simplexes.

---

[1]A *proper face* of a simplex τ is the convex hull of a non-empty proper subset of the vertices of τ.

3 Homotopy Properties of Metric Analogs

When (M, q) is a metric analog of X the homotopy properties of M are closely related to the homotopy properties of X. In the present section we give two theorems which make this vague statement precise.

THEOREM 3.1 *Let (M, q) be a metric analog of X, and let F_1 and F_2 be maps from a metric space A into M. Then qF_1 and qF_2 are homotopic if and only if F_1 and F_2 are homotopic.*

For suppose $H : A \times [0,1] \to X$ is a fixed base point homotopy of qF_1 to qF_2. Then H factors through a map $H_1 : (A \times [0,1])/(\{a\} \times [0,1]) \to X$, where a is the base point of A. Since A is a metric space the quotient space $(A \times [0,1])/(\{a\} \times [0,1])$ is metrizable (e.g., by the Hanai-Morita-Stone Theorem [Engelking 1977, Theorem 4.4.17]). As (M, q) is a metric analog of X, it follows that H_1 in turn factors through a map $H_2 : (A \times [0,1])/(\{a\} \times [0,1]) \to M$. By composing H_2 with the quotient map of $A \times [0,1]$ onto $(A \times [0,1])/(\{a\} \times [0,1])$, we obtain a fixed base point homotopy $H' : A \times [0,1] \to M$ of F_1' to F_2', where F_1' and F_2' are respectively homotopic to F_1 and F_2 (by homotopies that are ignored by q). Hence F_1 and F_2 are homotopic. The converse implication that if F_1 and F_2 are homotopic then so are qF_1 and qF_2 is a consequence of the continuity of q.

COROLLARY 3.2 *If (M, q) is a metric analog of X then q induces a bijection between the path components of M and the path components of X and induces isomorphisms between the homotopy groups of M and those of X. In other words, q is a weak homotopy equivalence.*

This corollary generalizes one of McCord's main results in [McCord 1966], which asserts that the quotient map $q : |K(X)| \to X$ we defined earlier is a weak homotopy equivalence. The proof given in [McCord 1966] is rather different from ours — it depends on Dold and Thom's results in [Dold and Thom 1958] on quasifibrations.

Now let X and Y be locally finite topological spaces, and let (M_X, q_X) and (M_Y, q_Y) be metric analogs of X and Y respectively. Let $f : X \to Y$ be any map. Then $fq_X : M_X \to Y$ is a map from the metric space M_X to

Y, so by the definition of a metric analog there is a map $F : M_X \to M_Y$ such that $fq_X = q_Y F$.

In the case when X is a locally finite T_0 space, $M_X = |K(X)|$, $M_Y = |K(Y)|$ and q_X, q_Y are the open quotient maps we defined earlier there is in fact a *natural* map with this property, which we now define. Let $\sigma = (v_1, v_2, ...v_n)$ be a simplex in $K(X)$. Then the v_i are pairwise connected in X. So by continuity of f the points $f(v_i)$ are pairwise connected in Y and are therefore the vertices of some simplex τ (which need not be n-dimensional) in $K(Y)$. We define $S_f|\sigma$ to be the affine map from σ onto τ that extends f. If we do this for all simplexes σ in $K(X)$ then the map S_f will satisfy $fq_X = q_Y F$, as required.

One consequence of this is the following theorem.

THEOREM 3.3 *Let X and Y be locally finite T_0 spaces, and let $f : X \to Y$ be any map. Let A be a metric space. Then f induces an injection (a surjection) of the homotopy classes of maps from A to X to the homotopy classes of maps from A to Y if and only if S_f induces an injection (a surjection) of the homotopy classes of maps from A to $|K(X)|$ to the homotopy classes of maps from A to $|K(Y)|$.*

This has an immediate corollary relating to the homotopy groups of locally finite spaces.

COROLLARY 3.4 *Let X and Y be locally finite T_0 spaces, and let $f : X \to Y$ be any map. Then the homomorphism of $\pi_n(X)$ to $\pi_n(Y)$ induced by f is an isomorphism (respectively, an epimorphism, a monomorphism) if and only if the homomorphism of $\pi_n(|K(X)|)$ to $\pi_n(|K(Y)|)$ induced by S_f is an isomorphism (respectively, an epimorphism, a monomorphism).*

4 Acknowledgement

We are grateful to Ralph Kopperman and Paul Meyer for useful discussions on the material presented here. This paper incorporates a number of their suggestions.

References

[Alexandroff 1937] P. S. Alexandroff, Diskrete Räume, *Math. Sbornik* **2**, 1937, 501 - 518.

[Deutsch 1972] E. S. Deutsch, Thinning algorithms on rectangular, hexagonal and triangular arrays, *Communications of the ACM* **9**, 1972, 827 - 837.

[Dold and Thom 1958] A. Dold and R. Thom, Quasifaserungen und undendliche symmetrische Produkte, *Annals of Mathematics* (2) **67**, 1958, 239 - 281.

[Engelking 1977] R. Engelking, *General Topology*, Państwowe Wydawnictwo Naukowe (PWN — Polish Scientific Publishers), Warsaw, 1977.

[Hilditch 1983] C. J. Hilditch, Comparison of thinning algorithms on a parallel processor, *Image and Vision Computing* **1**, 1983, 115 - 132.

[Khalimsky 1986] E. Khalimsky, Pattern analysis of N-dimensional digital images, *Proceedings of the IEEE International Conference on Systems, Man and Cybernetics*, CH2364-8/86, 1986, 1559 - 1562.

[Khalimsky 1987] E. Khalimsky, Motion, deformation and homotopy in finite spaces, *Proceedings of the 1987 IEEE International Conference on Systems, Man and Cybernetics*, 87CH2503-1, 1987, 227 - 234.

[Khalimsky 1988] E. Khalimsky, Finite, primitive and Euclidean spaces, *Journal of Applied Mathematics and Simulation*, **1**, 1988, 177 - 196.

[Khalimsky et al. 1988] E. Khalimsky, R. Kopperman and P. Meyer, Computer graphics and connected topologies on finite ordered sets. To appear in *Topology and its Applications*.

[Kong 1988] T. Y. Kong, A digital fundamental group. To appear in *Computers and Graphics*.

[Kong and Roscoe 1985] T. Y. Kong and A. W. Roscoe, Continuous analogs of axiomatized digital surfaces, *Computer Vision, Graphics and Image Processing* **29**, 1985, 60 - 86.

[Kopperman et al. 1988] R. D. Kopperman, P. R. Meyer and R. G. Wilson, A Jordan surface theorem for three-dimensional digital spaces. To appear in *Discrete and Computational Geometry.*

[Kovalevsky 1986] V. A. Kovalevsky, On the topology of discrete spaces, *Studientexte Digitale Bildverarbeitung*, Heft 93/86, 00 0012 93 0, Technische Universität Dresden, Sektionen Mathematik und Informationsverarbeitung, 1986, 56 – 77.

[Lobregt et al. 1980] S. Lobregt, P. W. Verbeek and F. C. A. Groen, Three-dimensional skeletonization, principle and algorithm, *IEEE Trans.* **PAMI 2**, 1980, 75 -77.

[McCord 1966] M. C. McCord, Singular homology groups and homotopy groups of finite topological spaces, *Duke Mathematical Journal* **33**, 1966, 465 – 474.

[Morgenthaler and Rosenfeld 1981] D. G. Morgenthaler and A. Rosenfeld, Surfaces in three-dimensional digital images, *Information and Control* **51**, 1981, 227 – 247.

[Reed 1984] G. M. Reed, On the characterization of simple closed surfaces in three-dimensional digital images, *Computer Vision, Graphics and Image Processing* **25**, 1984, 226 – 235.

[Reed and Rosenfeld 1982] G. M. Reed and A. Rosenfeld, Recognition of surfaces in three-dimensional digital images, *Information and Control* **53**, 1982, 108 – 120.

[Ronse 1986] C. Ronse, A topological characterization of thinning, *Theoretical Computer Science* **43**, 1986, 31 – 41.

[Rosenfeld 1970] A. Rosenfeld, Connectivity in digital pictures, *J. ACM* **17**, 1970, 146 – 160.

[Rosenfeld 1975] A. Rosenfeld, A characterization of parallel thinning algorithms, *Information and Control* **29**, 1975, 286 – 291.

[Rosenfeld 1979] A. Rosenfeld, Digital topology, *American Mathematical Monthly* **86**, 1979, 621 – 630.

[Rosenfeld 1981] A. Rosenfeld, Three-dimensional digital topology, *Information and Control* **50**, 1981, 119 – 127.

[Stefanelli and Rosenfeld 1971] R. Stefanelli and A. Rosenfeld, Some parallel thinning algorithms for digital pictures, *J. ACM* **18**, 1971, 255 – 264.

[Tamura 1978] H. Tamura, A comparison of line-thinning algorithms from a digital geometry viewpoint, *Proc. 4th International Joint Conference on Pattern Recognition*, 1978, 715 – 719.

[Tsao and Fu 1982] Y. F. Tsao and K. S. Fu, A 3D parallel skeletonwise thinning algorithm, *Proc. IEEE PRIP Conference*, 1982, 678 – 683.

[Toriwaki et al. 1982] J. I. Toriwaki, S. Yokoi, T. Yonekura and T. Fukumura, Topological properties and topology-preserving transformation of a three-dimensional binary picture, *Proc. 6th International Conference on Pattern Recognition*, 1982, 414 – 419.

Hereditary Properties and the Hodel Sum Theorem

John E. Mack and Marlon C. Rayburn

Department of Mathematics
The University of Kentucky
Lexington, Kentucky 40506, U.S.A.

Department of Mathematics
The University of Manitoba
Winnipeg, Manitoba R3T 2N2, Canada

1. INTRODUCTION

In [H-2], R. E. Hodel established the sum theorem which we state below as (3.2) Hodel's Lemma. He demonstrated that the sum theorem would hold for any closed hereditary property that satisfies the Locally Finite Sum Theorem, and noted that if P is any such property for which being open hereditary implies hereditary, then any totally normal space with P will have P hereditarily.

This paper generalizes those results. We shall refer to a regular-closed hereditary property P that satisfies the conclusion of Hodel's Lemma as an *Hodel* property and as an *Hodel open* property if additionally whenever the open sets have P, then X has P hereditarily.

Known examples of Hodel open properties (see [H-2]) include being normal, collectionwise normal, paracompact, metacompact, stratifiable, and as we shall note: countably paracompact, countably metacompact, sub-paracompact, sub-metacompact, paraLindelöf, and in spaces of non-measurable cardinality, almost realcompact. Being a cb space is not a Hodel property, but does satisfy a generalized condition called *weak Hodel*. On spaces of non-measurable cardinality, realcompactness is also weak Hodel. We consider the kind of open sets that inherit (weak) Hodel properties and observe the spaces where all open sets are of this kind. As one application, we find an easy proof of Nyikos' result that the Michael plane is hereditarily metacompact.

2 TOTAL SPACES

Let us begin with a consideration of some special open sets and the spaces whose open sets are all of these kinds.

Recollect that an open set G is a regular F_σ-set if $G = \bigcup_1^\infty F_n$, each F_n being a closed set, and $\{int(F_n)\}_1^\infty$ covers G. Every cozero set is a regular F_σ-set and each regular F_σ-set is an open F_σ-set.

2.1 DEFINITION: Let G be an open subset of X such that G is the union of a locally-finite-in-G family $\{U_\alpha: \alpha \in \Gamma\}$ of sets.

a) If each U_α is an open F_σ-set of X, then G is an *almost F_σ-set.*

b) If each U_α is a regular F_σ-set of X, then G is an *almost regular* F_σ*-set*.

c) If each U_α is a cozero set of X, then G is an *almost cozero set*.

Every cozero set of X is an almost cozero set. If D is the discrete space of cardinality c, then $X = D^*$ (the one point compactification) has D as an almost cozero set which is not cozero.

2.2 DEFINITION: a) If every open set of X is an almost F_σ-set, then X is *total*.

b) If every open set of X is an almost regular F_σ-set, then is *totally regular*.

c) If every open set of X is an almost cozero set, then X is *totally C.R.*

We may observe that every perfect space is total, every totally regular space is total and regular, and every totally C.R. space is completely regular and totally regular.

In [D], C. H. Dowker defined a totally normal space so as to be equivalent to total and normal. He showed that perfectly normal spaces are totally normal, that totally normal spaces are completely normal, and proved the "if" direction of the

DOWKER-HODEL THEOREM X is regular and hereditarily paracompact if and only if X is paracompact and totally normal.

The "only if" direction was shown by R. E. Hodel in [H-1]. As the referee remarks, by Urysohn's Lemma every totally normal space is totally C. R.

3. HODEL PROPERTIES

3.1 DEFINITION: a) A topological property P is an *Hodel property* if it is regular-closed hereditary and whenever X has a σ-locally finite open cover by sets whose closures have X has P, then X has P.

b) Property P is an *Hodel open property* if it is an Hodel property such that whenever the open sets of X have P, then X has P hereditarily.

A property P is said to satisfy the "Locally Finite Sum Theorem" if whenever X has a locally finite cover by closed sets, each of which has P, then X has P. The following result basically appears in [H-2] as "Sum Theorem I".

3.2 HODEL'S LEMMA: Any regular-closed hereditary property that satisfies the Locally Finite Sum Theorem is an Hodel property.

Proof: Let $\mathcal{V} = \bigcup_{n=1}^{\infty} \mathcal{V}_n$ be a σ-locally finite open cover of X such that for each n, each $V \in \mathcal{V}_n$, cl(V) has P. Let $V_n = \bigcup \{V: V \in \mathcal{V}_n\}$, so $\{V_n\ n \in \mathbb{N}\}$ is a countable open cover for which $cl(V_n)$ has P for each n. Let $W_1 = V_1$ and $W_n = V_n \setminus \bigcup_{j<n} cl(V_j)$ for $n > 1$. Then W_n is open and $cl(W_n)$ is a regular-closed subset of $cl(V_n)$, hence has P. Since $\{V_n: n \in \mathbb{N}\}$ is a cover and $W_k \cap V_i = \emptyset$ if $k > i$, $\{W_n\ n \in \mathbb{N}\}$ is locally finite, whence $cl(\bigcup W_n)$ has P.

It remains to show that for each n, $\bigcup_1^n cl(V_i) = \bigcup_1^n cl(W_i)$. Since $\bigcup_1^{n+1} cl(V_i) = \bigcup_{i=1}^n cl(V_i) \cup [cl(V_{n+1}) \setminus \bigcup_1^n cl(V_i)]$, let us take $x \in cl(V_{n+1})$. Let G be an open neighborhood of x, and $G_1 = G \setminus \bigcup_1^n cl(V_i)$. Since $G_1 \cap V_{n+1} \neq \emptyset$, we have $G \cap W_{n+1} \neq \emptyset$, and $x \in cl(W_{n+1})$.∎

3.3 THEOREM: Let P be an Hodel property. If a space X has P, then every almost regular F_σ-subset of X has P.

Proof: Let G be an almost regular F_σ-set. Then G is the union of a locally-finite-in-G family $\{U_\alpha: \alpha \in \Gamma\}$ of regular F_σ-sets of X. Each regular F_σ-set U_α is the union of a countable family $\{V_{\alpha,n}: n \in \mathbb{N}\}$ of regular open sets of X such that for each n, $cl(V_{\alpha,n}) \subseteq U_\alpha$. Let $\mathcal{V}_n = \{V_{\alpha,n}: \alpha \in \Gamma\}$. Then $\mathcal{V}_n$ is a locally-finite-in-G family and for each α, $cl(V_{\alpha,n})$ has P. Hence $\mathcal{V} = \bigcup_1^\infty \mathcal{V}_n$ is a σ-locally finite (in G) open cover of G by sets whose closures have P.■

3.4 COROLLARY: Let P be an Hodel open property and X be totally regular. If X has P, then X has P hereditarily.

It has been observed [S, Theorem 4.4] that the properties of (discretely) θ-expandablity and almost (discretely) θ-expandibility satisfy the Closed Subset Theorem, the Locally Finite Sum Theorem, and they are hereditary if every open subset has the property. Hence these properties are Hodel open properties.

In conjunction with identifying Hodel properties, it is useful to note:

3.5 THEOREM [Mo]: Let P be a property such that

a) whenever X_α has P for each $\alpha \in \Gamma$, then the free union $\oplus_\Gamma X_\alpha$ has P, and

b) whenever $f: X \longrightarrow Y$ is a finite-to-one closed map (continuous surjection), then X has P implies Y has P.

Then P satisfies the Locally Finite Sum Theorem.

Since it is easily seen that each of the following properties: subparacompact, submetacompact (also known as θ-refinability), and paraLindelöf, satisfy a) of Theorem 3.5, and are possessed hereditarily by any space where every open set has these properties, that they are Hodel open properties follows respectively from:

3.6 <u>Proposition</u> [B1, Theorem 3.1] Subparacompactness is preserved by closed maps.

3.7 <u>Proposition</u> [J, Proposition 3.4] Submetacompactness is preserved by pseudo-open, finite-to-one maps.

3.8 <u>Proposition</u> [B2, Corollary 3.2] ParaLindelöfness is preserved by closed maps f such that for each $y \in Y$, $f^{\leftarrow}(y)$ is Lindelöf.

Thus these three properties are seen to be Hodel open properties.

For metacompactness, we can improve on Corollary 3.4 slightly. Recollect that a "precise refinement" of an indexed cover $\{E_\alpha: \alpha \in \Gamma\}$ is a cover $\{A_\alpha: \alpha \in \Gamma\}$ with the same index set such that for each $\alpha \in \Gamma$, $A_\alpha \subseteq E_\alpha$. It is well known that any open cover of an arbitrary space that has a point finite open refinement has a precise point finite open refinement. An open shrinkage of an open cover $\mathcal{E} = \{E_\alpha: \alpha \in \Gamma\}$ is an open cover $\{A_\alpha: \alpha \in \Gamma\}$ such that $\{cl(A_\alpha): \alpha \in \Gamma\}$ is a precise refinement of $\mathcal{E}$. In [W], Theorem 15.10, we find the

SHRINKAGE THEOREM: A space is normal if and only if every point finite open cover has an open shrinkage.

3.9 THEOREM: If X is metacompact and every open set is the union of a point finite family of open F_σ-sets, then X is hereditarily metacompact.

Proof: We shall use the fact that X is hereditarily metacompact if and only if every open set of X is metacompact [H-2].

Let G be open in X and $G = \bigcup_\Gamma U_\alpha$, where $\{U_\alpha: \alpha \in \Gamma\}$ is a point finite family of open F_σ-sets of X. Since metacompactness is F_σ-hereditary [C], each U_α is metacompact. Now let $\mathcal{V} = \{V_\gamma: \gamma \in \Lambda\}$ be an open cover of G. For each α, $\{V_\gamma \cap U_\alpha: \gamma \in \Lambda\}$ is an open cover of metacompact U_α. Hence it has a precise point finite open refinement $\{W_{\gamma\alpha}: \gamma \in \Lambda\}$. Let $\mathcal{W} = \{W_{\gamma\alpha}: \gamma \in \Lambda, \alpha \in \Gamma\}$. This is an open refinement of $\mathcal{V}$. Let $x \in G$. Then x is in only finitely many U_α and for each U_α, only finitely many $W_{\gamma\alpha}$. Hence $\mathcal{W}$ is point finite, and G is metacompact.■

This theorem can be used to obtain an easy proof of the following result, first obtained by Nyikos in [Ny].

3.10 EXAMPLE: The Michael plane is hereditarily metacompact.

Let $\mathbb{I}$ be the irrational subspace of the reals, and $\mathbb{R}_\mathbb{Q}$ be the Michael line, obtained from $\mathbb{R}$ by additionally requiring every subset of irrationals to be open. The plane, $\mathbb{R}_\mathbb{Q} \times \mathbb{I}$, has been shown in [Mi] not to be normal.

Let G be any (non-empty) open set of $\mathbb{R}_\mathbb{Q} \times \mathbb{I}$. Then $\text{int}_{R\times\mathbb{I}}G \subseteq G$ is a cozero set in $\mathbb{R} \times \mathbb{I}$, hence in $\mathbb{R}_\mathbb{Q} \times \mathbb{I}$. If $p = (a,b) \in G \setminus \text{int}_{\mathbb{R}\times\mathbb{I}}G$, then a is irrational, so consider $(\{a\} \times \mathbb{I}) \cap G$, open in metric space $\{a\} \times \mathbb{I}$. Hence it is cozero, thus F_σ in $\{a\} \times \mathbb{I}$, hence in $\mathbb{R}_\mathbb{Q} \times \mathbb{I}$. Therefore G is the union of a point-finite family of open F_σ-sets.

A modification of the same reasoning shows the Michael plane to be metacompact. Let $\mathcal{U} = \{U_\alpha: \alpha \in \Gamma\}$ be an open cover of $\mathbb{R}_\mathbb{Q} \times \mathbb{I}$ and for each α, let V_α be the $\mathbb{R} \times \mathbb{I}$-interior of U_α. Let $Y = \bigcup_\Gamma V_\alpha \subseteq \mathbb{R} \times \mathbb{I}$. Then Y is metacompact, so there is a point finite open refinement $\mathcal{W}$ of $\mathcal{V} = \{V_\alpha: \alpha \in \Gamma\}$ covering Y. Let $p \in \mathbb{R}_\mathbb{Q} \times \mathbb{I} \setminus Y$. Then $p = (a,b)$ with a irrational, and $\{U_\alpha \cap [\{a\} \times \mathbb{I}: \alpha \in \Gamma\}$ is an open cover of metric $\{a\} \times \mathbb{I}$. Hence there is a point finite open refinement $\mathcal{W}_a$ covering $\{a\} \times \mathbb{I}$. Then $\mathcal{W} \cup \bigcup \mathcal{W}_a$ is a point-finite open refinement of $\mathcal{U}$ covering $\mathbb{R}_{\mathbb{Q}} \times \mathbb{I}$, and the plane is metacompact.

It is an easy exercise to show:

3.11 LEMMA: Let G be an open subset of X, and let V be a cozero set of G. If $cl_X(V) \subseteq G$, then V is a cozero set of X.

Notice that regular and hereditarily paracompact implies hereditarily normal, i.e. completely normal. So we may compare the following to the Dowker-Hodel Theorem. (N.b., in the absence of Hausdorff, normal need not imply regular.)

3.12 THEOREM: Let X be completely normal. Then X is regular and hereditarily metacompact if and only if X is metacompact and every open set is the union of a point finite family of cozero sets of X.

Proof: "If" follows from Theorem 3.9 and that completely regular (the cozero sets form a base) implies regular. "Only if" Let G be open. By regularity, for each $p \in G$, there is an open V_p with $p \in V_p \subseteq cl_X(V_p) \subseteq G$. Then $\{V_p: p \in G\}$ is an open cover of metacompact G, so there is a point finite open refinement $\mathcal{U} = \{U_\alpha: \alpha \in \Gamma\}$. Now G is normal, so by the Shrinkage Theorem, there is an open cover $\{W_\alpha: \alpha \in \Gamma\}$ such that for each α, $cl_G(W_\alpha) \subseteq U_\alpha$. Since G is normal, for each α there is a G-cozero set C_α with $cl_G(W_\alpha) \subseteq C_\alpha \subseteq U_\alpha \subseteq cl_X(V_p) \subseteq G$ for some $p \in G$. Hence $cl_X(C_\alpha) \subseteq G$, and C_α is a cozero set of X. Thus $\{C_\alpha: \alpha \in \Gamma\}$ is a cover of G by cozero sets of X, which is point finite because $\mathcal{U}$ is.∎

We may now observe that countable paracompactness and countable metacompactness are Hodel open properties:

It is easily checked that both properties are closed hereditary. It is also quickly verified that

a) If for each $\alpha \in \Gamma$, X_α is countably paracompact (countably metacompact), then $\oplus_\Gamma X_\alpha$ is countably paracompact (countably metacompact).

b) Let $f: X \longrightarrow Y$ be a perfect map. Then X is countably paracompact (countably metacompact) if and only if Y is countably paracompact (countably metacompact). Hence by Theorem 3.5, countable paracompactness and countable metacompactness satisfy the Locally Finite Sum Theorem. Therefore both properties are Hodel properties.

3.13 LEMMA: Let $A \subseteq X$. If for each open neighborhood U of A, there is a countably paracompact (countably metacompact) set S such that $A \subseteq S \subseteq U$, then A is countably paracompact (countably metacompact).

3.14 COROLLARY: X is hereditarily countably paracompact (countably metacompact) if and only if every open subset is countably paracompact (countably metacompact).

Thus both properties are Hodel open properties.

Next, let us consider almost realcompactness. A space is almost realcompact if every open ultrafilter Φ for which $\{cl(U): U \in \Phi\}$ has the countable intersection property, has $\bigcap \{cl(U): U \in \Phi\} \neq \emptyset$. Every realcompact space is almost realcompact.

3.15 THEOREM: If X has a σ-locally finite open cover $\mathcal{U}$ such that $|\mathcal{U}|$ is non-measurable, and each set in $\mathcal{U}$ has almost realcompact closure, then X is almost realcompact.

Proof: Let $\mathcal{U} = \bigcup_1^\infty \mathcal{U}_n$ be the open cover with each $\mathcal{U}_n$ locally finite. Suppose Φ to be a free open ultrafilter on

X. We shall show that Φ must contain a sequence $(U_n)_1^\infty$ for which $\bigcap_1^\infty \mathrm{cl}(U_n) = \emptyset$ by considering four cases:

1. $\mathcal{U} \cap \Phi \neq \emptyset$.
2. for each n, $\bigcup \mathcal{U}_n \notin \Phi$.
3. there is an n such that, all k, $\{x: \mathrm{ord}(x, \mathcal{U}_n) \geq k\} \in \Phi$.
4. $\mathcal{U} \cap \Phi = \emptyset$ and there exist integers n and k with $\{x: \mathrm{ord}(x, \mathcal{U}_n) \geq k\} \in \Phi$, while $\{x: \mathrm{ord}(x, \mathcal{U}_n) > k\} \notin \Phi$.

In case 1, let $G \in \mathcal{U} \cap \Phi$. By hypothesis, G is open and cl(G) is almost realcompact. Hence $\{U \cap \mathrm{cl}(G): U \in \Phi\}$ is a free open ultrafilter on cl(G), so there must be a sequence $(V_n)_1^\infty$ in Φ for which $\bigcap_1^\infty \mathrm{cl}[V_n \cap \mathrm{cl}(G)] = \emptyset$. Let $U_n = G \cap V_n \in \Phi$, and note $\bigcap_1^\infty \mathrm{cl}(U_n) = \emptyset$.

For case 2, for each n we can find some $G_n \in \Phi$ with $G_n \cap \bigcup \mathcal{U}_n = \emptyset$. Since $\mathcal{U}$ is a cover, $\bigcap_1^\infty \mathrm{cl}(G_n) = \emptyset$.

With case 3, it suffices to observe that since $\mathcal{U}_n$ is locally finite, $\bigcap_{k=1}^\infty \mathrm{cl}\{x: \mathrm{ord}(x, \mathcal{U}_n) \geq k\} = \emptyset$.

Finally in case 4, let $\Gamma = \{\gamma \subseteq \mathcal{U}_n: |\gamma| = k\}$. For each $\gamma \in \Gamma$, define $G_\gamma = \bigcap \{U: U \in \gamma\} \cap \mathrm{int}\{x: \mathrm{ord}(x, \mathcal{U}_n\} \leq k\}$. It is easy to see that $\{G_\gamma: \gamma \in \Gamma\}$ is a pairwise disjoint family of open subsets of X. Now if $\{x: \mathrm{ord}(x, \mathcal{U}_n) > k\} \notin \Phi$, then $\mathrm{int}\{x: \mathrm{ord}(x, \mathcal{U}_n) \leq k\} = X \setminus \mathrm{cl}\{x: \mathrm{ord}(x, \mathcal{U}_n) > k\} \in \Phi$. So $\bigcup_\Gamma G_\gamma \in \Phi$, while for each $\gamma \in \Gamma$, $G_\gamma \notin \Phi$ by hypothesis. Let $\mathcal{F} = \{ \Lambda \subseteq \Gamma: \bigcup_\Lambda G_\gamma \in \Phi\}$. Then $\mathcal{F}$ is a free ultrafilter on the discrete space Γ. Since $|\mathcal{U}|$ is non-measurable, so is $|\Gamma|$, hence we can find $\{\Lambda_k\}_1^\infty \subseteq \mathcal{F}$ with $\Lambda_{k+1} \subseteq \Lambda_k$ for all k, and $\bigcap_1^\infty \Lambda_k = \emptyset$. For each $k \in \mathbb{N}$, let $V_k = \bigcup \{G_\gamma: \gamma \in \Lambda_k\}$. Then $\{V_k\}_1^\infty \subseteq \Phi$. For $x \in X$, there is a neighborhood N(x) which meets only finitely many $U \in \mathcal{U}_n$, so N can meet only finitely many G_γ's, say $G_{\gamma j}$, $1 \leq j \leq m$. Since $\bigcup_1^m \gamma_j$ is finite, $\bigcap_1^\infty \Lambda_k = \emptyset$, and $\Lambda_{k+1} \subseteq \Lambda_k$, there must be some l_0 with $(\bigcup_1^\infty \gamma_j) \cap \Lambda_{l_0} = \emptyset$. Thus $N \cap V_{\ell_0} = \emptyset$ and $\bigcap_1^\infty \mathrm{cl}(V_k) = \emptyset$.■

3.16 THEOREM [F]: The intersection of any family of almost realcompact subspaces of X is an almost realcompact subspace.

3.17 COROLLARY: Let X be T_1. Then X is hereditarily almost realcompact if and only if every open subset is almost realcompact.

Hence, among T_3 spaces of non-measurable cardinality, almost realcompactness is an Hodel open property, and we may conclude:

3.18 COROLLARY: Suppose X is a totally regular T_1-space such that $|X|$ is non-measurable. If X is almost realcompact, then X is hereditarily almost realcompact.

3.19 EXAMPLE: The cardinality restriction is necessary. Let D be a discrete space of measurable cardinal and let X = $D \cup \{p\}$ be the one point realcompactification of D obtained by collapsing $cl_{\beta X}(\upsilon D \setminus D)$ to a point p. Then X is realcompact and totally C. R., but D is not almost realcompact.

4. WEAK HODEL PROPERTIES

4.1 Definition: A topological property P is a *weak Hodel property* if it is inherited by the closures of cozero sets and whenever X has a σ-locally finite cover by cozero sets whose closures have P, then X has P.

Every Hodel property is a weak Hodel property. Let us see that we may omit the "σ" in the definition. The following lemmas are well known (for example, see pages 4-10 in [Na]):

4.2 LEMMA: The union of any locally finite family of cozero sets is a cozero set.

4.3 LEMMA: Every countable cozero cover of an arbitrary space has a locally finite cozero refinement.

4.4 THEOREM: Every σ-locally finite cozero cover of an arbitrary space has a locally finite cozero refinement.

Proof: Let $\mathcal{U} = \bigcup_1^\infty \mathcal{U}_n$ be a cozero cover of X with each $\mathcal{U}_n = \{U_\alpha : \alpha \in \Gamma\}$ locally finite. By Lemma 4.2, $G_n = \bigcup_{\Gamma_n} U_\alpha$ is a cozero set, and $\{G_n\}_1^\infty$ is a countable cozero cover. By Lemma 4.3, there is a precise locally finite cozero refinement $\{W_n\}_1^\infty$. Let $\mathcal{W}_n = \{W_n \cap U_\alpha : \alpha \in \Gamma_n\}$, and $\mathcal{W} = \bigcup_1^\infty \mathcal{W}_n$. Then $\mathcal{W}$ is a cozero refinement of $\mathcal{U}$. Let $p \in X$ and V_0 be an open neighborhood of p with $V_0 \cap W_n = \emptyset$ except for at most a finite number of n's. For each n with $V_0 \cap W_n \neq \emptyset$, let V_n be an open neighborhood of p which intersects at most finitely many $U_\alpha \in \mathcal{U}_n$. Then $V = V_0 \cap \bigcap V_n$ is an open neighborhood of p which intersects at most finitely many sets of $\mathcal{W}$, and $\mathcal{W}$ is locally finite.■

4.5 COROLLARY: A topological property P is a weak Hodel property if and only if whenever X has a locally finite cover by cozero sets whose closures have P, then X has P.

Notice that if a space has a weak Hodel property, then every almost cozero subset of that space will also have the property. In particular, if the property is a weak Hodel *open* property, then any totally C. R. space having the property will have it hereditarily.

In [Ho], Horne defines a space to be a cb-space if each locally bounded, real valued function h on X is bounded above ($|h| \leq f$) by a continuous function f. These spaces, which lie between countably compact and countably paracompact, were further studied in [M].

4.6 THEOREM: Being a cb-space is a weak Hodel property.

Proof: Let $\{U_\alpha : \alpha \in \Gamma\}$ be a locally finite cover of X by cozero sets. For each α, let U_α be the cozero set for $g_\alpha \in C(X,I)$, where $I = [0,1]$ is the closed unit interval. Let $g = \Sigma_\Gamma g_\alpha$, so $g \in C(X)$ and $g(p) > 0$ for all $p \in X$. Let f be any locally bounded real- valued function on X. Then $F(p) = f(p)/g(p)$ is locally bounded. For each α, let F_α be the restriction of F to $cl(U_\alpha)$. Let $h_\alpha \in C(cl(U_\alpha))$ such that $h_\alpha \geq |F_\alpha|$. Define $H_\alpha : X \longrightarrow \mathbb{R}$ by: $H_\alpha | cl(U_\alpha) = h_\alpha$, and $H_\alpha = 0$ on $X \setminus cl(U_\alpha)$. Then $g_\alpha \circ H_\alpha \in C(X)$. Put $h = \Sigma(g_\alpha \circ H_\alpha) \in C(X)$. For $p \in X$, enumerate the finite set $\{\alpha : p \in cl(U_\alpha)\}$ so that $\{\alpha_j : 1 \leq j \leq n\} = \{\alpha : p \in cl(U_\alpha)\}$ $[\alpha_j \neq \alpha_k$ if $1 \leq j < k \leq n]$. Then $h(p) = \Sigma g_{\alpha_j} \circ h_{\alpha_j}(p) \geq \Sigma g_{\alpha_j}(p)|F(p)| = g(p)|F(p)| = |f(p)|$, whence $h \geq |f|$, and X is a cb-space.■

4.7 EXAMPLE: Being a cb-space is not an Hodel property. Let $\Omega = W^* \times W^* \setminus \{(\omega_1, \omega_1)\}$ be the space of [GJ: 8L]. Let $\Phi: \Omega \times \mathbb{N} \longrightarrow T$ be the quotient map which is at most two-to-one, with Φ being one-to-one on $W \times W \times \mathbb{N}$, while $\Phi(\alpha, \omega_1, 2n) = \Phi(\alpha, \omega_1, 2n+1)$ and $\Phi(\omega_1, \alpha, 2n+1) = \Phi(\omega_1, \alpha, 2n)$. (See also [MJ]). Then T is a locally compact, non-cb, non-realcompact space such that $|\upsilon T \setminus T| = 1$. To see that T is not cb, note every $g \in C(T)$ is constant on a "corner stairwell", $\Phi[([\alpha, \omega_1] \times [\alpha, \omega_1]) \cap \Omega \times \mathbb{N}]$. If f is a real-valued function on T such that $f(t) = \min\{\kappa : \Omega \times \{\kappa\} \cap \overleftarrow{\Phi}(t) \neq \emptyset\}$, then f is locally bounded, but unbounded on every corner stairwell. Put $V_n = int(\overleftarrow{f}[n-1, n+1])$. Then $\{V_a\}_1^\infty$ is a countable locally finite open cover of T such that $cl(V_n)$ is countably compact (hence cb). Let $A = \Phi[\{(\alpha, \beta, n) : \alpha \leq \beta,\ n \in \mathbb{N}\}]$ and $B = \Phi[\{(\alpha, \beta, n) : \beta \leq \alpha,\ n \in \mathbb{N}\}]$. Then A and B are closed and cover T. Moreover, each of these sets is a cb-space since they are each disjoint unions of countably compact spaces.

Mayhew's Theorem is an unpublished result, originally done in a seminar run by one of the authors at the University of Kentucky in the summer of 1967. It appears now with the permission of Mary Mayhew. A form of Lemma 4.8 appears as Theorem 4.7 in [Dy].

4.8 LEMMA: (Dykes): Let Tichonov X have non-measurable cardinality. Then X is realcompact if and only if for each free z-ultrafilter A^p, the cover $\{X \setminus Z: Z \in A^p\}$ contains a locally finite subcover.

4.9 THEOREM: (Mayhew): Let Tichonov X have non-measurable cardinality. If X is realcompact, then every almost cozero subset of X is realcompact.

Proof: Let X be realcompact and $G = \bigcup_\Gamma C_\alpha$ be an almost cozero set. Let $F: \beta G \longrightarrow \beta X$ be the Stone extension of $id: G \longrightarrow X$. Note $F^{\leftarrow}(G) = G$, so $\upsilon G \subseteq G \cup F^{\leftarrow}(X \setminus G)$. Now each $C_\alpha = C(f_\alpha)$ for some $f_\alpha \in C(X, [0, \infty))$. Let $V_{\alpha_n} = f_\alpha^{\leftarrow}(1/n, \infty)$. Choose any $p \in F^{\leftarrow}(X \setminus G)$. Then $\mathcal{V} = \{V_{\alpha_n} : n \in \mathbb{N}, \alpha \in \Gamma\}$ is a subfamily of $\{G \setminus Z: Z \in A^p\}$. Since $\mathcal{V}$ is a σ-locally finite cozero cover of G, it has a locally-finite-in-G cozero refinement $\mathcal{W}$. So $\mathcal{W}$ is a locally finite refinement of $\{G \setminus Z: Z \in A^p\}$. Since A^p is a Z-ultrafilter, $\mathcal{W} \subseteq \{G \setminus Z: Z \in A^p\}$ and the result follows from 4.8.∎

4.10 LEMMA: Let X be Tichonov and V be a cozero set of X. Then V is a cozero set of υX if and only if V is real-compact.

Proof: "Only if" Realcompactness is cozero hereditary. "If" Let $V = V(f)$ for some $f \in C(X)$. Extend f to $f^\upsilon \in C(\upsilon X)$ and let W be the cozero set of f^υ in υX. Then V is dense in W and $W \cap X = V$. By [GJ: 8G1], V is C-embedded in W. Since V is real compact, $V = W$.∎

4.11 THEOREM: Let Tichonov X have non-measurable cardinality. If X is the union of a locally finite family of realcompact cozero sets of X, then X is realcompact.

Proof: By Lemma 4.10, the realcompact cozero sets of X are cozero sets of υX, so X is an almost cozero subset of υX. By Mayhew's Theorem, X is realcompact.■

4.12 COROLLARY: Realcompactness is a weak Hodel property on Tichonov spaces of non-measurable cardinality. Hence every realcompact, totally C. R. space of non-measurable cardinality is hereditarily realcompact.

Realcompactness does not satisfy the Locally Finite Sum Theorem, as Mrowka's example in [Mr] shows. Note, however, that the interiors of Mrowka's two closed realcompact subsets in that example fail to cover the space. We do not know whether realcompactness is an Hodel property.

The authors would like to express particular thanks to the referee for encouragement and helpfulness far beyond the level they have come to expect. The constructiveness of the criticism and the attitude in which it was offered were highly appreciated.

BIBLIOGRAPHY

[B1] D. K. Burke, On subparacompact spaces, *Proc. Amer. Math. Soc.* (1969) 655-663.

[B2] ———————— , ParaLindelöf spaces and closed mappings, *Top. Proc.* (1980) 47-57.

[C] M. M. Coban, On the theory of p-spaces, *Sov. Math. Dokl.* *11* (1970), 1257-1260.

[D] C. H. Dowker, Inductive dimension of completely normal spaces *J. Math. Oxford* (2) *4* (1953) 267-281.

[Dy] N. Dykes, Generalizations of realcompact spaces, *Pac. J. Math 15* (1970) 571-581.

[F] Z. Frolík, A generalization of realcompact spaces, *Czec. Math. J. 13* (1963) 127-138.

[GJ] L. Gillman and M. Jerison, "Rings of Continuous Functions", D. Van Nostrand, New York, 1960.

[H-1] R. E. Hodel, Total normality and the hereditary property, *Proc. Amer. Math. Soc. 17* (1966) 462-465.

[H-2] ——————, Sum theorems for topological spaces, *Pac. J. Math. 30* (1969) 59-65.

[Ho] J. G. Horne, Jr., Countable paracompactness and cb-spaces, *Notices, Amer. Math. Soc. 6* (1959) 629-630.

[J] H. J. K. Junnila, On submetacompactness, *Top. Proc. 3* (1978) 375-405.

[M] John Mack, On a class of countably paracompact spaces, *Proc. Amer. Math. Soc. 16* (1965) 467-472.

[Mi] E. Michael, The product of a normal space and a metric space need not be normal, *Bull. Amer. Math. Soc. 69* (1963) 375-376.

[Mo] K. Morita, On spaces having the weak topology with respect to closed coverings, *Proc. Japan Acad. 29* (1953) 537-543.

[Mr] S. Mrówka, Some comments on the author's example of a non-$\mathcal{R}$-compact space. *Bull. Acad. Pol. Sci. Ser. Math. 18* (1970) 443-448.

[Na] K. Nagami, "Dimension Theory", Academic Press, New York, 1970.

[Ny] P. Nyikos, On the product of metacompact spaces I, *Amer. J. Math. 100* (1978) 829-835.

[S] J. C. Smith, On θ-expandable spaces, *Glasnik Mat. Ser. III 11 (31)* (1976) 335-346.

[W] S. Willard, "General Topology", Addison-Wesley, Reading, Mass., 1970.

On the Yosida Representation of a Riesz Space

James J. Madden

Indiana University at South Bend
South Bend, Indiana 46634

I. INTRODUCTION

This paper describes the operative ideas in the "localic Yosida theorem" (Madden-Vermeer 1989) in a way which does not require any knowledge of locales and illustrates them in situations which are intended to be familiar to readers of these proceedings. The classical Yosida theorem states that any archimedean Riesz space (= vector lattice) is isomorphic to a Riesz space of continuous $\mathbb{R} \cup \{\pm\infty\}$ - valued functions defined on a certain compact Hausdorff space (the "Yosida space" of the Riesz space) which is constructed from the space of prime ℓ-ideals of the Riesz space, see Hager-Robertson (1975). In the past dozen years or so, it has shown itself to be an extraordinarily useful tool for studying certain aspects of the category of Riesz spaces, see Aaron et.al., Hager, and Ball-Hager. The localic Yosida theorem has certain technical advantages over the classical version which make it easier to use and more powerful: it eliminates the necessity of considering extended-real-valued functions and it leads to a functorial duality between countably closed weak-unital Riesz spaces and regular Lindelöf locales.

Many readers are probably unfamiliar with locales and they will be gratified to learn that everything that goes on in the localic Yosida representation can be phrased entirely in terms of the lattice of ℓ-ideals of the space to be represented. It is hoped that those who are comfortable with locales will find some value in the perspective afforded by the present rephrasing and that those who are learning about locales will appreciate seeing localic ideas formulated in more familiar language.

This paper is intended as a research announcement, so no proofs are given. Full details of all results quoted here may be found in Madden-Vermeer (1986) and (1989) and Madden (to appear). The version of the localic Yosida theorem presented in these papers is valid for any archimedean ℓ-group with weak unit. The restriction to Riesz spaces in the present note enables us to ignore certain minor technical points which would distract from the main ideas.

Notational conventions in this paper are as follows: A Riesz space is a vector space equipped with lattice operations which are preserved by translations and by positive scalar multiplication. If V is a Riesz space, an element $e \in V$ is a 'strong unit' if $e \geq 0$ and for each $v \in V$ there is a constant a such that $ae \geq |v|$; e is called a 'weak unit' if $e \geq 0$ and $v \wedge e = 0$ implies $v = 0$. An ℓ-ideal of V is a subspace which contains w whenever it contains v and $|w| \leq |v|$. Alternatively, the ℓ-ideals of V are the kernels of the Riesz space maps with domain V . An ℓ-ideal is prime if it is the kernel of a Riesz space map to a totally-ordered Riesz space. Finally, V is archimedean if $v, w \in V$ and $n|w| \leq v$ for all $n = 1, 2, \ldots$ implies $w = 0$.

II. STRONG UNITS

Let V be a Riesz space. The spectral space of V, denoted Spec V is the topological space whose elements are the prime ℓ-ideals of V and whose topology has as its opens the sets $D(I) = \{p \in \text{Spec } V \mid I \not\subseteq p\}$, where I ranges over all ℓ-ideals of V (the hull-kernel or Zariski topology). The spectrum is useful because it keeps track of relations between the ℓ-ideals of V in a concrete way. The map $I \to D(I)$ is an order-preserving bijection from the lattice of ℓ-ideals of V to the lattice of open subsets of Spec V, as is well known and easily verified. Note that the opens determined by principal ℓ-ideals (the sets $D(f) = \{p \mid f \notin p\} \subseteq \text{Spec } V$), form a basis for the topology on Spec V because every ℓ-ideal is a union of principal ℓ-ideals and in a Riesz space an intersection of two principal ℓ-ideals is principal.

Only for unusual V is Spec V Hausdorff, so it is not a terribly attractive space. The Yosida theorem is useful because it enables one to view an arbitrary archimedean Riesz space V as a Riesz space of functions defined on a topological space with nice properties. When V has a strong unit e, the space chosen in the representation is not all of Spec V, but only the subspace whose points are the maximal ℓ-ideals. This is denoted Max V, and is always a compact Hausdorff space when a strong unit is present. Given any $m \in \text{Max } V$, there is a unique Riesz homomorphism with kernel m $\phi_m : V \to \mathbb{R}$ with the property that the image of the strong unit e is $\phi_m(e) = 1$. Any $f \in V$ defines a real-valued function $\hat{f} : \text{Max } V \to \mathbb{R}$ via the rule $\hat{f}(m) = \phi_m(f)$. It turns out that $\hat{f}$ is continuous and that the map $V \to C(\text{Max } V)$; $f \to \hat{f}$ is a Riesz homomorphism and is one-to-one if V is archimedean. This is the Yosida representation in the strong unit case.

In order to present the localic generalization of this, we consider the meaning of the topology of Max V more carefully. Consider first the situation when $V = C(X)$, with X compact Hausdorff, where $C(X)$ = the Riesz space of continuous real-valued functions on X. In this case, the maximal ℓ-ideals of $C(X)$ may be identified with the points of X. The topology on X induced by its inclusion in Spec C(X) is identical with the original topology on X , because X has a basis of cozero sets. Clearly every open in X is $D(I) \cap X$ for some ℓ-ideal I , but $I \to D(I) \cap X$ is many-to-one. What we want to emphasize at this juncture is that, just as the opens of Spec C(X) correspond one-to-one to *all* ℓ-ideals of C(X) , the opens of X = Max C(X) correspond one-to-one to the *uniformly closed* ℓ-ideals of C(X) . We make this more precise in the next paragraph.

Let V be an arbitrary Riesz space and let $s \in V^+$ be any positive element. A sequence $\{v_i\} \subseteq V$ is said to *converge s-uniformly* to $v \in V$ if for any $n = 1, 2, \ldots$, $|v - v_i| \leq \frac{1}{n}s$ for sufficiently large i . An ℓ-ideal $I \subseteq V$ is said to be *s-uniformly closed* if whenever $\{v_i\} \subseteq I$ converges s-uniformly to v then $v \in I$. See [7], for a complete discussion of these concepts.

PROPOSITION 1. Let X be a compact Hausdorff space and let 1 denote the constant function on X with value the real number one. Then the 1-uniformly closed ℓ-ideals of C(X) correspond one-to-one to the opens of X via $I \to \bigvee\{\text{coz } f \mid f \in I\}$. More generally, let V be any Riesz space with strong unit e and let Max $V \subseteq$ Spec V be the subspace whose points are the maximal ℓ-ideals of V . The map $I \to D(I) \cap$ Max V provides a bijective order-preserving correspondence between the e-uniformly closed ℓ-ideals of V and the open sets of Max V .

The statement about C(X) in the proposition is essentially exercise 40 of Gillman and Jerison. A proof of the remaining part may be extracted from Madden (to appear), but the reader should not find it difficult to provide his or her own proof.

We can use the e-uniformly closed ℓ-ideals to give an alternate description of the Yosida representation. Again, we will formulate the main ideas first in the case of C(X) , X compact. Suppose $f \in C(X)$ and $(a,b) \subseteq \mathbb{R}$. Then the 1-uniformly closed

ℓ-ideal of C(X) which corresponds to the open $f^{-1}(a,b)$ is generated as a 1-uniformly closed ℓ-ideal by any $g \in C(X)$ which satisfies coz $g = f^{-1}(a,b)$. (Whenever we say that an s-uniformly closed ℓ-ideal is generated as such by an element g, we mean that it is the smallest s-uniformly closed ℓ-ideal which contains g.) In particular, the 1-uniformly closed ℓ-ideal corresponding to $f^{-1}(a,b)$ is generated as such by the function $((f - a) \vee 0) \wedge \wedge ((b - f) \vee 0)$. If we did not already know f, we could recover f from a knowledge of what $f^{-1}(a,b)$ is for each open interval $(a,b) \subseteq \mathbb{R}$ --or equivalently, from a knowledge of the 1-uniformly closed ℓ-ideals generated by the functions $((f - a) \vee 0) \wedge ((b - f) \vee 0)$. Admittedly, there is not much point in considering these things when C(X) is sitting in front of us explicitly, but now consider an abstract archimedean Riesz space V with strong unit e, and let $v \in V$. Let $\Phi_o(v)$ be the function from the open intervals in $\mathbb{R}$ to the lattice of e-uniformly closed ℓ-ideals of V which assigns to (a,b) the e-uniformly closed ℓ-ideal generated as such by $((v - ae) \vee 0) \wedge ((be - v) \vee 0)$. The function $\Phi_o(v)$ is a complete description of the function $\hat{v}$ occuring in the Yosida representation. We shall return to this way of viewing things shortly, but first we shall attempt to generalize our remarks about uniformly closed ℓ-ideals to the context of Riesz spaces lacking strong units.

III. WEAK UNITS

We shall begin with a concrete example. Consider the Riesz space C(X) of all continuous real-valued functions on a non-compact Lindelöf space X. As was the case when X was compact, it is still possible to identify X with a subspace of Spec C(X), namely that whose points are the maximal ℓ-ideals of C(X) (i.e., the kernels of non-zero Riesz homomorphisms from C(X) to $\mathbb{R}$). Every open subset $U \subseteq X$ produces an ℓ-ideal, namely $\{f \in C(X) \mid \text{coz } f \subseteq U\}$. It is certainly not the case, however, that the opens of X correspond bijectively to the 1-uniformly closed ℓ-ideals of C(X) because, e.g., the bounded functions are a proper ℓ-ideal of $C(X)$ which is 1-uniformly closed. Moreover, those 1-uniformly closed ℓ-ideals whose elements are bounded functions (i.e., the 1-uniformly closed ℓ-ideals contained in $C^*(X) \cong C(\beta X)$) correspond, by proposition 1, with the opens of the Stone-Čech compactifcation βX, not with the opens of X.

There are too many 1-uniformly closed ℓ-ideals, so we consider a stronger completeness property. A sequence $\{v_i\}$ of elements of a Riesz space V is said to *converge relatively uniformly* to $v \in V$ if there is some $s \in V^+$ such that $\{v_i\}$ converges s-uniformly to v. An ℓ-ideal $I \subseteq V$ is said to be *relatively uniformly closed* if whenever $\{v_i\} \subseteq I$ converges relatively uniformly to v then $v \in I$. It is easy to check that for any open $U \subseteq X$, the ℓ-ideal $\{f \mid \operatorname{coz} f \subseteq U\} \subseteq C(X)$ is relatively uniformly closed. Also, a little algebra shows that any relatively uniformly closed ℓ-ideal of $C(X)$ which contains a nowhere vanishing function is all of $C(X)$. Indeed, we have:

PROPOSITION 2. If X is a regular Lindelöf space, then the lattice of relatively uniformly closed ℓ-ideals of $C(X)$ is isomorphic with the lattice of opens of X via $I \longrightarrow D(I) \cap X = \bigcup\{\operatorname{coz} f \mid f \in I\}$.

The proof of this is left to the reader. This proposition does not generalize beyond Lindelöf spaces. If X is realcompact but *not* Lindelöf, then $C(X)$ has relatively uniformly closed ℓ-ideals $I \neq J$ such that $D(I) \cap X = D(J) \cap X$, and *there is no topological space whose lattice of opens is order-isomorphic with the lattice of relatively uniformly closed ℓ-ideals* of $C(X)$, see Madden-Vermeer (1986). On the other hand, the relatively uniformly closed ℓ-ideals of an arbitrary Riesz space are highly significant from an algebraic point of view: a quotient V/I of a Riesz space V is archimedean if and only if I is relatively uniformly closed.

We showed previously that an element of $C(X)$, X compact, gives rise to a map from the open intervals of $\mathbb{R}$ to the lattice of

1-uniformly closed ℓ-ideals of $C(X)$. This generalizes. If X is Lindelöf, $f \in C(X)$ and $(a,b) \subseteq \mathbb{R}$, then the relatively uniformly closed ℓ-ideal of functions vanishing outside $f^{-1}(a,b)$ is precisely the same as the relatively uniformly closed ℓ-ideal of $C(X)$ generated as such by $((f-a) \vee 0) \wedge ((b-f) \vee 0)$. If V is a Riesz space with weak unit e, let $\mathcal{Y}^*(V,e)$ denote the lattice of all relatively uniformly closed ℓ-ideals of V which are contained in the relatively uniformly closed ℓ-ideal generated as such by e. (In many important cases, e is contained in no proper relatively uniformly closed ℓ-ideal, and $\mathcal{Y}^*(V,e)$ is the lattice of all relatively uniformly closed ℓ-ideals.) This lattice might or might not occur as the topology of some space, as we have mentioned, but regardless of whether it does or not each $v \in V$ gives rise to a map $\Phi_o(v)$ from the open intervals of $\mathbb{R}$ to $\mathcal{Y}^*(V,e)$ by means of the following rule:

$$\Phi_o(v)(a,b) = \langle\langle((v - ae) \vee 0) \wedge (be - v) \vee 0\rangle\rangle \cap \langle\langle e\rangle\rangle$$

where $\langle\langle v\rangle\rangle$ denotes the relatively uniformly closed ℓ-ideal generated as such by v. The map $v \mapsto \Phi_o(v)$ is the main ingredient of the "localic Yosida representation". Its usefulness springs from remarkable properties possessed by $\mathcal{Y}^*(V,e)$ and by Φ_o, which we now present.

THEOREM. Let V be an archimedean Riesz space with weak unit e and let $\mathcal{Y}^*(V,e)$ and Φ_o be as described above. Then:

A) $\mathcal{Y}^*(V,e)$ is a complete lattice which satisfies the infinite distributive law:

$$I \cap \vee\{J_\lambda\} = \bigvee(I \cap J_\lambda),$$

the regularity axiom:

$$I = \bigvee\{J \mid \exists J' \text{ s.t. } J' \vee I = \langle\langle e\rangle\rangle \text{ and } J' \cap J = \{0\}\}$$

and the Lindelöf axiom:

If $\langle\langle e\rangle\rangle = \bigvee\{J_\lambda \mid \lambda \in \Lambda\}$, then there is countable $\Lambda_o \subseteq \Lambda$ such that $\langle\langle e\rangle\rangle = \bigvee\{J_\lambda \mid \lambda \in \Lambda_o\}$.

B) Moreover, for each $v \in V$, the map $\Phi_o(v)$ from the open intervals of $\mathbb{R}$ to $\mathcal{Y}^*(V,e)$ extends uniquely to a map $\Phi(v)$ from the opens of $\mathbb{R}$ to $\mathcal{Y}^*(V,e)$ which satisfies

$$\Phi(v)(\emptyset) = \{0\}, \quad \Phi(v)(\mathbb{R}) = \langle\langle e\rangle\rangle,$$
$$\Phi(v)(U_1 \cap U_2) = \Phi(v)(U_1) \cap \Phi(v)(U_2)$$
$$\text{and} \quad \Phi(v)(\bigvee\{U_\lambda \mid \lambda \in \Lambda\}) = \bigvee\{\Phi(v)(U_\lambda) \mid \lambda \in \Lambda\}.$$

A proof can be found in Madden (to appear). To get some idea of what this all means, observe that part A) of the theorem is saying essentially that $\mathcal{Y}^*(V,e)$ --though it might not be the topology of a space--at least possesses the formal properties which are enjoyed by the lattice of opens of a regular Lindelöf space. As for part B), this says that the map $\Phi(v)$ possesses the formal properties which the map $U \mapsto f^{-1}(U)$ from the opens of $\mathbb{R}$ to the opens of some topological space has whenever f is a continuous real-valued function on that space. Indeed, if $\mathcal{Y}^*(V,e)$ happens to be the topology of some space, the conditions stated for $\Phi(v)$ are necessary and sufficient for $\Phi(v)$ to induce a continuous real-valued function $\hat{v}$ on that space with the property that $\hat{v}^{-1}(a,b) = \Phi(v)(a,b)$.

Consider the collection of all maps γ from the opens of $\mathbb{R}$ to $\mathcal{Y}^*(V,e)$ satisfying the conditions:

$$\gamma(\emptyset) = \{0\}, \quad \gamma(\mathbb{R}) = \langle\langle e\rangle\rangle,$$
$$\gamma(U_1 \cap U_2) = \gamma(U_1) \cap \gamma(U_2),$$
$$\gamma(\bigcup\{U_\lambda\}) = \bigvee\{\gamma(U_\lambda)\}.$$

It would be quite reasonable to call this $C(\mathcal{Y}^*(V,e))$ in analogy with the notation $C(X)$ for the continuous real-valued functions on X. It can be shown that the order and arithmetic operations on $\mathbb{R}$ induce on $C(\mathcal{Y}^*(V,e))$ the structure of a Riesz space (and a ring). The basic idea involved in defining sums, for instance, is the following. If f and g are real valued functions on a space X, then $(f+g)^{-1}(a,b) = \bigvee\{f^{-1}(a_1,b_1) \cap g^{-1}(a_2,b_2) \mid a < a_1 + a_2 \ \& \ b_1 + b_2 < b\}$. Using this as a clue, one expects that $\gamma_1 + \gamma_2$ ought to satisfy $(\gamma_1+\gamma_2)(a,b) = \bigvee\{\gamma_1(a_1,b_1) \cap \gamma_2(a_2,b_2) \mid a < a_1 + a_2 \ \& \ b_1 + b_2 < b\}$. In fact, it can be shown that there is always exactly one element of $C(\mathcal{Y}^*(V,e))$ which does satisfy this. The details are somewhat technical, so we don't go any further into this here. To the theorem above, we can add the following, the proof of which can also be found in Madden (to appear).

THEOREM (continued) C) The function $\Phi:V \rightarrow C(\mathcal{Y}^*(V,e))$ is an injective Riesz-space homomorphism.

To fully appreciate the generality of this result, the reader should note that there are Riesz spaces V with weak unit which admit no non-zero Riesz homomorphisms whatsoever to $\mathbb{R}$. The measurable real functions on a non-atomic measure space provide an example of such a Riesz space (see [LZ], example 27.8, for details). Obviously such V --and all quotients of such V --cannot be represented as Riesz spaces of real-valued functions. Thus, the introduction of the Riesz space $C(\mathcal{Y}^*(V,e))$ (which is in fact a Riesz space of continuous real-valued functions on a locale) is really necessary. We cannot get away with using just C(X) , X a space.

We shall bring our discussion to a close now by mentioning a few applications of the localic Yosida theorem. A result of Isbell (see Madden-Vermeer 1989) says essentially that the homomorphism Φ is an isomorphism if and only if V is "closed under countable composition". This term is introduced and fully explained in Henriksen-Isbell-Johnson (1961). The embedding $\Phi:V \rightarrow C(\mathcal{Y}^*(V,e))$ turns out to be equivalent to the "countably closed hull" of V which was first constructed, and its reflectivity proved, in Aron-Hager (1981). See also Hager (1985). (A reasonable project, which has yet to be carried out, would be to attempt to reprove the functorial characterization of closure under countable composition given in Hager (1985) by using the localic Yosida theorem.)

A regular Lindelöf frame is a lattice which has the properties attributed to $\mathcal{Y}^*(V,e)$ in the theorem above. These lattices form a category many features of which are well understood, see Madden-Vermeer (1986). In Madden-Vermeer (1989) it is shown that the category of regular Lindelöf frames is equivalent to the category of countably closed Riesz spaces with weak unit, and this fact is used to study epimorphisms of archimedean Riesz spaces.

Within the category of regular Lindelöf frames, it is relatively easy to make constructions which are analogous to the projective cover, the quasi-F cover and the many quasi-F_k covers in between as well as covers with more exotic properties. In combination with the localic Yosida theorem, this gives an alternate approach to the results of Ball-Hager-Neville which appear in these proceedings. This will be described fully in a future joint paper of R. Ball and the author.

REFERENCES

1. E. R. Aron and A. W. Hager, Convex vector-lattices and ℓ-algebras, *Top. and its Applic.*, 12(1981), 1-10.

2. E. R. Aron, A. W. Hager and J. Madden, Extensions of ℓ-homormorphisms, *Rocky Mtn. J. Math.*, 12(1982), 482-490.

3. R. N. Ball and A. W. Hager, Epimorphisms in archimedean ℓ-groups and vector-lattices I: characterization and examples. In *Proc. Bowling Green conference on ordered algebraic structures.* Marcel Dekker, New York, to appear.

4. L. Gillman and M. Jerison, *Rings of continuous functions.* Van Nostrand, Princeton, 1960.

5. A. W. Hager, Algebraic closures of ℓ-groups of continuous functions. In *Rings of continuous functions*, Lecture notes in pure and applied math. vol. 95, 164-194. Marcel Dekker, New York, 1985.

6. A. W. Hager and L. Robertson, Representing and ringifying a Riesz space. *Symposia Math.* 21(1977), 411-431.

7. Luxemburg and Zancen, Riesz spaces, North Holland.

8. J. Madden and J. Vermeer, Lindelöf locales and realcompactness. *Math. Proc. Camb. Phil. Soc.* 99(1986) 473-480.

9. J. Madden and J. Vermeer, Epicomplete archimedean ℓ-groups via a localic Yosida theorem. *J. Pure and Appl. Alg.* to appear 1989.

10. J. Madden, Lattices and frames associated with an abelian ℓ-group, to appear.

Recent Results and Open Problems on the Countability Index and the Density Index of S(X)

K.D. Magill, Jr.

Department of Mathematics
106 Diefendorf Hall
SUNY at Buffalo
Buffalo, New York 14214–3093

1. INTRODUCTION AND BACKGROUND

S(X) denotes the semigroup of all continuous selfmaps of the topological space X. We will assume throughout that the spaces are Hausdorff and when we topologize S(X) it will always be with the compact–open topology. The topics we are going to consider here were motivated by a fundamemental paper [12] published by Schreier and Ulam in 1934. They showed that $S(I^N)$ contains a dense subsemigroup generated by only five elements where I^N denotes the Euclidean N–cell. In other words, there exist five continuous selfmaps of I^N such that any continuous selfmap of I^N can be approximated as closely as one wishes just by composing these five functions. Naturally, it was reasonable to ask if fewer functions will suffice. Recall first of all that $S(X)$ is a topological semigroup whenever X is locally compact and note that if a topological semigroup has a dense subsemigroup generated by only one element then it will be commutative. Since $S(X)$ is commutative only when X has one point, it must take at least two functions to generate a dense subsemigroup of $S(I^N)$. In the same year that Schreier and Ulam published their paper, Sierpinski [13] showed that four functions sufficed for $S(I)$. In fact Sierpinski's paper appeared in the same volume of the Fundamenta Mathematica immediately following Schreier and Ulam's paper. Then the following year in 1935 Jarnik and Knichal [2] succeeded in showing that two functions did, in fact, suffice for $S(I)$.

For some reason the problem of determining the cardinality of a minimal generating set remained open for the higher dimensions for quite a few years. Then in 1969 Cook and Ingram [1] completely settled the problem. They produced a class of spaces which properly included all Euclidean N–cells such that for any X from the class, $S(X)$ contains a dense subsemigroup generated by two elements. Six years later Subbiah [14] independently rediscovered their result and proved something in addition. She proved that $S(R^N)$ contains a dense subsemigroup generated by three elements where R^N is the Euclidean N–space.

It's appropriate at this point to discuss how the result is obtained for $S(I^N)$. What one actually does is to show that given any countable subset $\mathcal{F}$ of $S(X)$ there exist two functions which generate a subsemigroup containing $\mathcal{F}$. The desired result is then an easy consequence of this. Since $S(I^N)$ is separable in the compact–open topology, one can choose $\mathcal{F}$ to be dense. All this led to the formation of two related concepts. One can be defined for any semigroup while the other is defined for topological semigroups.

DEFINITION (1.1). The *countability index* or *C–index* of a semigroup S, denoted by $C(S)$, is the smallest positive integer n (if such an integer exists) such that every countable subset of S is contained in a subsemigroup with n generators. If no such integer exists, we define $C(S) = \infty$.

DEFINITION (1.2). The *density index* or *D–index* of a topological semigroup

S, denoted by D(S), is the smallest positive integer n (if such an integer exists) such that S contains a dense subsemigroup generated by n elements. If no such integer exists, we define $D(S) = \infty$.

Note that if D(S) is finite then S is separable and $D(S) \leq C(S)$. It can happen that C(S) is finite and D(S) is infinite. Of course, S cannot be separable in a situation like this. We will discuss examples of this later.

As the title of the paper indicates, this is a survey paper and it will contain no formal proofs. Our main purpose here is to discuss some of the more recent results concerning the C–index and the D–index of S(X) and to formulate some of the problems in these areas which are still open.

2. THE C–INDEX OF S(X)

It is an easy matter to verify that if $C(S) = 1$ for some semigroup S then S must be commutative. The semigroups in which we are interested are various transformation semigroups and they are commutative only in trivial situations. Consequently, for us, two is really the minimal number. The surprising thing is that two turns out to be the C–index of a natural transformation semigroup as often as it does. We do have sufficient conditions for the C–index of a transformation semigroup to be two and the result (Theorem 2.6 of [7]) is useful in that it applies not only to semigroups of continuous selfmaps but other natural transformation semigroups like the semigroup of all endomorphisms of a vector space. Nevertheless the conditions are not (likely) necessary and they are also rather complicated which is one reason we didn't state them here. We also have a necessary condition for rather general transformation semigroups (Theorem 2.2 of [7]) which we state here only for S(X).

THEOREM (2.1). [7, p. 351] *Suppose there are an infinite number of homeomorphisms from* X *onto* X *and* $C(S(X)) = 2$. *Then* X *is homeomorphic to a proper retract of itself.*

This necessary condition is not sufficient and this leads to

OPEN PROBLEM (2.2). Find useful necessary and sufficient conditions on a transformation semigroup T(X) (and on S(X) in particular) for

$C(T(X)) = 2$.

We mentioned previously that Theorem 2.6 of [7] applies not only to semigroups of continuous selfmaps but to other semigroups as well. For example, we used it to prove the following result.

THEOREM (2.3). [7, p. 354] *Let* V *be a vector space over a field* F *and let* End V *denote the semigroup of all endomorphisms of* V. *Then* $C(\text{End } V) = 2$ *if and only if*

(2.3.1) V *is infinite dimensional or*

(2.3.2) $\text{Dim } V = 1$ *and* F *is finite.*

In that same paper, we also proved

THEOREM (2.4). [7, p.356] $C(\text{End } V) = \infty$ *if and only if* F *is infinite and* Dim V *is an integer* $N \geq 1$.

It is immediate from Theorem (2.1) that $C(S(S^N)) \geq 3$ where S^N denotes the Euclidean N–sphere. It turns out that, in fact, $C(S(S^N))$ is quite a bit larger than three.

THEOREM (2.5). [5, p. 442] $C(S(S^N)) = \infty$.

This latter result can be verified quite easily. Choose any infinite family of prime numbers and for each of these primes, choose a continuous selfmap of S^N whose degree is that particular prime. The resulting family of functions will not be contained in any finitely generated subsemigroup of $S(S^N)$. This is an immediate consequence of the fact that $\deg(f \circ g) = (\deg f)(\deg g)$ for any two continuous selfmaps f and g of S^N.

We then looked at $S(R^N)$ where R^N denotes the Euclidean N–space but thus far we have made very little progress with the problem of determining $C(S(R^N))$. All we have is the following result which, although it appears in [4], is the consequence of a joint effort by S. Subbiah and me.

THEOREM (2.6). [4, p. 92] $C(S(R^N)) \geq 4$.

OPEN PROBLEM (2.7). determine $C(S(R^N))$. We conjecture that $C(S(R^N)) = \infty$.

Until recently, we didn't know if there existed any spaces X with the property that $2 \neq C(S(X)) \neq \infty$. We now know and there are some but we really don't know how extensive they are. Before we can state our next result, we need some terminology. We say that a space X has the *internal extension property* if each continuous map from a closed subspace of X into X can be extended to a continuous selfmap of X.

THEOREM (2.8). [9] *Let* X *consist of a finite number of components, each of which is a compact* N–*dimensional subspace of Euclidean* N–*space and has the internal extension property. Then* C(S(X)) *is finite.*

This result was then used to prove

THEOREM (2.9). [9] *Let* X *consist of two components, each of which is a compact* N–*dimensional subspace of Euclidean* N–*space and has the internal extension property. Then* $3 \leq C(S(X)) \leq 8$ *and if the two components are homeomorphic then* $C(S(X)) \leq 7$.

OPEN PROBLEM (2.10). Let X be the discrete union of N copies of the closed unit interval. Determine C(S(X)).

OPEN PROBLEM (2.11). Determine if, for each positive integer N there exists a topological space X such that C(S(X)) = N.

OPEN PROBLEM (2.12). Characterize those spaces X for which C(S(X)) is finite.

OPEN PROBLEM (2.13). For each positive integer $N \geq 2$, characterize those spaces X for which C(S(X)) = N. In particular, characterize those spaces X for which C(S(X)) = 2.

It may shed some light on things if we were able to determine C(S(T)) where T is a triod. If C(S(T)) happens to be finite, we won't be able to prove it by using the techniques used in proving the other results. Although T has a lot of nice properties (including that of being an absolute retract) there is one which it does not have which is used in all the other proofs in which we verified that the C–index is finite. That is the property of containing a countably infinite mutually disjoint family of subspaces each homeomorphic to the superspace.

OPEN PROBLEM (2.14). Determine C(S(T)).

We now turn to the problem of determining if, given any semigroup T, one can always embed T into another semigroup whose C–index is two. The answer is easily seen to be yes if no further conditions are imposed. Simply embed T into $\mathcal{T}_X$, the semigroup of all transformations on a set X of sufficiently high cardinality, taking care to make sure that it is at least infinite. It then follows immediately from results in both [1] and [14] that $C(\mathcal{T}_X) = 2$. Of course, $\mathcal{T}_X$ coincides with S(X) whenever X is discrete and this suggested the problem of determining if, given any semigroup T, there exists a less trivial topological space X such that T can be embedded in S(X) and C(S(X)) = 2. Our first step in this direction was the following

THEOREM (2.15). [8, p. 94] *Let* T *be any semigroup whatsoever. Then there exists a compact Hausdorff space* X *such that* T *can be embedded in* S(X) *and* C(S(X)) = 2.

In order to get the latter result, we first proved

THEOREM (2.16). [8, p. 95] *Let* X *be any infinite discrete space. Then* $C(S(\beta X)) = 2$.

In order to derive Theorem (2.15) from Theorem (2.16), take X to be a discrete space of sufficiently high cardinality and embed T into S(X). Then simply observe that the map which sends $f \in S(X)$ into its extension over βX is an isomorphism from S(X) into $S(\beta X)$. One cannot hope to completely avoid the Stone–Cech compactification in deriving a result such as Theorem (2.15). For example if S(R) can be embedded in S(X) and X is compact, then X must contain a copy of βN. More generally, we proved

THEOREM (2.17). [3, p. 355] *Let* X *be a pathwise connected normal space, let* Y *be compact and suppose that* S(X) *can be embedded in* S(Y). *Then* Y *contains a copy of* βZ *for each discrete closed subspace* Z *of* X.

After deriving Theorem (2.15), it seemed reasonable to try to get spaces with still nicer properties. In particular, can one always take X to be a continuum (generally not metrizable, of course)? The answer in this case also is yes. We first proved

THEOREM (2.18). [8] *Let* Y *be any space with more than one point, let* X *be any infinite power of* Y *and suppose that* X *has the internal extension property. Then* C(S(X)) = 2.

An immediate corollary to this is the following

COROLLARY (2.19). [8] *Let* Y *be any nondegenerate absolute retract and let* X *be any infinite power of* Y. *Then* C(S(X)) = 2.

One then shows that given any semigroup T, one can embed T into S(X) where X is the product of an arc with itself a sufficient number of times (at least infinite). The following generalization of Theorem (2.15) is then a consequence of this and Corollary (2.19).

THEOREM (2.20). [8] *Let* T *be any semigroup whatsoever. Then there exists a continuum* X *such that* T *can be embedded in* S(X) *and* C(S(X)) = 2.

As we have already noted, one cannot in general expect the continuum in the previous theorem to be metrizable. However if T is countable then T can be embedded in S(H) where H is the Hilbert cube and, of course, C(S(H)) = 2.

3. THE D–INDEX OF S(X)

It has been known for a long time that if X is locally compact then S(X) with the compact–open topology is a topological semigroup. The converse isn't quite true but it's almost true in the sense that there is a very extensive class of spaces with the property that for any space X within the class, S(X) with the compact–open topology is a topological semigroup if and only if X is locally compact.

DEFINITION (3.1). A topological space X is said to be *quasi–completely regular* if there exists an element $a \in X$ and an open set G containing a such that for each compact subset K of X and each point $x \notin K$, there is a function $f \in S(X)$ such that $f(y) = a$ for all $y \in K$ and $f(x) \notin G$.
Quasi–completely regular spaces are quite abundant containing, as they do, all 0–dimensional spaces (which here means possesses a clopen basis) and every completely regular hausdorff space which contains an arc. The following is then a portion of Subbiah's Theorem (2.3) [16].

THEOREM (3.2). [16, p. 30] *Let* X *be a quasi–completely regular space. Then* S(X) *with the compact–open topology is a topological semigroup if and only if* X *is locally compact.*

Hereafter in this section it will be assumed, without further mention, that S(X) has the compact–open topology.

As we stated previously, Subbiah showed in [14, p. 230] that $D(S(R^N) \leq 3$ for all N. Her technique there was different than the technique she used in proving that $D(S(I^N)) = 2$. In the latter case she proved first that $C(S(I^N)) = 2$ and then used the fact that $S(I^N)$ is separable. She definitely did not prove that $C(S(R^N)) \leq 3$ and then resort to the separability of $S(R^N)$ in order to prove that $D(S(R^N)) \leq 3$. Indeed, we found out later that the result cannot be proved in this manner (Theorem (2.6)). She simply produced three functions which generate a dense subsemigroup of $S(R^N)$. For a long time, those of us who are interested in these things felt that $D(S(R^N)) = 3$. Subbiah and I actually conjectured this in print in [10]. But apparently Subbiah didn't quite believe it as much as I did because three years later in 1983 she proved

THEOREM (3.3). [15, p. 85] $D(S(R)) = 2$.

Her methods do not seem to carry over at all to the higher dimensions and this leads to

OPEN PROBLEM (3.4). Determine $D(S(R^N))$ for $N > 1$.

We now have reason to suspect that the answer may be two for all N but still, there are many instances in semigroups of continuous selfmaps where one asks a question and there is a common answer for R^N when $N > 1$ and a different answer for R.

It follows from the theorem in [11] that if X is second countable and regular, then S(X) is separable (in fact, hereditarily separable) so the following result is an immediate consequence of this fact and Theorem (2.8).

COROLLARY (3.5). [9] *Let* X *consist of two components, each of which is a compact* N*–dimensional subspace of Euclidean* N*–space and has the internal extension property. Then* D(S(X)) *is finite.*

DEFINITION (3.6). Let A be a subset of a space X. We say that S(X) is *doubly transitive* on A if for $a,b,x,y \in A$ with $x \neq y$, there exists an $f \in S(X)$ such that $f(a) = x$ and $f(b) = y$.

THEOREM (3.7). [9] *Let* X *be a compact disconnected space with a finite*

number of components at least one of which contains more than one point and suppose also that S(X) *is doubly transitive on its components. Then* $D(S(X)) \geq 3$.

The following result is an immediate consequence of Theorems (2.9), (3.7) and the fact that S(X) is separable.

COROLLARY (3.8). [9] *Let* X *consist of two components, each of which is a compact* N*–dimensional subspace of Euclidean* N*–space and has the internal extension property.* Then $3 \leq D(S(X)) \leq 8$ *and if the two components are homeomorphic then* $D(S(X)) \leq 7$.

In section 1 we remarked that the C–index of a topological semigroup could be finite while its D–index is infinite. Let X be an uncountable discrete space. Then $C(S(X)) = 2$ by results in both [1] and [14] but $D(S(X)) = \infty$ since S(X) is not separable. Furthermore $C(S(\beta X)) = 2$ by Theorem (2.16) but $D(S(\beta X)) = \infty$. $S(\beta X)$ cannot be separable because of the fact that each $\langle x \rangle$, $x \in X$ (where $\langle x \rangle$ denotes the constant function which maps everything into x) is isolated in $S(\beta X)$ and open subsets of separable spaces must be separable. In our closing result the semigroups are separable but have infinite D–index nevertheless.

THEOREM (3.9). [5, p. 442] $D(S(S^N)) = \infty$.

OPEN PROBLEMS (3.10–3.12). Solve problems (2.11–2.13) with C(S(X)) replaced by D(S(X)).

REFERENCES

1. H. Cook and W. T. Ingram, Obtaining AR–like continua as inverse limits with only two bonding maps, *Glasnik Math.* 4 (24) 2 (1969) 309–311.

2. V. Jarnik and V. Knichal, Sur l'approximation des fonctions continues par les superpositions de deux fonctions, *Fund. Math.* 24 (1935) 206–208.

3. K.D. Magill, Jr., Embedding S(X) into S(Y) whenever Y is compact and X is not, *Semigroup Forum*, 12 (1976) 347–366.

4. K.D. Magill, Jr., Recent results and open problems in semigroups of continuous selfmaps, *Russian Math Surveys* 35:3 (1980) 91–97.

5. K.D. Magill, Jr., Some open problems and directions for further research in semigroups of continuous selfmaps, *Univ. Alg. and App.*, *Banach Center*

Pub., PWN–Polish Sci. Pub. Warsaw 9 (1982) 439–454.

6. K.D. Magill, Jr., The countability indices of certain transformation semigroups, *Semigroups and their Applications*, D. Reidel Pub. Co. Boston (1987) 91–97.

7. K.D. Magill, Jr., The countability index of the endomorphism semigroup of a vector space, *Linear and Multilin. Alg.* 22 (1988) 349–360.

8. K.D. Magill, Jr., Embedding semigroups into semigroups with countability index two, Semigroup Forum 39 (1989) 51–57.

9. K.D. Magill, Jr., More on the countability index and the density index of S(X), Proc. Edinburgh Math. Soc. (to appear).

10. K.D. Magill, Jr. and S. Subbiah, Solution of problem 6244 proposed by John Myhill, *Amer. Math. Monthly*, (8) 87 (1980) 676–678.

11. E. Michael, On a theorem of Rudin and Klee, *Proc. Amer. Math. Soc.* 12 (1961) 921.

12. J. Schreier and S. Ulam, Über topologische Abbildungen der euklidschen Sphäre, *Fund. Math.* 23 (1934) 102–118.

13. W. Sierpinski, Sur l'approximation des fonctions continues par les superpositions de quatre fonctions, *Fund. Math.* 23 (1934) 119–120.

14. S. Subbiah, Some finitely generated subsemigroups of S(X), *Fund. Math.* 86 (1975) 221–231.

15. S. Subbiah, A dense subsemigroup of S(R) generated by two elements, *Fund. Math.* 117 (1983) 85–90.

16. S. Subbiah, The compact–open topology for semigroups of continuous selfmaps, *Semigroup Forum*, 35 (1987) 29–33.

A Survey of Regular $\mathcal{J}$ - Classes of S(X)

Prabudh R. Misra

Department of Mathematics
College of Staten Island
City University of New York
Staten Island, NY 10301

1. INTRODUCTION

The collection of all continuous selfmaps on a topological space X forms a semigroup under composition and is denoted by S(X). Regular $\mathcal{J}$ - classes of S(X) first surfaced in [15] where Magill and Subbiah showed that, for X a Peano continuum with no cut points, they form a chain if and only if X is homeomorphic to a simple closed curve. This was an interesting algebraic characterization of a simple closed curve within Peano continua with no cut points. In another paper [16] the same authors characterized those plane Peano continua, without having any condition about cut points, whose semigroups of continuous selfmaps are such that their regular $\mathcal{J}$ - classes form a finite chain. These spaces turned out to be essentially five different kind of plane Peano continua. Later Magill [9] showed that this result holds good for any Peano continuum and not just the plane Peano continuum. In the present paper we will survey the results obtained as a consequence of attempts to generalize and extend these results of Magill and Subbiah and our attention for the most part will be on the regular $\mathcal{J}$ - classes of S(X).

2. BASIC DEFINITIONS AND RESULTS

Two elements a and b of a semigroup S are called $\mathcal{J}$ - related if they generate the same two sided ideal, that is, if $S^1aS^1 = S^1bS^1$. They are said to be $\mathcal{L}$-($\mathcal{R}$-) related if they generate the same left(right) ideal in S(X). All these relations are equivalence relations. Moreover, $\mathcal{L} \circ \mathcal{R} = \mathcal{R} \circ \mathcal{L}$ and hence this composition is also an equivalence relation, denoted by $\mathcal{D}$. These along with the equivalence relation $\mathcal{H} = \mathcal{L} \cap \mathcal{R}$ were defined by Green and are known as Green's relations on the semigroup S. An element a of a semigroup S is called regular if there exists an element b in S such that aba = a. A $\mathcal{D}$ - class or a $\mathcal{J}$ - class is called regular if it contains a regular element. In the case of a $\mathcal{D}$ - class, if it contains one regular element then all of its elements will be regular. Each $\mathcal{D}$ - class is contained in a $\mathcal{J}$ - class. The $\mathcal{J}$ - classes of S can be partially ordered in the usual manner: $J_a \le J_b$ if and only if $S^1aS^1 \subseteq S^1bS^1$, where J_a and J_b are $\mathcal{J}$ - classes containing elements a and b respectively. It is easily seen that $\mathcal{H} \subseteq \mathcal{L} \subseteq \mathcal{D} \subseteq \mathcal{J}$ and $\mathcal{H} \subseteq \mathcal{R} \subseteq \mathcal{D} \subseteq \mathcal{J}$. All of this and related material can be found in [2]. Let $\mathcal{D}_R(S)$ and $\mathcal{J}_R(S)$ denote the collections of all regular $\mathcal{D}$ - classes and all regular $\mathcal{J}$ - classes of S respectively. Clearly $\mathcal{J}_R(S)$ is a partially ordered set.

For X a topological space, the regular elements of S(X) have been characterized by Magill and Subbiah in [13]. An element f of

S(X) is regular if and only if range of f, denoted by ran f in the sequal, is a retract of X and f maps some retract of X homeomorphically on to its range. In the same paper it was shown that two regular elements f and g of S(X) are $\mathcal{D}$ - related if and only if ran f and ran g are homeomorphic and that they are $\mathcal{J}$ - related if and only if range of each contains a retract of X which is homeomorphic to the range of the other. In the passing we will like to mention that such characterizations are not yet known for elements of S(X) which are not necessarily regular. This appears as Project (4.6) in [5]. An excellent source for background material and results on S(X) is the survey article of Magill [3].

3. LOCAL DENDRITES WITH FINITE BRANCH NUMBERS

Local dendrites with finite branch number come up quite naturally when one considers finiteness of $\mathcal{D}_R(S(X))$ or that of $\mathcal{J}_R(S(X))$ within Peano continua. In fact, Magill and Subbiah [13], [14] characterized the Peano continua X for which their $\mathcal{D}_R(S(X))$ contain exactly two or exactly three elements. These continua are arc, in the first case, and a simple closed curve or a triode (i.e., a space homeomorphic to $\perp$ which is often referred to as a T as well) in the second. Now, this prompted Magill to pose (and solve) the problem of characterizing those Peano continua X for which $\mathcal{D}_R(S(X))$ is finite [5] and [6], and this is one place where local dendrites with finite branch number show up in a natural way. We need to define them before we can give Magill's result.

A dendrite is a Peano continuum which contains no simple closed curves and a local dendrite is any Peano continuum with the property that each point has a neighborhood which is a dendrite [17, p.88]. Let X be any local dendrite and let D be a dendrite neighborhood of an element x of X. The number of components of

D - {x} does not depend upon the choice of D [6] and thus this number is called the rank of x in X. The point x is defined to be an end point if the rank of x in X is one, a local cut point if the rank of x in X is > 1, and a branch point if the rank of x in X is > 2. The branch number of a local dendrite X is the sum of the ranks of all the branch points of X. Certainly this number is finite if and only if X has only finitely many branch points and the rank of each branch point is finite.

Theorem (3.1). (Magill [6]) Let X be a Peano continuum. Then $\mathcal{D}_R(S(X))$ is finite if and only if X is a local dendrite with finite branch number.

This theorem has been suitably extended in [9]. Let $\mathfrak{U}$ be a nonempty family of subcontinua of any topological space X with the property that if $A \in \mathfrak{U}$, $f \in S(X)$ and f maps A homeomorphically on to B, then $B \in \mathfrak{U}$. That is, the family $\mathfrak{U}$ is a unifying family of X [3] all of whose members are subcontinua of X. The unifying congruence $\sigma(\mathfrak{U})$, obtained by identifying two functions in S(X) if whenever one of them is injective on any $A \in \mathfrak{U}$, then they both agree on A, will be called a continuum congruence and $\mathrm{Con}_C(S(X))$ will denote the partially ordered collection of all continuum congruences on S(X).

Theorem (3.2). (Magill [9]) Let X be any Peano continuum. The following are equivalent.

(i) $\mathrm{Con}_C(S(X))$ is finite.

(ii) $\mathcal{D}_R(S(X))$ is finite.

(iii) $\mathcal{J}_R(S(X))$ is finite.

(iv) X is a local dendrite with finite branch number.

It seems appropriate to digress here to recall that the relations $\mathcal{D}$ and $\mathcal{J}$ are most often distinct for S(X). For example, these are differrent relations whenever X is a completely regular Hausdorff space containing an arc [13]. However, there are instances when these two coincide as well. It is proved in [2] that $\mathcal{D} = \mathcal{J}$ for $\mathcal{T}_X$, the full transformation semigroup on X, which is really S(X) if we consider the discrete topology on X. This result has been generalized by Magill for 0 - dimensional metric spaces. Some parts of the following theorem have credits to Cezus [1] and Magill and Subbiah.

Theorem (3.3). (Magill [4]) Let X be a 0 - dimensional metric space. The following are equivalent.

(i) $\mathcal{D} = \mathcal{J}$ in S(X).

(ii) Every element of S(X) is regular, i.e., S(X) is a regular semigroup.

(iii) X is either discrete or is the one - point compactification of the countably infinite discrete space.

Since the local dendrites with finite branch number are important for us, we would like to mention that the relations $\mathcal{D}$ and $\mathcal{J}$ also coincide for the regular elements of the semigroup of selfmaps of such continua and it is easy to see that this property is equivalent to each $\mathcal{J}$ - class contain at most one regular $\mathcal{D}$ - class [7].

4. REGULAR $\mathcal{J}$ - CLASSES

Once an appropriate equivalence relation is defined on the collection of all the retracts of X, the characterization of regular elements of S(X) enables one to get without much difficulty the following general and very useful lemma for regular $\mathcal{J}$ - classes of S(X),

Let $\mathcal{R}(X)$ be the collection of all the retracts of X. For R_1, $R_2 \in \mathcal{R}(X)$ define $R_1 \approx R_2$ whenever R_1 is homeomorphic to a retract of X contained in R_2 and also R_2 is homeomorphic to a retract of X contained in R_1. Let $\mathcal{R}^*(X)$ denote the equivalence classes of $\mathcal{R}(X)$ with respect to the equivalence relation $\approx$. Let $R_1^* \leq R_2^*$ if and only if R_1 is homeomorphic to a retract of X contained in R_2 where R_i^* are elements of $\mathcal{R}^*(X)$ containing R_i $(i = 1, 2)$. Then $\leq$ is a partial ordering on $\mathcal{R}^*(X)$.

Lemma (4.1). (Magill and Subbiah [16]) for any topological space X, $\mathcal{J}_R(S(X))$ is order isomorphic to $\mathcal{R}^*(X)$.

We will like to mention another result about $\mathcal{J}_R(S(X))$ before we go on to discuss the consequences of Lemma (4.1). The theorem given below holds for many spaces beside local dendrites with finite branch number and it parallels results in representation theory of inverse and regular semigroups. Observe that if X contains more than one point then S(X) can not be an inverse semigroup and that it is rarely regular either (Theorem (3.3)). We first need some definitions.

Any semigroup of the form vSv, where v is an idempotent of a semigroup S, is known as a <u>local subsemigroup</u> of S. Two local subsemigroups vSv and wSw are called equivalent if each can be algebraically embedded in the other. This equivalence relation is denoted by $\mathcal{E}$, the equivalence class containing vSv is denoted by $\mathcal{E}(vSv)$ and the family of all these equivalence classes is denoted by $\mathcal{E}Loc(S)$. $\mathcal{E}(vSv) \leq \mathcal{E}(wSw)$ in $\mathcal{E}Loc(S)$ means that vSv can be algebraically embedded in wSw. It is easily checked that $\leq$ is a partial ordering on $\mathcal{E}Loc(S)$.

We also need the definition of an L-admissible space.

A topological space X is a <u>strong S^*- space</u> if it is Hausdorff and for each pair of nonempty disjoint closed subsets A and B of X, there exist two distinct points a and b of X, such that $f(x) = a$ for $x \in A$ and $f(x) = b$ for $x \in B$. A strong S^*- space X is called <u>strongly conformable</u> if it is first countable and for each pair of compact countable subspaces A and B, each having exactly one limit point, there exists a continuous selfmap of X mapping A into B such that B - f[A] is finite. Next, S(X) is said to be <u>doubly transitive</u> on X if for each quadruple of points a,b,c,d from X with a ≠ b, there exists an f ∈ S(X) such that f(a) = c and f(b) = d. A space X is called <u>accessible</u> if every continuous map from a closed, connected subspace of X into X can be extended to a continuous selfmap of X. Finally, A space X is called <u>L - admissible</u> if it is connected, second countable, strongly conformable, accessible and if S(X) is doubly transitive on X. Any second countable absolute retract is L - admissible. Thus all Euclidean N - spaces, Euclidean N - cell, Hilbert space, the Hilbert cube and all dendrites are L - admissible. Local dendrites are examples of L - admissible spaces which are not absolute retracts.

Theorem (4.2). (Magill and Misra [10]) Let X be an L - admissible space. Then $\mathcal{E}$Loc(S(X)) is order isomorphic to $\mathcal{J}_R$(S(X)).

As we mentioned in the introduction, first Magill and Subbiah [16] for plane Peano continua and then Magill [9] for any Peano continuum, characterized X whose $\mathcal{J}_R$(S(X)) form a finite chain. Of course, by Theorem (3.2), the finiteness of $\mathcal{J}_R$(S(X)) implies that such Peano continua are necessarily local dendrites with finite branch number. The chain condition on $\mathcal{J}_R$(S(X)) forces these local dendrites to be either a point, a simple closed curve, a triple T, an N - star or a bed of nails. By a triple T is meant any space homeomorphic to

$$\{(x,0) \in \mathbb{R}^2 : -2 \le x \le 2\} \cup \{(-1,y) \in \mathbb{R}^2 : -1 \le y \le 0\} \cup$$
$$\{(1,y) \in \mathbb{R}^2 : -1 \le y \le 0\} \cup \{(0,y) \in \mathbb{R}^2 : 0 \le y \le 1\} \cup$$
$$\{(x,1) \in \mathbb{R}^2 : -1 \le x \le 1\}.$$

and it can be represented by the figure . By an N - star we mean any space homeomorphic to

$$\cup \{(n,y) \in \mathbb{R}^2 : y = (1/n)x \text{ and } 0 \le x \le 1\}_{n=1}^{N}.$$

Finally, by a bed of nails with N spikes is meant any space homeomorphic to

$\{(n,y) \in \mathbb{R}^2 : 0 \le y \le 1\}_{n=1}^{N} \cup \{(x,0) \in \mathbb{R}^2 : 0 \le x \le N + 1\}$.

Thus, for example, a 1 - star and a 2 - star are homeomorphic to the closed unit interval, a 3 - star is homeomorphic to  i.e., a triode, a 4 - star is homeomorphic to , etc. Further, a bed of nails with 5 spikes, for example, can be represented by the figure . Lemma (4.1) has been extensively used in obtaining these results.

More specifically, the following theorem was proved.

Theorem (4.3). (Magill [9]) Let X be a Peano continuum. Then the following are equivalent.

(i) $Con_c(S(X))$ is a finite chain.

(ii) $\mathcal{J}_R(S(X))$ is a finite chain.

(iii) $Con_c(S(X))$ and $\mathcal{J}_R(S(X))$ are finite and have the same number of elements.

(iv) $Con_c(S(X))$ and $\mathcal{D}_R(S(X))$ are finite and have the same number of elements.

(v) X is either a point, a simple closed curve, a triple T, an N - star or a bed of nails.

Thus by Theorem (4.2) each of these statements is equivalent to

(vi) $\mathcal{E}Loc(S(X))$ forms a finite chain.

For local dendrites with finite branch number, the family $Con_c(S(X))$ has a "nice" characterization. A collection $\mathcal{K}_c$ of subcontinua of X is called a <u>characteristic collection</u> if no two distinct elements of $\mathcal{K}_c$ are homeomorphic to each other and each subcontinuum of X is homeomorphic to some element of $\mathcal{K}_c$. Let $Ind(\mathcal{K}_c)$ be the collection of all independent subcollections of $\mathcal{K}_c$ where a subcollection $\mathfrak{U}$ of $\mathcal{K}_c$ is called <u>independent</u> if no $A \in \mathfrak{U}$ is the union of copies of elements of $\mathfrak{U} - \{A\}$. The relation $\mathcal{A} \le \mathcal{B}$ if and only if each B $\in \mathcal{B}$ is the union of copies of elements of $\mathcal{A}$ is reflexive and transitive on Ind($\mathcal{K}_c$). This relation is also antisymmetric if X is a local dendrite with finite branch number.

Theorem (4.4). (Magill [8]) Let X be a local dendrite with finite branch number. Then $\text{Con}_c(S(X))$ is order isomorphic to Ind($\mathcal{K}_c$).

Example (4.5). Let $X = \{(x,0) \in \mathbb{R}^2 : -1 \le x \le 2\} \cup$
$\{(0,y) \in \mathbb{R}^2 : -1 \le y \le 1\} \cup$
$\{(1,y) \in \mathbb{R}^2 : 0 \le y \le 1\}$.

Then X can be represented by 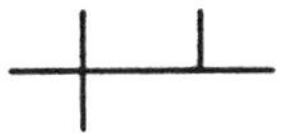 . The characteristic collection $\mathcal{K}_c$ of subcontinua of X contains elements which can be represented by the following figures.

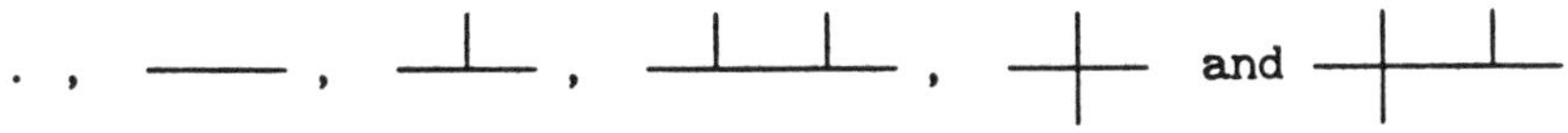

1 2 3 4 5 6

Thus $\mathcal{K}_c$ = {1,2,3,4,5,6} and Ind($\mathcal{K}_c$) = {1,2,3,4,5,6,{4,5}}. The diagram of Ind($\mathcal{K}_c$) can be given as follows.

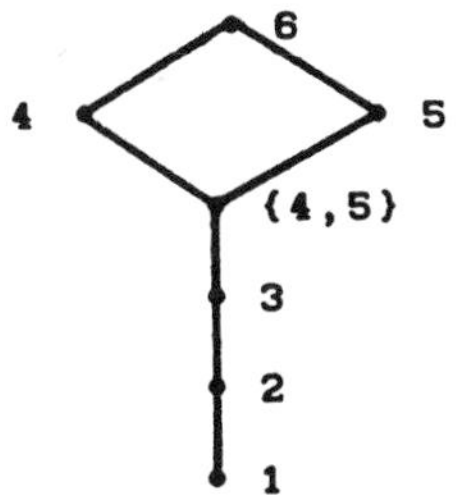

By Theorem (4.4) this is also the diagram of Con_c(S(X). Note that it is a lattice in this case. An example of a continuum X for which $\text{Con}_c(S(X))$ is not a lattice will be given shortly.

As a matter of fact, for any topological space X, one can define two relations on its characteristic collection $\mathcal{K}_c$ itself in natural ways. For A, B $\in \mathcal{K}_c$, let $A \le_e B$ if and only if A can be embedded in B and let $A \leq_u B$ if and only if B is the union of copies of A. Both these relations are reflexive and transitive. They are antisymmetric if X is a local dendrite with finite branch number. Clearly $\leq_u \subseteq \leq_e$ but they need not be equal in general.

Example (4.6). Let $X = \{(x,0) \in \mathbb{R}^2 : -2 \leq x \leq 2\} \cup$
$\{(-1,y) \in \mathbb{R}^2 : -1 \leq y \leq 1\} \cup$
$\{(0,y) \in \mathbb{R}^2 : 0 \leq y \leq 1\} \cup$
$\{(1,y) \in \mathbb{R}^2 : -1 \leq y \leq 1\}$.

Thus X can be represented by 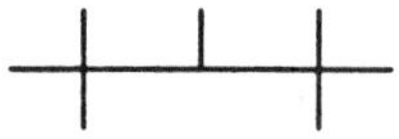. Now the elements of a characteristic collection $\mathcal{K}_c$ of X can be represented by the following figures.

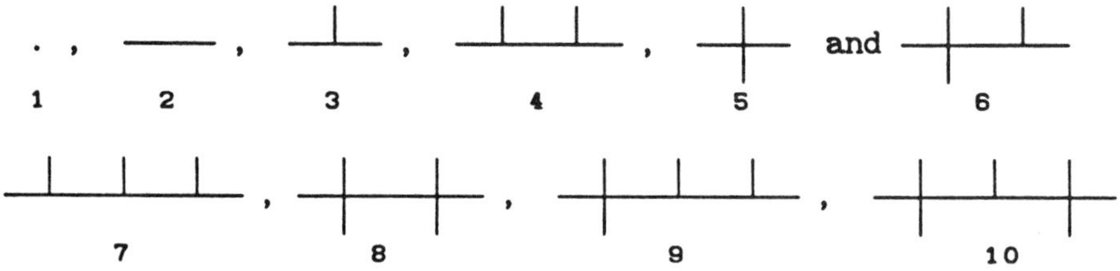

We draw the diagrams for $(\mathcal{K}_c, \leq_e)$ and $(\mathcal{K}_c, \leq_u)$. They are

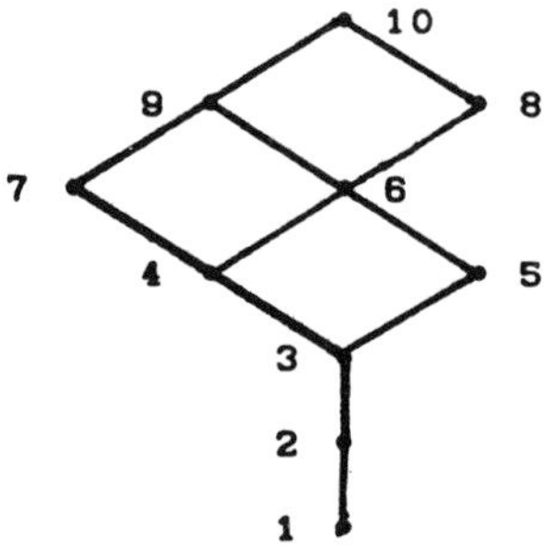

and

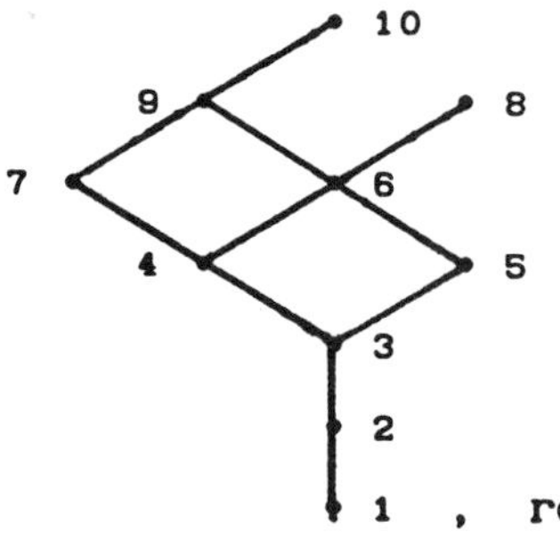

, respectively.

Observe that $(\mathcal{K}_c, \leq_u)$ is not a lattice in this case and also that $\leq_u$ and $\leq_e$ are different relations here. Ind($\mathcal{K}_c$) for this X is {1,2,3,4,5,6,7,8,9,10,{4,5},{5,7},{6,7},{7,8},{8,9},{8,10}} which is easily seen to be not a lattice. Thus $\mathrm{Con}_c(S(X))$ is not a lattice when X is homeomorphic to 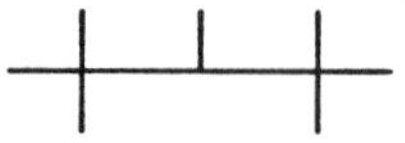.

Since any subcontinuum of a local dendrite is a retract of the local dendrite, the following lemma can be easily established.

Lemma (4.7). (Magill and Misra [12]) Let X be a local dendrite with finite branch number. Then $\mathcal{J}_R(S(X))$ is order isomorphic to $(\mathcal{K}_c, \leq_e)$.

Next, denote by $\text{Con}_{PC}(S(X))$ the partially ordered subfamily of $\text{Con}_C(S(X))$ consisting of all <u>principal</u> <u>continuum</u> <u>congruences</u> $\sigma(K)$ defined for each subcontinuum K of X and obtained by identifying two functions in S(X) if whenever one of them is injective on any subspace of X homeomorphic to K, then they both agree on that subspace. For local dendrites with finite branch number, we have

Lemma (4.8). (Magill and Misra [11]) Let X be a local dendrite with finite branch number. Then $\text{Con}_{PC}(S(X))$ is order isomorphic to $(\mathcal{K}_c, \leq_u)$.

The last two lemmas show us that, for local dendrites with finite branch number, $\mathcal{J}_R(S(X))$ and $\text{Con}_{PC}(S(X))$ contain same number of elements and also that these two families are order isomorphic if and only if $(\mathcal{K}_c, \leq_e)$ and $(\mathcal{K}_c, \leq_u)$ are order isomorphic. This brings us to the next theorem.

Theorem (4.9). (Magill and Misra [11]) Let X be a Peano continuum. The following are equivalent.

(i) $\text{Con}_{PC}(S(X))$ and $\mathcal{J}_R(S(X))$ are isomorphic finite partially ordered sets.

(ii) $\text{Con}_{PC}(S(X))$ and $\mathcal{E}\text{Loc}(S(X))$ are isomorphic finite partially ordered sets.

(iii) X is a local dendrite with finite branch number such that for any two subcontinua H and K of X, if H can be embedded in K then K is the union of copies of H.

A continuum arising in (iii) of the above theorem has been called an <u>accordant</u> <u>space</u> in [11] and it has been shown there that an accordant space is either a simple closed curve or a dendrite. Note that not every dendrite is an accordant space. In example (4.6) the continuum 8 is a subcontinuum of 10 but 10 is not the union of copies of 8. Theorem (4.9) is extremely useful in characterizing the Peano continua whose $Con_{PC}(S(X))$ and $\mathcal{J}_R(S(X))$ are isomorphic finite lattices. To help save space, these continua will be described below by their figures only as we did for most of the continua arising in examples (4.5) and (4.6). Moreover, an N - star will be represented by  and a bed of nails with N spikes will be represented by , without any explicit mention of N.

Theorem (4.10). (Magill and Misra [11]) Let X be a Peano continuum. The following are equivalent.

(i) $Con_{PC}(S(X))$ and $\mathcal{J}_R(S(X))$ are isomorphic finite lattices.

(ii) $Con_{PC}(S(X))$ and $\mathcal{E}Loc(S(X))$ are isomorphic finite lattices.

(iii) X is an accordant space such that for each pair of subcontinua H and K of X, among all those subcontinua which can be embedded in both H and K there is one which contains copies of all the others.

(iv) X is either a simple closed curve or is homeomorphic to a subcontinuum of one of the following six continua.

(a) 

(b) 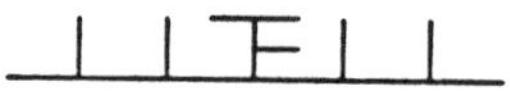

(c) 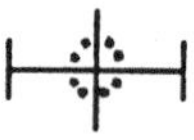

(d) 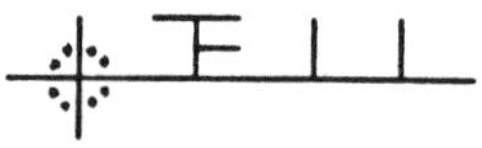

(e) 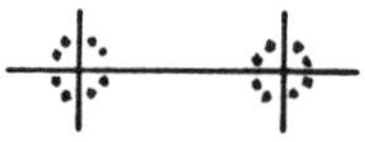

(f) 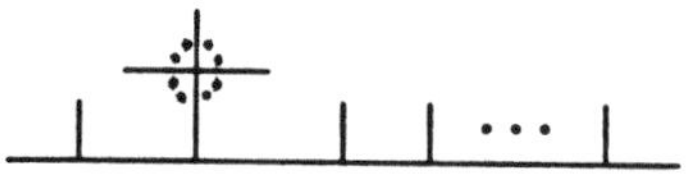

This is a considerable generalization of Theorem (4.3). In the spirit of one of the first results obtained for regular $\mathcal{J}$ - classes of S(X) by Magill and Subbiah [15], we observe here that if a Peano continuum X has no cut points, then any of the first two conditions of the above theorem characterizes simple closed curves among such continua. This is corollory (3.11) of [11].

In order to characterize the Peano continua whose $\mathcal{J}_R$(S(X)) form a finite lattice, one has to consider local dendrites with finite branch number which are not necessarily an accordant space. Such spaces could be very involved. However, the following theorem does characterize the dendrites X whose $\mathcal{J}_R$(S(X)) are finite lattices.

Theorem (4.11). (Magill and Misra [12]) Let X be a dendrite. Then $\mathcal{J}_R$(S(X)) is a finite lattice if and only if X is homeomorphic to a subcontinuum of one of the following continua.

(a)

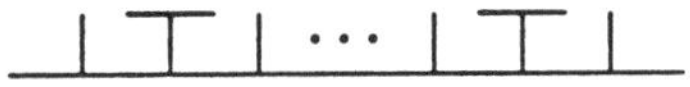

(b)

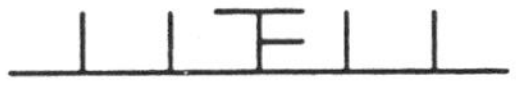

(c)

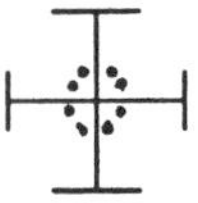

(d)

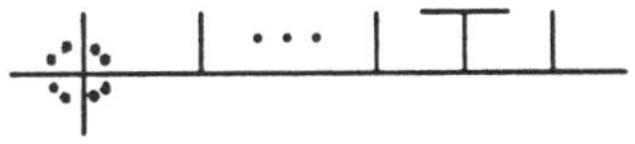

(e)

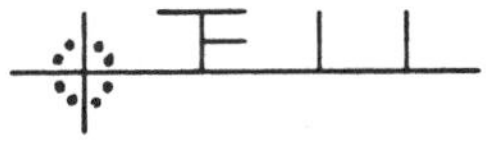

(f)

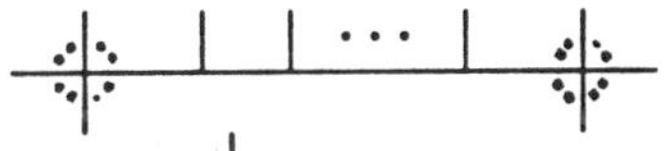

(g)

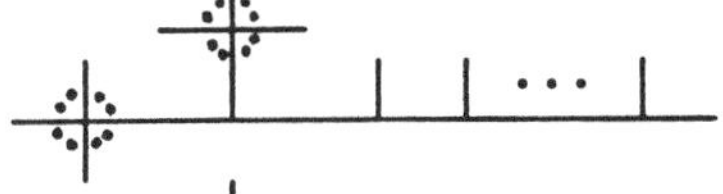

(h)

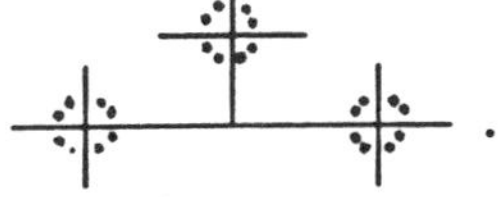

In view of Theorem (3.2) this result can be extended to Peano continua as soon as one proves it for local dendrites.

REFERENCES

1. F. A. Cezus, Green's relations in semigroups of functions, Ph.D. thesis at Australian National University, Canberra, Australia (1972).

2. A. H. Clifford and G. B. Preston, The algebraic theory of semigroups, Math. Surveys, No. 7, Vol.I, Amer. Math. Soc., Providence, Rhode Island (1964).

3. K. D. Magill, Jr., A survey of semigroups of continuous selfmaps, Semigroup Forum, 11 (3) (1975/76) 189-282.

4. K. D. Magill, Jr., Semigroups of continuous selfmaps for which Green's $\mathcal{D}$ and $\mathcal{J}$ relations coincide, Glasgow Math. J., 20 (1979) 25-28.

5. K. D. Magill, Jr., Some open problems and directions for further research in semigroups of continuous selfmaps, Univ. Alg. and Appl., Banach Center Pub., PWN-Polish Sci. Publ., Warsaw 9 (1982) 439-454.

6. K. D. Magill, Jr., Semigroups with only finitely many regular $\mathcal{D}$ - classes, Semigroup Forum, 25 (3/4) (1982) 361-377.

7. K. D. Magill, Jr., Semigroups in which $\mathcal{D}$ and $\mathcal{J}$ coincide for regular elements, Semigroup Forum, 25 (3/4) (1982) 383-385.

8. K. D. Magill, Jr., Congruences on semigroups of continuous selfmaps, Semigroup Forum, 29 (1984) 159-182.

9. K. D. Magill, Jr., Semigroups for which the continuum congruences form a finite chain, Semigroup Forum, 30 (1984) 221-230.

10. K. D. Magill, Jr. and P. R. Misra, ordering the local subsemigroups of a semigroup, Semigroup Forum, 35 (1987) 101-117.

11. K. D. Magill, Jr. and P. R. Misra, Principal continuum congruences, regular $\mathcal{J}$ - classes and local subsemigroups, to appear in Semigroup Forum.

12. K. D. Magill, Jr. and P. R. Misra, Semigroups whose regular $\mathcal{J}$ - classes form a lattice, to be submitted.

13. K. D. Magill, Jr. and S. Subbiah, Green's relations for regular elements of semigroups of endomorphisms, Canad. J. Math., 26 (1974) 1484-1497.

14. K. D. Magill, Jr. and S. Subbiah, Semigroups with exactly three regular $\mathcal{D}$ - classes, Semigroup Forum, 17 (1979) 279-281.

15. K. D. Magill, Jr. and S. Subbiah, Another semigroup characterization of the simple closed curve, Semigroup Forum, 21 (1980) 89-90.

16. K. D. Magill, Jr. and S. Subbiah, Regular $\mathcal{J}$ - classes of semigroups of continua, Semigroup Forum, 22 (1981) 159-179.

17. G. T. Whyburn, Analytic topology, colloq. Pub., Vol.28., Amer. Math. Soc., Providence, Rhode Island (1963).

A Localic Construction of Some Covers of Compact Hausdorff Spaces

A. T. Molitor, Department of Mathematics, Wesleyan University, Middletown, CT

This paper describes a simple construction, using the theory of locales, of a class of covering spaces related to the quasi-F cover. Let $\mathcal{P}$ be a property of locales satisfying the condition that every compact Hausdorff space has a smallest dense sublocale with $\mathcal{P}$. The Stone-Čech compactification of this sublocale is an irreducible preimage of the original space with some properties similar to those of the quasi-F cover. This generalizes a construction used in [J] to produce the Gleason cover, and in [MV] to produce the quasi-F cover.

I am pleased to thank A.W. Hager, A.J. Macula and C. Neville for various suggestions and valuable conversations concerning this work.

§0 PRELIMINARIES

A frame is a complete lattice in which finite meets distribute across arbitrary joins, a frame homomorphism is a map between frames which commutes with finite meets and all joins, and preserves the top and bottom elements. This defines the category of frames, **Frm**. The category **Loc** of locales is the formal opposite of **Frm**.

Sublocales correspond to quotient frames, and the latter are conveniently described by nuclei. A nucleus on a frame is a function from the frame to itself which is idempotent, increasing and commutes with finite meets. The image of a nucleus is, in the order it inherits by virtue of being a subset of a frame, itself a frame. Furthermore, every quotient frame is isomorphic to the image of some nucleus. Thus, sublocales may be identified with nuclei. Given a nucleus j, on a frame $\mathbf{A}$, the image of j is denoted $\mathbf{A}_j$. To compute joins in $\mathbf{A}_j$, one computes the join in $\mathbf{A}$ (since $\mathbf{A}_j$ is a subset of $\mathbf{A}$) and applies j to the result.

A locale is said to be *compact* if the associated frame has the property that whenever a family of elements has join $\mathbf{1}$, there is a finite subfamily with join $\mathbf{1}$. A sublocale (i.e. quotient frame) given by a nucleus j is *dense* if $j(\emptyset) = \emptyset$. An element, x, of a frame is said to be *well below* an element y if there is a third element z with $x \wedge z = \mathbf{0}$ and $y \vee z = \mathbf{1}$. If the frame is the frame of open sets of some space, this means exactly that $cl(x) \subseteq y$. A locale is said to be *regular* if every element of the associated frame is the join of elements well below it.

The category of compact regular locales is equivalent to the category of compact Hausdorff spaces, i.e. the compact regular frames are exactly the topologies of compact Hausdorff spaces (assuming the axiom of choice). Thus, without loss of generality, we will not distinguish between compact regular locales and compact Hausdorff spaces.

The category of compact regular locales is reflective in **Loc**, with the reflection being denoted β, and called the Stone-Čech compactification. This is an extension of the usual Stone-Čech compactification of Tychonov spaces, and behaves very much like it. In particular, if L is a sublocale of a compact Hausdorff space $\mathbf{X}$, then $\mathbf{X} = \beta(L)$ if and only if L is C*-embedded, i.e. every **Loc** morphism from L to $[0, 1]$ (or any compact Hausdorff space) lifts over $\mathbf{X}$.

All of the above is to be found, in much greater detail, in [J].

§1 THE QUASI-$\mathcal{P}$ COREFLECTION

Let $\mathcal{P}$ be any (fixed) property of locales satisfying the condition the any compact Hausdorff space, $\mathbf{X}$, has a minimum dense sublocale with $\mathcal{P}$, $\mathbf{X}_{\mathcal{P}}$. That is, $\mathbf{X}_{\mathcal{P}}$ is a dense sublocale of $\mathbf{X}$, has $\mathcal{P}$, and is a sublocale of any other such sublocale of $\mathbf{X}$. For example, if $\mathcal{P}$ is the property "compactness", then $\mathbf{X}_{\mathcal{P}}$ is simply $\mathbf{X}$. If $\mathcal{P}$ is some vacuous property (say, the property of being a locale), then $\mathbf{X}_{\mathcal{P}}$ is the minimum dense sublocale of $\mathbf{X}$ (i.e. maximum dense quotient of the topology of $\mathbf{X}$, which is the complete Boolean algebra of regular open sets. See [J]). In the next section, the property "κ-Lindelöf" is shown to be such a property.

DEFINITION 1.0. A continuous map $f: \mathbf{X} \to \mathbf{Y}$ between compact Hausdorff spaces is said to be *$\mathcal{P}$-skeletal* if the map $f|_{\mathbf{X}_{\mathcal{P}}}$ factors through $\mathbf{Y}_{\mathcal{P}}$, i.e. if f "drops" to a map $\hat{f}: \mathbf{X}_{\mathcal{P}} \to \mathbf{Y}_{\mathcal{P}}$. If $\hat{f}$ is a homeomorphism, we say f is *$\mathcal{P}$-irreducible.*

DEFINITION 1.1. A space, $\mathbf{X}$, is said to be *quasi-$\mathcal{P}$* if $\mathbf{X}_{\mathcal{P}} \subseteq \mathbf{X}$ is the Stone-Čech compactification of $\mathbf{X}_{\mathcal{P}}$.

LEMMA 1.2. $[\beta(\mathbf{X}_{\mathcal{P}})]_{\mathcal{P}} = \mathbf{X}_{\mathcal{P}}$ for any compact Hausdorff space $\mathbf{X}$, thus $\beta(\mathbf{X}_{\mathcal{P}})$ is quasi-$\mathcal{P}$.

Proof: this follows immediately from the condition on $\mathcal{P}$.

■

PROPOSITION 1.3. A space $\mathbf{X}$ is quasi-$\mathcal{P}$ if and only if every dense sublocale with $\mathcal{P}$ is C*-embedded.

Proof: By the definition of $\mathbf{X}_{\mathcal{P}}$ every dense sublocale with $\mathcal{P}$ contains $\mathbf{X}_{\mathcal{P}}$. Since the latter is C*-embedded, every dense sublocale with $\mathcal{P}$ is C*-embedded. The converse is trivial, since $\mathbf{X}_{\mathcal{P}}$ has $\mathcal{P}$.

■

LEMMA 1.4. A $\mathcal{P}$-irreducible map is irreducible.

Proof: A continuous map $f:\mathbf{X} \to \mathbf{Y}$ is irreducible if and only if f^{-1} induces an isomorphism $\mathcal{R}(\mathbf{X}) \cong \mathcal{R}(\mathbf{Y})$ of the complete Boolean algebras of regular open sets (see, for example, [M]). These algebras are the frames associated with the minimum dense sublocales of $\mathbf{X}$ and $\mathbf{Y}$. Since f, by hypothesis, restricts to a homeomorphism $\mathbf{X}_{\mathcal{P}} \to \mathbf{Y}_{\mathcal{P}}$, it certainly restricts further to a homeomorphism of minimum dense sublocales, which means precisely that f induces an isomorphism between the corresponding frames. ∎

For the following, note that for any compact Hausdorff space $\mathbf{X}$, $\beta(\mathbf{X}_{\mathcal{P}})$ is quasi-$\mathcal{P}$.

DEFINITION 1.5. For any compact Hausdorff space $\mathbf{X}$, the Stone-Čech extension, $q:\beta(\mathbf{X}_{\mathcal{P}}) \to \mathbf{X}$, of the inclusion $\mathbf{X}_{\mathcal{P}} \subseteq \mathbf{X}$ is the *quasi-$\mathcal{P}$ cover* of $\mathbf{X}$.

As we will see later, the two extreme cases, $\mathbf{X}_{\mathcal{P}} = \mathbf{X}$ and $\mathbf{X}_{\mathcal{P}}$ the minimum dense sublocale of $\mathbf{X}$, result in the smallest and largest irreducible preimages of $\mathbf{X}$ ($\mathbf{X}$ and the Gleason cover of $\mathbf{X}$) respectively. We can now state the main theorem:

THEOREM 1.6. For any space $\mathbf{X}$, the quasi-$\mathcal{P}$ cover is:

1) the minimum $\mathcal{P}$-irreducible preimage of $\mathbf{X}$

2) the quasi-$\mathcal{P}$ coreflection of $\mathbf{X}$ in the category of compact Hausdorff spaces and $\mathcal{P}$-skeletal maps.

Proof: (1) That the covering map is $\mathcal{P}$-irreducible follows from lemma 1.2. Let $g:\mathbf{Z} \to \mathbf{X}$ be any $\mathcal{P}$-irreducible map, where $\mathbf{Z}$ is any compact Hausdorff space. Let $h:\mathbf{X}_{\mathcal{P}} \to \mathbf{Z}$ be the composition

$$\mathbf{X}_{\mathcal{P}} \cong \mathbf{Z}_{\mathcal{P}} \subseteq \mathbf{Z}$$

Let $\hat{h}:\beta(\mathbf{X}_{\mathcal{P}}) \to \mathbf{Z}$ be the Stone-Čech extension of h. A routine diagram chase shows the $g \circ \hat{h} = q$ (See figure 1). Note that the inclusion $i:\mathbf{X}_{\mathcal{P}} \subseteq \beta(\mathbf{X}_{\mathcal{P}})$ is epimorphic.

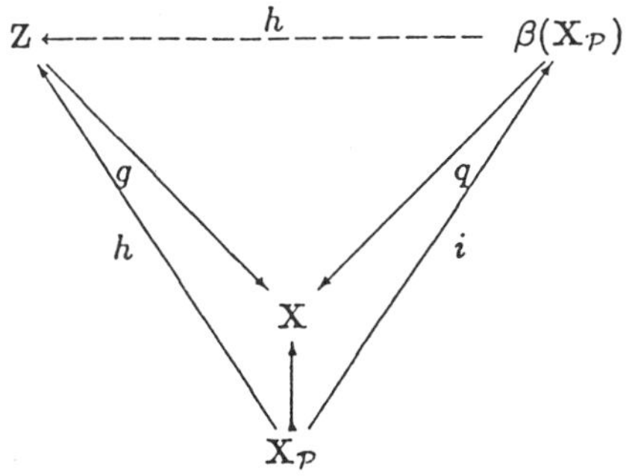

figure 1

(2) Let Z be a quasi-$\mathcal{P}$ space, and $g:\mathbf{Z} \to \mathbf{X}$ be $\mathcal{P}$-skeletal. By hypothesis, $\mathbf{Z} \cong \beta(\mathbf{Z}_{\mathcal{P}})$. In figure 2, $\hat{g}$ is the restriction map given by the hypothesis on g and f is the Stone-Čech extension of $i \circ \hat{g}$. Another routine diagram chase, noting that the inclusion $\mathbf{Z}_{\mathcal{P}} \subseteq \mathbf{Z}$ is epimorphic, shows that $q \circ f = g$. It is clear that f is $\mathcal{P}$-skeletal, $\hat{g}$ being the required restriction. In order that $\beta(\mathbf{X}_{\mathcal{P}})$ be a coreflection, it is necessary that f be the unique $\mathcal{P}$-skeletal map satisfying $q \circ f = g$. To see this, note that any other such map must agree with $i \circ \hat{g}$, and therefore with f, on the dense sublocale $\mathbf{Z}_{\mathcal{P}}$ (this will follow from the fact that the inclusion $\mathbf{X}_{\mathcal{P}} \subseteq \mathbf{X}$ is monomorphic). Uniqueness of f then follows from the fact the if two maps between regular locales agree on a dense sublocale, they are equal.

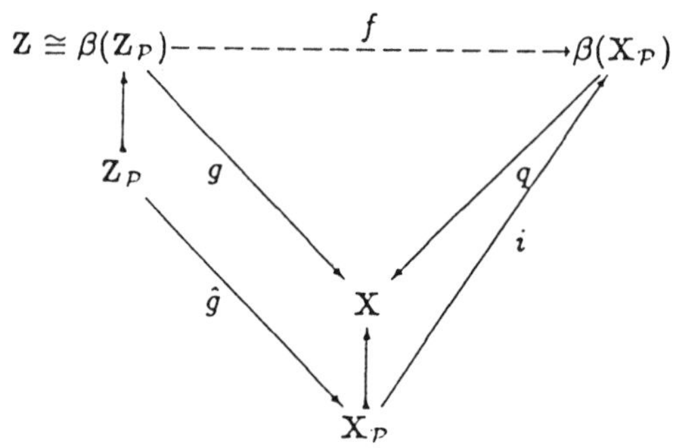

figure 2

■

PROPOSITION 1.7. The quasi-$\mathcal{P}$ cover of X is the unique (up to homeomorphism) quasi-$\mathcal{P}$, $\mathcal{P}$-irreducible preimage of X.

Proof: Routine.

■

§2 THE QUASI-F_κ COVER

Throughout this section, κ is a fixed regular cardinal number. Following the conventions of [M] we a define a *κ-cozero set* to be the union of fewer than κ cozero sets. A locale is *κ-Lindelöf* if every open cover admits a subcover of cardinality strictly less than κ. Thus, Lindelöf is, in this notation, ω_1-Lindelöf.

The object of this section is to show that the property "κ-Lindelöf" is a property $\mathcal{P}$ as described in the previous section. In this case the quasi-$\mathcal{P}$ cover turns out to be the quasi-F_κ cover introduced in [N] and investigated so thoroughly in [BHN]. In the case $\kappa = \omega_1$, the cover is the quasi-F, see [DHH]. In the case $\kappa = \infty$, i.e. no cardinal bound, the Gleason cover results.

The contents of this section can be, and first were, deduced from section 7 of [M].

Let X be a compact Hausdorff space.

DEFINITION 2.0. Let $U \subseteq X$ be an open set. Define j by $j(U) = \{intcl(C) : C \subseteq U, \text{and } C \text{ a } \kappa\text{-cozero set}\}$.

LEMMA 2.1. The map j is a nucleus on the topology of X.

Proof: Routine.

■

Let X_κ be the sublocale of X given by the above nucleus. Since $j(\emptyset) = \emptyset$, X_κ is a dense sublocale. We will now show that X_κ is the minimum dense κ-Lindelöf sublocale of X.

LEMMA 2.2. For any open set $U \subseteq X$, $j(U) = X$ if and only if U contains a dense κ-cozero set.

Proof: If U contains a dense κ-cozero set, it is evident that $j(U) = X$. Conversely, if $j(U) = X$, then

$$\{intcl(C) : C \subseteq U, C\ \kappa\text{-cozero}\}$$

is an open cover of X. Choose $C_1, ..., C_n$ such that $X = \bigcup_{i=1}^n intcl(C_i)$. Then $\bigcup_{i=1}^n C_i$ is a dense κ-cozero set contained in U.

■

LEMMA 2.3. X_κ is κ-Lindelöf.

Proof: It suffices to show that if $\mathcal{U}$ is a family of open subsets of X with $j(\bigcup\mathcal{U}) = X$, then there is a subfamily $\mathcal{V}$, of cardinality less than κ with $j(\bigcup\mathcal{V}) = X$. This follows from lemma 2.2 applied to $\mathcal{U}$, and the fact that in compact Hausdorff spaces κ-cozero sets are κ-Lindelöf.

∎

We now show that X_κ is minimum for being dense and κ-Lindelöf. This is most conveniently proved by applying the following result, 2.3 (1) of [J].

THEOREM 2.4. In a regular locale any sublocale is the intersection of open sublocales.

Note that the open sublocales of a space are the open subspaces.

THEOREM 2.5. If X is a compact Hausdorff space, then X_κ is the minimum dense κ-Lindelöf sublocale.

Proof: Let L be a dense κ-Lindelöf sublocale of X. We must show that $X_\kappa \subseteq L$. In view of 2.4, we need only show that if $L \subseteq U$, where U is some open subset of X, then $X_\kappa \subseteq U$. In terms of nuclei, we must show that $L \subseteq U$ implies that $j(U) = X$. Applying lemma 2.2 we see that is suffices to show that $L \subseteq U$ implies that U contains a dense κ-cozero set. This last implication follows readily from the fact that L is dense and κ-Lindelöf.

∎

The existence of X_κ shows that "κ-Lindelöf" is a property $\mathcal{P}$ as described in the previous section. In this case, the quasi-$\mathcal{P}$ cover can be identified as the quasi-F_κ cover of [BHN] and [N].

Before establishing this, we first indicate some of the basic properties of the quasi-F_κ cover.

DEFINITION 2.6. a compact Hausdorff space, X, is *quasi*-F_κ if every dense κ-Lindelöf subspace is C^*-embedded.

DEFINITION 2.7. A continuous map f:X → Y is *κ-irreducible* if, for every κ-cozero set U in X, there is a κ-cozero set V in Y with

$$intcl(U) = intcl(f^{-1}[V])$$

THEOREM 2.8. (Theorem 4.5 (a) of [BHN]) Every compact Hausdorff space has a unique, up to homeomorphism, quasi-F_κ, κ-irreducible preimage, the quasi-F_κ cover.

Proof: See [BHN]

∎

To show that the quasi-"κ-Lindelöf" cover is this preimage, we must establish that $\beta(X_\kappa)$ is quasi-F_κ, and that the covering map is κ-irreducible. The first follows from proposition 1.3, since subspaces are sublocales. To establish the second we use the following:

PROPOSITION 2.9. "κ-Lindelöf"-irreducible maps are κ-irreducible.

To prove this, it is convenient to use the following lemma, which is a corollary to lemma 4.2 of [M]. For completeness we give a version of Madden's proof as well as the statement.

LEMMA 2.10. If $C \subseteq X$ is a κ-cozero set, $\mathcal{U}$ a collection of open sets with

$$j(C) \subseteq j(\bigcup\mathcal{U})$$

then there is a subfamily $\mathcal{V} \subseteq \mathcal{U}$, of cardinality less than κ, with $j(C) \subseteq j(\bigcup\mathcal{V})$.

Proof: write $C = \bigcup C_i (i \in I)$ where each C_i is a cozero set with $cl(C_i) \subseteq C$, and $|I| < \kappa$. Denote by V_i the complement of $cl(C_i)$ for each i. Let $\mathcal{U}$ be as hypothesized. Since $cl(C_i) \subseteq C$, $V_i \cup C = \mathbf{X}$ for each i.

Thus:

$$\mathbf{X} = V_i \cup C \subseteq j(V_i) \cup j(\bigcup \mathcal{U})$$

Since the sublocale $\mathbf{X}_\kappa$ is κ-Lindelöf, and the right hand side of the above expression corresponds to an open cover of $\mathbf{X}_\kappa$, we may select a collection $\mathcal{V}_i \subseteq \mathcal{U}$ for each i, with $|\mathcal{V}_i| < \kappa$ and

$$j[j(V_i) \cup j(\bigcup \mathcal{V}_i)] = \mathbf{X}$$

Furthermore:

$$\begin{aligned} j(C_i) &= j(C_i) \cap \mathbf{X} \\ &= j(C_i) \cap j[j(V_i) \cup j(\bigcup \mathcal{V}_i)] \end{aligned}$$

by computation of joins in quotient frames:

$$\begin{aligned} &= j(C_i) \cap j[V_i \cup \bigcup \mathcal{V}_i] \\ &= j[(C_i \cap V_i) \cup (C_i \cap \bigcup \mathcal{V}_i)] \\ &= j(C_i \cap \bigcup \mathcal{V}_i) \qquad \text{(since } C_i \cap V_i = \emptyset) \\ &= j(C_i) \cap j(\bigcup \mathcal{V}_i) \end{aligned}$$

So $j(C_i) \subseteq j(\bigcup \mathcal{V}_i)$ for each i. Thus:

$$j(C) = j(\bigcup_{i \in I} C_i) \subseteq j(\bigcup_{i \in I} j(\bigcup \mathcal{V}_i)) = j(\bigcup_{i \in I} \bigcup \mathcal{V}_i)$$

by nucleic arithmetic. So $\mathcal{V} = \bigcup \mathcal{V}_i (i \in I)$ is a subfamily of $\mathcal{U}$ with the required property.

∎

Proof: (of 2.9.) Since a "κ-Lindelöf"-irreducible map $f : \mathbf{X} \to \mathbf{Y}$ restricts to a homeomorphism of the locales $\mathbf{X}_\kappa \cong \mathbf{Y}_\kappa$, f evidently satisfies:

a) if U and V are open in $\mathbf{Y}$ and $j(U) = j(V)$ then $j(f^{-1}[U]) = j(f^{-1}[V])$ (this is the condition permitting f to be restricted), and

b) for each open $U \subseteq \mathbf{X}$ there is some V open in $\mathbf{Y}$ with $j(f^{-1}[V]) = j(U)$. This makes the frame map corresponding to the homeomorphism a surjection (it is, of course, also one-to-one).

Let C be any κ-cozero set in $\mathbf{X}$. There is, by (b), an open set $V \subseteq \mathbf{Y}$ with $j(f^{-1}[V]) = j(C)$. Writing V as a union of cozero sets, $V = \bigcup_{i \in I} C_i$, we have $j(\bigcup_{i \in I} f^{-1}[C_i]) = j(C)$. Applying lemma 2.10 gives a set $J \subseteq I$ with $|J| < \kappa$ and

$$j(\bigcup_{j \in J} f^{-1}[C_j]) = j(C)$$

So $W = \bigcup_{j \in J} C_j$ is a κ-cozero set in $\mathbf{Y}$ with $j(f^{-1}[W]) = j(C)$. Noting that, for a κ-cozero set C, $j(C) = intcl(C)$, one sees that f is κ-irreducible.

∎

§3 REMARKS

(a) The major defect of the theory outlined above is a lack of examples. The only interesting examples discovered to date are the quasi-F_κ covers, including the extreme case of the Gleason cover. In [MV] it was pointed out that paracompactness is a property $\mathcal{P}$ as above, unfortunately the minimum dense paracompact sublocale is simply the minimum dense sublocale (since in a complete Boolean algebra, any family of elements can be refined by a family of pairwise disjoint elements with the same join, assuming the axiom of choice). Thus the quasi-"paracompact" cover is simply the Gleason cover.

(b) To attempt an extension of the theory, it is natural to drop the requirement that $\mathbf{X}_\mathcal{P}$ actually have $\mathcal{P}$. For example, one might define $\mathbf{X}_\mathcal{P}$ to be the localic intersection of all dense sublocales or subspaces with $\mathcal{P}$. Then $\mathbf{X}_\mathcal{P}$ is a dense sublocale, and $\beta(\mathbf{X}_\mathcal{P})$ with the usual covering map is an irreducible preimage of $\mathbf{X}$. However, lemma 1.2 fails, and with it a substantial portion of the theory.

(c) The theory of §1 resembles the quasi-$\mathcal{P}$ covers of Vermeer [V] and the $\mathcal{P}^3(-)$ covers of Hager [H1]. It is clearly a special case of these, but it is probably not equivalent, even to the compact case. However, the construction given here is, modulo the theory of locales, more elementary than that of [V] and [H1].

(d) It is clear that two spaces, $\mathbf{X}$ and $\mathbf{Y}$, have the same quasi-$\mathcal{P}$ cover if and only if $\mathbf{X}_\mathcal{P}$ is homeomorphic to $\mathbf{Y}_\mathcal{P}$. This appears to be, in some sense, a generalization of theorems of the following form:

THEOREM 3.0. Two Čech-complete spaces, $\mathbf{X}$ and $\mathbf{Y}$, which have G_δ diagonals (i.e. the diagonal in $\mathbf{X} \times \mathbf{X}$ is a G_δ set in $\mathbf{X} \times \mathbf{X}$, similarly $\mathbf{Y}$) have the same Gleason cover if and only if they have homeomorphic dense G_δs.

Proof: see [H2], [MS]

∎

A natural question is whether or not similar theorems exist for the quasi-F_κ covers, and if they do, what are they?

REFERENCES

[BHN] R.N. Ball, A.W. Hager, C. Neville, The κ-ideal completion of an archimedean ℓ-group and the quasi-$F_\kappa$cover of a compact Hausdorff space, to appear in this volume.

[BM] B. Banaschewski and C.J. Mulvey, Stone-Čech compactification of locales I, HoustonJ.Math. 6 (1980), 301-312.

[DHH] F. Dashiell, A. Hager, M. Henriksen, Order-Cauchy completions of rings and vector lattices of continuous functions., Canad. J. Math. (32) 1980, 657-685.

[DS] C.H. Dowker and D. Strauss, Sums in the category of frames, HoustonJ.Math. 3 (1976), 17-32.

[H1] Anthony .W. Hager, Minimal Covers of Topological Spaces, **Papers on General Topology and Related Category Theory and Topological Algebra**, Vol. 552 of the *Annals of the New York Academy of Sciences*, March 15, 1989, 44-59.

[H2] A.W. Hager, Isomorphisms of Some Completions of C(**X**), Topology Proceedings 4 (1979), 407-435.

[I] J. Isbell, Atomless Parts of Spaces, *Math. Scand.* 31(1972), 5–32.

[J] P.T. Johnstone, StoneSpaces, Cambridge Studies in Advanced Math. no. 3. (Cambridge University Press, 1982).

[M] J. Madden, κ-frames, Manuscript, 1988.

[MS] D. Maharam and A.H. Stone, Realizing Isomorphisms of Category Algebras, Bull. Austral. Math. Soc. 19 (1978), 5-10.

[MV] J. Madden, J. Vermeer, Lindelöf Locales and Realcompactness, Math.Proc.Camb.Phil.Soc. 99 (1986), 473-480.

[N] C.W. Neville, Quasi-$F_{\aleph}$ Spaces as Projectives, Manuscript, (1979).

[V] J. Vermeer, On Perfect Irreducible Preimages, Topology Proceedings 9 (1984), 173-189.

H-Closed Spaces of Large Local π-Weight

Jack R. Porter and R. Grant Woods*

Dept. of Mathematics
The University of Kansas
Lawrence, Kansas 66045
U.S.A.

Dept. of Mathematics and Astronomy
University of Manitoba
Winnipeg, Manitoba R3T 2N2
Canada

* The research of the second-named author was supported by Grant No. A7592 from the Natural Sciences and Engineering Research Council of Canada.

1. INTRODUCTION

B. D. Shapirovskii [1980] proved the following result.

1.1 *THEOREM* The following are equivalent.

(a) $2^{\aleph_0} = \aleph_1$

(b) Each compact Hausdorff space contains either a copy of βN or a point with a countable π-base.

This theorem answered affirmatively a question posed in Rudin (1975). (Recall that a *π-base* at a point p of a space X is a collection $\mathfrak{B}$ of nonempty open subsets of X such that each neighborhood of p contains a member of $\mathfrak{B}$).

As H-closed spaced (defined below) play much the same role with respect to Hausdorff spaces as compact spaces do with respect to Tychonoff spaces, Shapirovskii's result suggests the following question.

1.2 *QUESTION* Is it consistent with ZFC that each H-closed space contain either a copy of βN or a point with a countable π-base?

Our principal results are the following; they assume no special set-theoretic axioms beyond ZFC.

1.3 *THEOREM* There exists an H-closed Urysohn space in which no point has a countable π-base and which contains no infinite separable H-closed subspaces.

1.4 *THEOREM* There exists an H-closed semiregular space in which no point has a countable π-base and which contains no H-closed extension of N in which N is C*-embedded.

We now give a brief review of the basic properties of H-closed spaces. The reader is advised to consult chapter 7 of Porter and Woods (1987) for more extensive background on this topic. Throughout this paper, all hypothesized topological spaces are assumed to be Hausdorff. Hence the word "space" will mean "Hausdorff topological space".

A space is called *H-closed* if it is closed in each space in which it is embedded. Each compact space is H-closed; an H-closed space is compact iff it is regular (Porter and Woods (1987), 4.8 (c)). There are lots of non-compact H-closed spaces. A space is *Urysohn* if distinct points can be put inside disjoint closed neighborhoods; it is *semiregular* if its regular open sets form a base for its open sets. There are H-closed Urysohn spaces, and H-closed semiregular spaces, that are not compact; however, an H-closed space is simultaneously Urysohn and semiregular iff it is compact (Porter and Woods (1987), 7.5 (a), (b)). Hence if we attempted to find a simultaneous improvement of the examples in 1.3 and 1.4 by considering H-closed Urysohn semiregular spaces, we would be doomed to failure; such spaces are compact, and hence their properties are determined by 1.1.

Each space X can be densely embedded in its Katetov H-closed extension κX. The following theorem (which is part of 4.8(p) of Porter and Woods (1987)) summarizes the properties of κX that we will need later.

1.5 *THEOREM* Each space X is a dense subspace of an H-closed space κX with these properties.

(a) $\kappa X \backslash X$ is the set of free open ultrafilters on X (an open ultrafilter α is free if $\cap\{c\ell_X V : V \in \alpha\} = \phi$).

(b) $\kappa X \backslash X$ is a closed discrete subspace of κX.

(c) if $\alpha \in \kappa X \backslash X$, then $\{\{\alpha\} \cup A : A \in \alpha\}$ is a neighborhood base at α.

(d) if A is a closed nowhere dense subset of X, then A is closed in κX.

Finally, we consider weak P-points and weak P-spaces. Weak P-points were introduced by Kunen (1978).

1.6 *DEFINITION* (a) A point p of a space X is called a *weak P-point* of X if p is not a limit point of any countable subset of X. The set of all weak P-points of X is denoted by $wP(X)$.

(b) A space X is called a *weak P-space* if $wP(X) = X$.

The properties of weak P-spaces that we will subsequently need are summarized in the next result.

1.7 *THEOREM* (a) A space is a weak P-space iff its countable subsets are closed and discrete.

(b) A space X is a weak P-space if κX is a weak P-space. If X is a weak P-space without isolated points, then κX is a weak P-space.

(c) Let X be an H-closed, weak P-space without isolated points. Then X contains no infinite separable H-closed subspace and contains no point with a countable local π-base.

PROOF (a) If A is a countable subset of the weak P-space X, then no point of X is a limit point of A. Hence A is closed and discrete. Conversely, if X is not a weak P-space, it has a point x that is a limit point of a countable set B. Then $B \setminus \{x\}$ is countable but not closed in X.

(b) Being a weak P-space is obviously hereditary, so if κX is a weak P-space then X is. Conversely, if X is a weak P-space without isolated points, let J be a countably infinite subset of κX. Then by (a) $J \cap X$ is a closed discrete subspace of X. As X has no isolated points, $\mathrm{int}_X(J \cap X) = \phi$ and so $J \cap X$ is closed in κX by 1.5 (d). By 1.5 (b) $J \setminus X$ is also closed in κX, so

$$\begin{aligned} c\ell_{\kappa X} J &= c\ell_{\kappa X}(J \cap X) \cup c\ell_{\kappa X}(J \setminus X) \\ &= (J \cup X) \cap (J \setminus X) \\ &= J. \end{aligned}$$

Thus J is closed in κX.

Since each countable subset of κX is closed, no point of κX is the limit point of a countable subset of κX. Hence κX is a weak P-space.

(c) Since countable subsets are closed and discrete and no infinite discrete space is H-closed, X contains no infinite separable H-closed subspace. Suppose $\{V_n : n \in N\}$ is a countable family of nonempty open sets. As X has no isolated points, there exists $v_n \in V \setminus \{p\}$ for each $n \in N$. So $B = \{v_n : n \in N\}$ is closed in X and $X \setminus B$ is a neighborhood of p that contains no V_n. Hence, there is no countable local π-base at p.

□

2. AN H-CLOSED URYSOHN SPACE

In this section we construct two examples with the properties described in 1.3. It is clear from 1.7 (b), (c) that if we can find a weak P-space X without isolated points such that κX is Urysohn, then κX will be the desired example.

Recall (see chapter 14 of Gillman and Jerison (1960), or chapter 1 of Walker (1974), or 6L of Porter and Woods (1987)) that a Tychonoff space is an *F-space* if its cozero-sets are C*-embedded. A space is *extremally disconnected* if its open sets have open closures. Each extremally disconnected space is an F-space. A space is *ccc* (ie., satisfies the countable chain condition) if it has no uncountable family of pairwise disjoint open sets.

Each space X has an *absolute*, which is a pair (EX, k_X) consisting of an extremally disconnected zero-dimensional space EX and a perfect irreducible θ-continuous surjection $k_X : EX \to X$ (Recall a closed surjection is called *irreducible* if proper closed subsets of the domain are not mapped onto the range. A function $g : S \to T$ is *θ-continuous at* $x \in S$ if, whenever $g(x) \in V$ and V is open in T, there exists an open subset U of S such that $x \in U$ and $g[c\ell_S U] \subseteq c\ell_T V$. A function that is θ-continuous at each point of its domain is called *θ-continuous*. See 4.8 of Porter and Woods (1987) for more information concerning θ-continuity).

We will need the following results. The first appears, among other places, as 6.6 (e) (6) of Porter and Woods (1987), the second is a special case of 6.5 (d) (3) of Porter and Woods (1987), and the third is 6.7 (a) of Porter and Woods (1987).

2.1 *THEOREM* The map $k_X : EX \to X$ is continuous iff X is regular.

We denote by $\mathcal{R}(X)$ the set of regular closed subsets of the space X.

2.2 *THEOREM* If $g : X \to Y$ is a perfect irreducible θ-continuous surjection, then $A \to g[A]$ is a bijection from $\mathcal{R}(X)$ onto $\mathcal{R}(Y)$.

2.3 *THEOREM* The absolute (EX,k_X) of a space X is unique in the following sense: if (F,g) is a pair consisting of a zero-dimensional extremally disconnected space F and a perfect irreducible θ-continuous surjection $g : F \to X$, then there exists a homeomorphism $h : EX \to F$ such that $g \circ h = k_X$.

We will also use the following result, which is 2.3 of Dow (1982). We denote $2^{\aleph}$ by c.

2.4 *THEOREM* Each compact ccc F-space of weight greater than c contains weak P-points.

2.5 *EXAMPLE* Let m be a cardinal greater than c, and let $K = E(\underline{2}^m)$, where $\underline{2}$ is the two-point discrete space. Since the weight of $\underline{2}^m$ is m, and since the weight of $E(\underline{2}^m)$ is at least as great as the weight of $\underline{2}^m$, we see that $E(\underline{2}^m)$ is an extremally disconnected compact space of weight greater than c. It is well-known that $\underline{2}^m$ is ccc, and so $E(\underline{2}^m)$ is also (Porter and Woods (1987), 6B (4)). Hence by 2.4 $E(\underline{2}^m)$ contains weak P-points. Let us denote the map $k_{\underline{2}^m}$ by k.

In fact, $E(\underline{2}^m)$ contains a dense set of weak P-points. To see this, note that if U and W are non-empty open subsets of $E(\underline{2}^m)$, and if $c\ell\, U \subseteq W$, than by 2.1 $k[c\ell U]$ must contain a basic open set of $\underline{2}^m$. Evidently basic open subsets of $\underline{2}^m$ are homeomorphic to $\underline{2}^m$ and clopen in $\underline{2}^m$, so there exists a clopen set A of $E(\underline{2}^m)$ such that $A \subseteq k\,[c\ell\, U]$. Since $\underline{2}^m$ is regular, it follows from 2.1 that k is continuous and so $k^{\leftarrow}[A]$ is a clopen subset of $E(\underline{2}^m)$ and hence regular closed. By 2.2 it follows that $k^{\leftarrow}[A] \subseteq c\ell\, U \subseteq W$. It is easily checked that $k \mid k^{\leftarrow}[A]$ is a perfect irreducible continuous function from the extremally disconnected space $k^{\leftarrow}[A]$ onto A, so by 2.3 $k^{\leftarrow}[A]$ is homeomorphic to $E(\underline{2}^m)$. Hence $k^{\leftarrow}[A]$ has a weak P-point q, and because $k^{\leftarrow}[A]$ is clopen in $E(\underline{2}^m)$, q is a weak P-point of $E(\underline{2}^m)$. Hence $wP(E(\underline{2}^m))$ is dense in $E(\underline{2}^m)$ as claimed.

Denote $wP(E(\underline{2}^m))$ by X. As X is dense in $E(\underline{2}^m)$, X is extremally disconnected (Porter and Woods (1987), 6.2 (b) or Gillman and Jerison (1960), lH). As $E(\underline{2}^m)$ has no isolated points, neither has X. It remains to show that κX is Urysohn; if it is, then by 1.7 (c) κX is an example satisfying 1.3. Observe that κX is extremally disconnected (see Porter and Woods (1987), 6.2(b)).

By 7B (2), (3) of Porter and Woods (1987) κX is Urysohn iff all boundaries of regular closed subsets of X are compact. Since X is extremally disconnected, its regular closed subsets are clopen and hence have empty boundaries. Thus κX is Urysohn, and is the desired example. □

Our second example is simpler.

Let X be a space and X(s) denote X with the topology generated by the regular open sets (recall that $U \subseteq X$ is regular open if $U = \mathrm{int}_X c\ell_X(U)$); X(s) is called *semiregularization* of X. We need two well-known results about the relationship between a space and its semiregularization.

2.6 *LEMMA* (i) (Porter and Woods (1987), 4.8(h)(8)) A space X is H-closed iff X(s) is H-closed.

(ii) Porter and Woods (1987), 4K(7) A space X is Urysohn iff X(s) is Urysohn.

2.7 *EXAMPLE* Let X be the unit interval I with the topology generated by $\{I\backslash C : C$ is countable$\} \cup \tau$, where τ is the usual topology on I. Observe that X is a weak P-space. Since $\{I\backslash C : C$ is countable$\}$ is a filter on $\{D : D$ dense in I$\}$, it follows by 7M in Porter and Woods (1987) that X(s) = I. By (i) and (ii), X is H-closed and Urysohn. Clearly, X has no isolated points. Thus, by 1.7(c), X is another example of a space with the properties of 1.3.

3. AN H-CLOSED SEMIREGULAR SPACE

In this section we construct an example of the sort described in 1.4. Our construction comes in two parts. First, for any compact F-space X without isolated points for which $|X| = 2^c$, we construct disjoint subsets A and B of X such that if k is any compact separable subset of X, then $K \cap A \neq \phi \neq K \cap B$. Second, we use a construction due to Vermeer (see Vermeer (1985) or Vermeer and Wattel (1981)) to identify each point of A with the corresponding point of B and thereby create a perfect irreducible θ-continuous image H of X. The space H will turn out to satisfy the requirements of 1.4.

3.1 *LEMMA* Let X be a compact F-space without isolated points, and assume $|X| = 2^c$. Then there exist disjoint subsets A and B of X such that $|A| = |B| = 2^c$ and such that if K is an infinite compact separable subspace of X, then $A \cap K \neq \phi \neq B \cap K$.

PROOF Evidently X has $(2^c)^{\aleph} = 2^c$ countable subsets, so the collection of separable infinite compact subsets of X can be indexed as $\{K_\alpha : \alpha < 2^c\}$. Now countable subsets of X are C*-embedded in X (see Porter and Woods (1987), 6L (6) or Gillman and Jerison (1960), 14N (5)) and so each K_2 contains a copy of βN. Hence $|K_\alpha| = 2^c$, since $|\beta N| = 2^c$ (see Gillman and Jerison (1960), 9.3).

Choose a_0, b_0 to be distinct points of K_0. Fix $\delta < 2^c$ and assume that for all $\alpha < \delta$ we have chosen $a_\alpha, b_\alpha \in X$ such that $\{a_\alpha, b_\alpha\} \subseteq K_\alpha$ and $(a_\alpha)_{\alpha<\delta} \cap (b_\alpha)_{\alpha<\delta} = \phi$. Furthermore, assume that if $\alpha < \delta < \delta$ then $a_\alpha \neq a_\delta$ and $b_\alpha \neq b_\alpha$. Then

$|(a_\alpha)_{\alpha<\delta} \cup (b_\alpha)_{\alpha<\delta}| < 2^c$, so $S_\delta = K_\delta \setminus [(a_\alpha)_{\alpha<\delta} \cup (b_\alpha)_{\alpha<\delta}] \neq \phi$. Choose distinct points a_δ, b_δ from S_δ. Our inductive hypotheses remain satisfied. Now let $A = (a_\alpha)_{\alpha<2^c}$ and $B = (b_\alpha)_{\alpha<2^c}$. Evidently A and B satisfy the conclusions of the lemma.

3.2 *LEMMA* Let A and B be as in 3.1. Then compact subspaces of $X \setminus A$, A, $X \setminus B$, and B are finite.

PROOF Suppose H were an infinite compact subspace of $X \setminus A$. Let C be a countably infinite subset of H; then there exists $\alpha < 2^c$ such that $c\ell_X C = K_\alpha$. But $K_\alpha \cap A \supset \{a_\alpha\} \neq \phi$, so H is not a subset of $X \setminus A$, which is a contradiction. If we replace $X \setminus A$ by A, and A by B, in the above argument, we show that A contains no infinite compact subset. Similar arguments apply to $X \setminus B$ and B. □

The following result appears in Vermeer (1983) and as 2.3 of Vermeer and Wattel (1981). It also appears in 7H of Porter and Woods (1987).

3.3 *THEOREM* Let X be a compact space, Y a set, and $f : X \to Y$ an irreducible surjection (ie. if A is closed in X and $X \setminus A \neq \phi$ then $Y \setminus f[A] \neq \phi$). Assume that if $y \in Y$ then $f^{\leftarrow}(y)$ is a compact subset of X. Then $\{f[A] : A \text{ closed in } X\}$ is a closed base for a semiregular Hausdorff topology on Y, and with respect to this topology f is a perfect irreducible θ-continuous surjection.

A slight modification of 4.3 of Vermeer (1985) and 7H(3) of Porter and Woods (1987) is needed; the proof in Vermeer (1985) also works for this modification.

3.4 *THEOREM* Let $f : Y \to X$ be a perfect, irreducible θ-continuous surjection from a compact space Y onto an H-closed space X. A subspace A of X is H-closed iff there is a compact subspace B of Y such that $f | B : B \to A$ is a perfect, irreducible θ-continuous surjection from B onto A.

We will also need a modification of two results in Dow (1982). The proofs of 4.2(i), (ii) of Dow (1982) can be used (with the following minor changes) to prove 3.5 below. Note that the c.c.c. hypothesis implicit in 4.2 of Dow (1982) is not necessary to its proof, as the "B_i"s constructed in that proof can be taken to be pairwise disjoint cozero-sets of X. The hypothesis that X is an F-space now allows us to conclude (in the notation of 4.2 of Dow (1982)) that $U \cap V = \phi$, and that either $X \setminus U$ or $X \setminus V$ is the "required neighborhood" of x (as in the proof of 4.2 (i) of Dow (1982)). Observe also that the proof of 4.2 (ii) of Dow (1982) uses only the conclusion of 4.2 (i) of Dow (1982).

3.5 *THEOREM* Suppose that X is a compact F-space without isolated points, and suppose that each open subset of X has uncountable π-weight. Then each finite subset of X has uncountable local π-weight. (The π-weight of a finite subset F of X is $\min\{|\mathcal{A}| :$ $\mathcal{A}$ is a collection of non-empty open subsets of X such that each open set containing F contains a member of $\mathcal{A}\}$).

The proof of 3.5 (being a straightforward modification of that of 4.2 of Dow (1982)) is not included.

3.6 *EXAMPLE* Let X be a compact zero-dimensional F-space without isolated points such that $|X| = 2^c$. Assume that X has no open subsets of countable π-weight. (The spaces $\beta N \setminus N$ or $E(\underline{2}^c)$ satisfy these hypotheses). Let A and B be as constructed in 3.1. Let $H = [X \setminus (A \cup B)] \cup \{ c_\alpha : \alpha < 2^c\}$, and define $f : X \to H$ by: $f(x) = x$ if $x \in A \cup B$, and $f(a_\alpha) = f(b_\alpha) = c_\alpha$ if $\alpha < 2^c$. Give H the topology described in 3.3. Then H is an H-closed semiregular space and $f : X \to H$ is a perfect irreducible θ-continuous surjection.

We claim that no point of H has a countable local π-base. For suppose that $y \in H$ and H has a countable local π-base at y. By definition of the topology on H (see 3.3) there exists a countable collection $\{G_n : n \in N\}$ of proper closed subsets of X such that $\{H \setminus f[G_n] : n \in N\}$ is a π-base at y in H. Since X is zero-dimensional, we may choose the G_n's to be clopen subsets of X. Suppose that V is open in X and $f^{\leftarrow}(y) \subseteq V$. As $f^{\leftarrow}(y)$ is compact and X is zero-dimensional there is a clopen subset C of X such that $f^{\leftarrow}(y) \subseteq X \setminus C \subseteq V$. Then $y \in H \setminus f[C]$, so there exists $k \in N$ such that $H \setminus f[G_k] \subseteq H \setminus f[C]$, i.e. $f[C] \subseteq f[G_k]$. Assume that $C \setminus G_k \neq \emptyset$. By the irreducibility property of f, there is some $x \in X$ such that $f^{\leftarrow}(f(x)) \subseteq C \setminus G_k$. But $f(x) \in f[C] \subseteq f[G_k]$ implies $f^{\leftarrow}(f(x)) \cap G_k \neq \emptyset$, a contradiction. Thus $C \subseteq G_k$ as claimed. Thus $X \setminus G_k \subseteq X \setminus C \subseteq V$, and it follows that $\{X \setminus G_n : n \in N\}$ is a countable π-base at $f^{\leftarrow}(y)$ in X, in contradiction to 3.5. Thus no point of H has a countable local π-base.

Assume that H contains an H-closed subspace hD which is an extension of a countable infinite discrete space D such that D is C*-embedded in hD. By 4M of Porter and Woods (1987), there is a continuous surjection $g : hD \rightarrow \beta D$ such that $g(d) = d$ for all $d \in D$. In particular, if $L \subseteq hD$ is closed and $g[L] = \beta D$, then $D \subseteq L$; so, $L = hD$. By 3.4, there is a compact subspace E of X such that $f \mid E : E \rightarrow hD$ is perfect, irreducible, θ-continuous and onto. Also, $g \circ f \mid E$ is perfect since E is compact (see 1.8 (b) of Porter and Woods (1987)) and is θ-continuous (see 4.8(h)(1) of Porter and Woods (1987)). If $K \subseteq E$ is closed and $g[f[K]] = \beta D$, then f[K] is closed and, hence, $hD = f[K]$; thus, $K = E$ as f is irreducible. This shows that $g \circ f \mid E$ is irreducible. By 6.5 (d) (4) of Porter and Woods (1987), $g \circ f \mid E$ is one-to-one as βD is extremally disconnected by 6.2(c) of Porter and Woods (1987). It follows that $f \mid E$ is one-to-one. Since E is an infinite compact space, E contains K_α for some α. Thus, $a_\alpha, b_\alpha \in E$ and $f(a_\alpha) = f(b_\alpha)$, contradicting the fact that $f \mid E$ is one-to-one. □

We do not know if the space H described in 3.6 has the stronger property that is contains no infinite separable H-closed subspaces (the space described in 1.3 has this property). So, it is natural to ask this question:

Does there exist an H-closed semiregular space in which no point has a countable π-base and which contains no infinite separable H-closed subspace? more particularly, is there an H-closed semiregular weak P-space?

REFERENCES

1. Dow, A. (1982). *Trans. Amer. Math. Soc. 269* : 557.
2. Gillman, L. and Jerison, M. (1960). *Rings of Continuous Functions*, Van Nostrand, Princeton.
3. Kunen, K. (1978). " Weak P-points in N*", Proceedings of Bolyäi Janos Soc. Colloquium on Topology 1978, Budapest, pp.741-750.
4. Porter, J. R. and Woods, R. G. (1987). *Extensions and Absolutes of Hausdorff spaces*, Springer-Verlag, New York.
5. Rudin, M. E. (1975). " Lectures on Set-theoretical Topology", Conference Board Series on Mathematics 23, American Mathematical Society, Providence, Rhode Island.
6. Shapirovskii, B. D. (1980). *Uspekhi Mat. Nauk. 35: 3* , 122.
7. Vermeer, J. (1983). *Expansions of H-closed Spaces*, Doctoral dissertation, Vrije Universiteit, Amsterdam.
8. Vermeer, J. (1985). *Pac. J. Math., 118* : 229.
9. Vermeer, J. and Wattel, E. *Canad. J. Math. 33* : 872.
10. Walker, R. C. (1974) *The Stone-Cech compactification*, Springer-Verlag, New York.

Ideal-Preserving Boolean Automorphisms

R. M. SHORTT Wesleyan University, Middletown, Connecticut

§0. Introduction

It is well-known that if I_1 and I_2 are the σ-ideals of Lebesgue measure zero and Baire first category sets in $B = \mathcal{B}(\mathbf{R})$, the Borel σ-algebra of the real line, then the quotients B/I_1 and B/I_2 are non-isomorphic Boolean algebras. One might ask other questions, however:

1) Can two such non-isomorphic Boolean algebras have isomorphic automorphism groups? This question can, of course, be phrased in terms of the homeomorphism groups of the corresponding Stone spaces.

2) For any σ-ideal I in B, let $\mathrm{Aut}(B,I)$ be the group of all automorphisms of B that preserve I. Does $\mathrm{Aut}(B,I_1) \cong \mathrm{Aut}(B,I_2)$ imply that $B/I_1 \cong B/I_2$?

3) The same as 2), assuming now that $Aut(B,I_1) = Aut(B,I_2)$.

In [1], R.D. Anderson demonstrated a method for showing that, in certain circumstances, one homeomorphism of a space onto itself is the product of conjugates of another such homeomorphism (or its inverse). A similar argument was used by Fathi [3] to prove that the group of all measure-preserving transformations of the interval is perfect, simple, and is generated by its involutions. Finally, S. Eigen [2], using the results of Fathi, showed that every automorphism of this group is inner, and that the same is true for the group of non-singular transformations.

Considering these proofs, one quickly sees that their basic ideas may be transferred to the more general context of automorphisms of Boolean σ-algebras respecting a given σ-ideal, especially those algebras satisfying a certain homogeneity condition. Following the trail of [1], we show in proposition 1.11 that under very general conditions, one automorphism in $Aut(B,I)$ is a product of conjugates of another (or its inverse). This result is used (proposition 1.13) to identify, under certain conditions, the maximal proper, normal subgroup of $Aut(B,I)$.

Generalisations of the results of Fathi [3] and Eigen [2] to the context of certain homogeneous Boolean algebras are stated in corollary 1.14 and propositions 2.1 and 2.2. Such algebras are characterised by their automorphism groups, and this result speaks to question 1). In particular, for the σ-ideals I_1 and I_2 of measure zero and first category subsets of **R**, the quotients B/I_1 and B/I_2 have non-isomorphic automorphism groups. It should be noted, however, that it is possible for non-isomorphic Boolean algebras to have isomorphic automorphism groups:

witness the existence of non-isomorphic rigid algebras: [5; page 99].

In §3, we specialise to the case of the Borel σ-field $B = \mathcal{B}(S)$ of a complete, separable metric (i.e. *Polish*) space S. In this case, automorphisms of B are in one-one correspondence with invertible Borel-measurable mappings of S onto itself. If I_1 and I_2 are "homogeneous" σ-ideals in B with B/I_1 and B/I_2 complete, then $\mathrm{Aut}(B,I_1) \cong \mathrm{Aut}(B,I_2)$ implies $B_1/I_1 \cong B_2\ I_2$, answering question 2) in a special case.

Lastly, §4 examines the yet more restrictive case, where $\mathrm{Aut}(B,I_1) = \mathrm{Aut}(B,I_2)$. In proposition 4.6, it is shown that whenever I_1 and I_2 contain all singleton sets, then, with one irregular exception, this equality implies that $I_1 = I_2$. A similar result holds for I_1 and I_2 homogeneous (corollary 4.2).

We assume familiarity with fundamental concepts of Boolean algebra, σ-algebra and σ-ideal. The notations a^*, $a - b$, and $a \Delta b$ indicate complement, difference, and symmetric difference. Elements a and b are *orthogonal* if $a \wedge b = 0$. If B is a Boolean algebra, then $\mathrm{Aut}(B)$ is the group of all automorphisms of B. The symbol id indicates the identity automorphism. If I is an ideal in B (all ideals are assumed to be proper), then B/I is the quotient algebra; for $a \in B$, the equivalence class of a *modulo* I is $[a]$.

If X is a topological space, then $\mathcal{B}(X)$ is its σ-field of Borel sets. If X and Y are spaces, then a one-one correspondence $f : X \to Y$ is a *Borel-isomorphism* if $A \in \mathcal{B}(X)$ if and only if $f(A) \in \mathcal{B}(Y)$.

§1. Automorphisms respecting an ideal

Let α be an automorphism of a Boolean algebra B. We say that α is supported in $a \in B$ if $\alpha(u) = u$ for all $u \in B$ with $u \wedge a = 0$. Note that if α is supported in $a \in B$, so also is α^{-1}. Automorphisms α and β are disjointly supported if there is some $a \in B$ such that α is supported in a and β is supported in a^*.

1.1 Lemma: Disjointly supported automorphisms of a Boolean algebra commute.

Proof: Let α and β be such automorphisms and $a \in B$ be as above. For each $u \in B$, write

$$\begin{aligned}(\alpha\beta)(u) &= \alpha(\beta((u\wedge a) \vee (u\wedge a^*))) \\ &= \alpha((u\wedge a) \vee (\beta(u)\wedge a^*)) \\ &= \alpha(u\wedge a) \vee (\beta(u)\wedge a^*) \\ &= (\alpha(u)\wedge a) \vee (\beta(u)\wedge a^*).\end{aligned}$$

By a parallel computation, we see that this is $(\beta\alpha)(u)$.

Q.E.D.

1.2 Lemma: Let α and β be automorphisms of a Boolean algebra B. Then α is supported in $a \in B$ if and only if $\beta\,\alpha\,\beta^{-1}$ is supported in $\beta(a)$.

Proof: We have the equivalent statements

i) $\beta\,\alpha\,\beta^{-1}$ is supported in $\beta(a)$;

ii) $(\beta\alpha\beta^{-1})(u) = u$ whenever $u \leq \beta(a)^* = \beta(a^*)$;

iii) $\alpha(\beta^{-1}(u)) = \beta^{-1}(u)$ whenever $\beta^{-1}(u) \leq a^*$;

iv) $\alpha(v) = v$ whenever $v \leq a^*$;

v) α is supported in a.

Q.E.D.

Let I_1 and I_2 be ideals in the Boolean algebras B_1 and B_2, respectively. Say that the pairs (B_1,I_1) and (B_2,I_2) are <u>isomorphic</u> if there is a Boolean isomorphism $\alpha : B_1 \to B_2$ such that $u \in I_1$ if and only if $\alpha(u) \in I_2$. In this case, the mapping $[a] \to [\alpha(a)]$ gives an isomorphism of B_1/I_1 with B_2/I_2. Given a pair (B,I) and $a \in B - I$, define

$$B(a) = \{b \in B : b \leq a\} \qquad I(a) = \{b \in I : b \leq a\}.$$

Now $B(a)$ may be considered as a Boolean algebra with a new maximal element <u>a</u>. Then $I(a)$ is an ideal in $B(a)$. A pair (B,I) is <u>homogeneous</u> if, for each $a \in B - I$, the pairs $(B(a),I(a))$ and (B,I) are isomorphic. A Boolean algebra B is homogeneous if the pair $(B,\{0\})$ is homogeneous.

<u>1.3 Lemma</u>: Let (B,I) be a homogeneous pair. Then $B_0 = B/I$ is homogeneous.

<u>Proof</u>: Given $a \in B - I$, we must prove that $B_0([a])$ and B_0 are isomorphic. Let $\alpha : B(a) \to B$ be an isomorphism with $\alpha(u) \in I$ if and

only if $u \in I$. Define $\alpha_0 : B_0([a]) \to B_0$ by $\alpha_0([b]) = [\alpha(b)]$. It is easy to see that α_0 is an isomorphism.

Q.E.D.

1.4 Lemma: A homogeneous Boolean algebra is either trivial or non-atomic.

Proof: Suppose that B is homogeneous. If a is an atom of B, then $B(a)$ and hence B must be trivial.

Q.E.D.

1.5 Lemma: Let I be a σ-ideal in a Boolean σ-algebra B such that the quotient B/I is non-atomic. Then each $a \in B - I$ may be written as

$$a = \sup\{a_n : n \geqq 0\},$$

where the a_n are pair-wise orthogonal elements of $B - I$.

Proof: Since B/I is non-atomic, we may choose $a > b_1 > b_2 > \cdots$ with $[a] > [b_1] > [b_2] > \cdots$. Put

$$a_0 = (a-b_1) \vee \inf\{b_n : n \geqq 1\}$$

$$a_n = b_n - b_{n+1}.$$

Verification is easy.

Q.E.D.

Suppose that I is an ideal in a Boolean algebra B. We denote by

Aut(B,I) the group of all automorphisms α of B such that $\alpha(u) \in I$ iff $u \in I$. Define N(B,I) as the set of all automorphisms α of B such that $\alpha(u) \Delta u \in I$ for all $u \in B$. It is easy to check that N(B,I) is a normal subgroup of Aut(B,I). Now define a mapping $\pi : Aut(B,I) \to Aut(B/I)$ by the rule $\pi(\alpha)[a] = [\alpha(a)]$. Then π is a well-defined group homomorphism whose kernel is precisely N(B,I).

The following lemma, which shows how to build automorphisms from pieces of other automorphisms, will be important in the sequel.

<u>1.6 Welding Lemma</u>: Let I be a σ-ideal in a Boolean σ-algebra B. Suppose that

1) $\alpha_0\ \alpha_1\ \alpha_2 \ldots$ is a sequence of automorphisms in Aut(B,I);

2) $a_0\ a_1 \ldots$ and $b_0\ b_1 \ldots$ are sequences of pair-wise orthogonal elements of B;

3) $\sup\{a_n : n \geq 0\} = \sup\{b_n : n \geq 0\} = 1$;

4) $\alpha_n(a_n) = b_n$ for $n \geq 0$.

Then define $\alpha : B \to B$ by

$$\alpha(u) = \sup\{\alpha_n(u \wedge a_n) : n \geq 0\}.$$

Then α is an automorphism in Aut(B,I) such that $\alpha(u) = \alpha_n(u)$ for $u \leq a_n$, $n \geq 0$.

Proof: We compute

$$\alpha(u\vee v) = \sup\{\alpha_n((u\vee v) \wedge a_n) : n \geq 0\}$$

$$= \sup\{\alpha_n(u\wedge a_n) \vee \alpha_n(v\wedge a_n) : n \geq 0\}$$

$$= \alpha(u) \vee \alpha(v).$$

Now write

$$\alpha(u) = \sup\{\alpha_n(u)\wedge b_n : n \geq 0\}$$

$$\alpha(u)^* = \sup\{\alpha_n(u)^*\wedge b_n : n \geq 0\}$$

$$= \sup\{\alpha_n(u^*)\wedge b_n : n \geq 0\}$$

$$= \alpha(u^*),$$

noting that the b_n form a "partition of unity". It is easy to check that

$$\alpha^{-1}(u) = \sup\{\alpha_n^{-1}(u\wedge b_n) : n \geq 0\},$$

that $\alpha \in \mathrm{Aut}(B,I)$, and that $\alpha(u) = \alpha_n(u)$ for $u \leq a_n$.

Q.E.D.

Let I be a σ-ideal in a Boolean σ-algebra B and suppose that $\alpha_1\ \alpha_2 \ldots$ is a sequence of automorphisms in $\mathrm{Aut}(B,I)$ supported on pairwise orthogonal elements $a_1\ a_2 \ldots$, respectively. Then we define the product map $\alpha = \Pi\alpha_n$ by

$$\alpha(u) = \sup\{\alpha_n(u\wedge a_n) : n \geq 1\} \vee (u \wedge \sup\{a_n : n \geq 1\}^*).$$

Then $\alpha \in \mathrm{Aut}(B,I)$ is such, by the welding lemma, that $\alpha(u) = \alpha_n(u)$ for $u \leqq a_n$.

Let I be an ideal in a Boolean algebra B. An element $a \in B$ is <u>I-proper</u> if a and a^* are elements of $B - I$. An element $a \in B$ is <u>proper</u> if a is I-proper for the trivial ideal $I = \{0\}$, i.e. $a \neq 0$ or 1.

<u>1.7 Lemma</u>: Let I be a σ-ideal in a Boolean σ-algebra B. Suppose that B/I is non-atomic. The following are equivalent:

1) The pair (B,I) is homogeneous.

2) Whenever a and b are I-proper elements of B, the pairs (B(a),I(a)) and (B(b),I(b)) are isomorphic.

3) Whenever a and b are I-proper elements of B, there is an automorphism $\alpha \in \mathrm{Aut}(B,I)$ such that $\alpha(a) = b$.

<u>Proof</u>: <u>$1 \Longrightarrow 2$</u>: Trivial.

<u>$2 \Longrightarrow 3$</u>: Let $\alpha_1 : B(a) \to B(b)$ and $\alpha_2 : B(a^*) \to B(b^*)$ be isomorphims such that

$\alpha_1(u) \in I(b)$ if and only if $u \in I(a)$

$\alpha_2(u) \in I(b^*)$ if and only if $u \in I(a^*)$.

Define $\alpha : B \to B$ by the formula

$$\alpha(u) = \begin{cases} \alpha_1(u) & u \leqq a \\ \alpha_2(u) & u \leqq a^*. \end{cases}$$

By the welding lemma (1.6), α is an automorphism in Aut(B,I). Clearly, $\alpha(a) = b$.

3⟹1: Given $a \in B - I$, we must prove that $(B(a), I(a))$ and (B,I) are isomorphic pairs. By lemma 1.5, we may write

$$a = \sup\{a_n : n \geqq 0\} \qquad 1 = \sup\{b_n : n \geqq 0\},$$

where the a_n and b_n are pair-wise orthogonal sequences of elements of $B - I$. Let $\alpha_0\ \alpha_1\ \alpha_2 \ldots$ be automorphisms in Aut(B,I) such that $\alpha_n(a_n) = b_n$. Define $\alpha : B(a) \to B$ by setting $\alpha(u) = \alpha_n(u)$ if $u \leqq a_n$. (The welding lemma (1.6) applies.) We see that α provides the desired isomorphism.

Q.E.D.

1.8 Lemma: Let I be an ideal in a Boolean algebra B and suppose that β is an automorphism in Aut(B,I) with $\beta(u) \Delta u \in B - I$ for some $u \in B$. Then there is some $a \in B - I$ with $a \leqq u$ and $\beta(a) \wedge a = 0$.

Proof: Suppose first that $u - \beta(u) \in B - I$. Then put $a = u - \beta(u)$. On the other hand, if $\beta(u) - u \in B - I$, put $a = \beta^{-1}(\beta(u)-u) = u - \beta^{-1}(u)$.

Q.E.D.

An automorphism $\alpha \in \mathrm{Aut}(B,I)$ is <u>properly supported</u> (with respect to I) if α is supported in some I-proper element of B.

<u>1.9 Lemma</u>: Let I be σ-ideal in a Boolean σ-algebra B and suppose that (B,I) is a homogeneous pair. Let a be an element of B - I and suppose that α is a properly supported automorphism in Aut(B,I). Then α is conjugate to some $\beta \in \mathrm{Aut}(B,I)$ supported in a.

<u>Proof</u>: There is some I-proper element $b \in B$ such that α is supported in b. Using homogeneousness (lemma 1.7.3), we may find $\gamma \in \mathrm{Aut}(B,I)$ with $\gamma(b) \leq a$. Put $\beta = \gamma \alpha \gamma^{-1}$ and apply lemma 1.2.

Q.E.D.

<u>1.10 Lemma</u>: Let I be a σ-ideal in a Boolean σ-algebra B with (B,I) a homogeneous pair and B/I non-trivial. If $a \in B - I$, then there is a sequence $a_0\ a_1\ a_2 \ldots$ of elements of B - I such that

i) the a_n are pair-wise orthogonal;

ii) $\sup\{a_n : n \geqq 0\} \leqq a$.

Also, there is an automorphism ρ in Aut(B,I) such that

iii) ρ is supported in a;

iv) $\rho(a_n) = a_{n+1}$ $(n \geqq 0)$.

<u>Proof</u>: By lemma 1.5, we may write the element <u>a</u> as the supremum of a double sequence $\ldots a_{-2}\ a_{-1}\ a_0\ a_1\ a_2 \ldots$ of pair-wise orthogonal elements in B - I. By homogeneousness (lemma 1.7.3), there are

automorphisms ρ_n $(n = 0, \pm 1, \pm 2 \ldots)$ in Aut(B,I) with $\rho_n(a_n) = a_{n+1}$. Using the welding lemma (1.6), one may define ρ in Aut(B,I) with

$$\rho(u) = \sup\{\rho_n(u \wedge a_n) : n = 0, \pm 1, \ldots\} \vee (u \wedge a^*).$$

Q.E.I.

Propositions 1.11 and 1.13 following are the main results of this section. The technique of proof is adapted from [1].

<u>1.11 Proposition</u>: Let I be a σ-ideal in Boolean σ-algebra B with (B,I) a homogeneous pair and B/I non-trivial. Let β be an automorphism in Aut(B,I) - N(B,I). Then every properly supported α in Aut(B,I) is a product of two conjugates of β with two conjugates of β^{-1}. Also, each such α is a commutator in Aut(B,I).

<u>Demonstration</u>: Since $\beta \in N(B,I)$, there is, by lemma 1.8, some $a \in B - I$ with $a \wedge \beta^{-1}(a) = 0$. Apply lemma 1.10 to $a \in B - I$ to find elements a_0 a_1 ... and an automorphism $\rho \in$ Aut(B,I) with the properties therein described. By lemma 1.9, we see that α is conjugate to some automorphism of (B,I) supported in a_0. Without loss of generality, we may assume that α itself is supported in a_0.

By lemma 1.2, we have that $\rho^n \alpha \rho^{-n}$ is supported in $\rho^n(a_0) = a_n$ for $n = 0, 1, \ldots$. Thus, we may define the product

$$\gamma = \prod_{n=0}^{\infty} \rho^n \alpha \rho^{-n}$$

as an automorphism of (B,I) supported in a. Likewise, γ^{-1} is supported in $\underline{a}$. The same holds true of $\rho\gamma^{-1}\rho^{-1}$. Now $\beta^{-1}\gamma\beta$ and $\beta^{-1}\gamma^{-1}\beta$ are supported in $\beta^{-1}(a)$. By lemma 1.1, we have that the pairs

$$\rho \text{ and } \beta^{-1}\gamma\beta \qquad \rho\gamma^{-1}\rho^{-1} \text{ and } \beta^{-1}\gamma^{-1}\beta$$

each commute. We calculate

$$(\rho\beta^{-1}\gamma\beta\gamma^{-1}\rho^{-1})\ (\beta^{-1}\gamma^{-1}\beta\gamma)$$

$$= (\rho\beta^{-1}\gamma\beta\rho^{-1})\ (\rho\gamma^{-1}\rho^{-1})\ (\beta^{-1}\gamma^{-1}\beta)\gamma$$

$$= (\rho\beta^{-1}\gamma\beta\rho^{-1})\ (\beta^{-1}\gamma^{-1}\beta)\ (\rho\gamma^{-1}\rho^{-1})\gamma$$

$$= (\rho\gamma^{-1}\rho^{-1})\gamma = \alpha.$$

We have written α as a commutator. Then

$$\alpha = (\rho\beta^{-1}\gamma\beta\gamma^{-1}\rho^{-1})\ (\beta^{-1}\gamma^{-1}\beta\gamma)$$

$$= (\rho\beta^{-1}\rho^{-1})\ (\rho\gamma\beta\gamma^{-1}\rho^{-1})\ (\beta^{-1})\ (\gamma^{-1}\beta\gamma),$$

the four desired conjugates.

Q.E.D.

1.12 Lemma: Let I be an ideal in a Boolean algebra B and suppose that B/I is non-atomic. Then every $\alpha \in \text{Aut}(B,I) - N(B,I)$ may be written $\alpha = \alpha_1 \circ \alpha_2$, where α_1 and α_2 are I-properly supported.

Proof: By lemma 1.8, there is some $a \in B - I$ with $\alpha(a) \wedge a = 0$. Since B/I is non-atomic, there is some $b \leq a$ with both b and $a - b$ in $B - I$. Define α_1 by

$$\alpha_1(u) = \begin{cases} \alpha(u) & \text{if } u \leqq b \\ \alpha^{-1}(u) & \text{if } u \leqq \alpha(b) \\ u & \text{if } u \leqq (b \vee \alpha(b))^*. \end{cases}$$

Then $\alpha_1 = \alpha_1^{-1}$ is I-properly supported (in $b \vee \alpha(b)$) . Put $\alpha_2 = \alpha_1 \circ \alpha$. We see that α_2 is supported in b^*.

Q.E.D.

1.13 Proposition: Let I be a σ-ideal in a Boolean σ-algebra B with (B,I) a homogeneous pair and B/I non-trivial. Then N(B,I) is the maximal proper, normal subgroup of Aut(B,I).

Demonstration: Clearly, N(B,I) is a proper, normal subgroup of Aut(B,I). Suppose that M is another such subgroup and that there is some $\beta \in M - N(B,I)$. By lemma 1.12, each $\alpha \in \text{Aut}(B,I) - N(B,I)$ may be written as $\alpha = \alpha_1 \circ \alpha_2$ with α_1 and α_2 properly supported. It follows from proposition 1.11 that α is in the normal closure of β. We have shown

$$\text{Aut}(B,I) - N(B,I) \subseteq M,$$

implying M = Aut(B,I). M could not have been proper.

Q.E.D.

From the preceding proof and the last two propositions one has

1.14 Corollary: Let B be a homogeneous Boolean σ-algebra. Then

1) Aut(B) is simple;

2) every element of Aut(B) is the product of at most eight involutions;

3) every element of Aut(B) is the product of two commutators.

Note: In [3], A. Fathi showed that in the group of all measure-preserving transformations of [0,1], every element is the product of at most ten involutions and five commutators. Our methods diverge somewhat, however.

Note: J.D. Monk has pointed out to the author that part 1) of corollary 1.14 is known. A proof may be found in Anderson's article [10].

§2. Algebras whose only automorphisms are inner

As noted in the introduction, the techniques of [2] and [3] can be adapted to the context of certain types of Boolean algebras. In particular, this applies to complete, homogeneous algebras.

Suppose that α is an automorphism of a Boolean algebra B. Where possible, we define the support of α as

$$\mathrm{supp}(\alpha) = \inf\{a \in B : \alpha \text{ is supported in } a\}$$

whenever this infimum exists. It is easy to see that when the infimum exists, then α is supported in $\mathrm{supp}(\alpha)$.

2.1 Proposition: Let B be a homogeneous Boolean σ-algebra such that every involution of B has a support. Then

1) every automorphism of Aut(B) is inner;

2) $\text{Aut}(\text{Aut}(B)) \cong \text{Aut}(B)$.

Indication: In [3], S. Eigen showed that the group of non-singular transformations of [0,1] is simple and that every one of its automorphisms is inner. The methods used there may be transcribed *mutatis mutandis* to the Boolean algebra setting. Here follows a very brief outline of a part of the argument. Let Γ be an automorphism of $\text{Aut}(B)$. Given $a \in B$, let α be an involution in $\text{Aut}(B)$ with $a = \text{supp}(\alpha)$. Define $F(a) = \text{supp}(\Gamma(\alpha))$. One can show that $F(a)$ does not depend on the particular choice of α, and moreover, that F is an automorphism of B. The next step is to show that $\Gamma(\alpha) = F \circ \alpha \circ F^{-1}$ for each $\alpha \in \text{Aut}(B)$. (By corollary 1.14, it is enough to check this for α on involution.) Thus Γ is an inner automorphism.

Since $\text{Aut}(B)$ is simple (corollary 1.14) and non-Abelian, it has a trivial centre. Thus $\text{Aut}(\text{Aut}(B)) \cong \text{Aut}(B)$.

Q.E.D.

2.2 Proposition: Let B_1 and B_2 be homogeneous Boolean σ-algebras such that every involution of B_1 or B_2 has a support. Then $\text{Aut}(B_1) \cong \text{Aut}(B_2)$ if and only if $B_1 \cong B_2$.

Sketch of proof: Once again, we may say that the methods of [2] apply. As in the previous proposition, suppose that $\Gamma : \text{Aut}(B_1) \to \text{Aut}(B_2)$ is an isomorphism. Given $a \in B_1$, choose an involution α of B_1 with $\text{supp}(\alpha) = a$. Define $F(a) = \text{supp}(\Gamma(\alpha))$. As in the previous proposition, one proves that $F : B_1 \to B_2$ is a well-defined Boolean isomorphism.

Q.E.D.

In particular, the preceding proposition applies to complete, homogeneous algebras. Transferring this to the context of Stone spaces, one has

2.3 Corollary: Consider the class of extremally disconnected, compact Hausdorff spaces X such that any two non-empty clopen subsets of X are homeomorphic. Then two such spaces are homeomorphic if and only if they have isomorphic homeomorphism groups.

Results such as corollary 2.3 have been obtained for other classes of topological spaces, such as compact manifolds. See the work of Whittaker [9] and Filipkiewicz [4].

§3. Automorphisms of Borel structures

This section approaches the issues raised by question 2) of the introduction. We specialise to the context of σ-ideals in the Borel σ-field of a Polish space. In this case, the map π defined in §1. is surjective. This fact, combined with the results of §2., enables us to characterise certain quotients B/I by the associated group $\mathrm{Aut}(B,I)$.

3.1 Lemma: Suppose that I is a σ-ideal of sets in the Borel σ-field $B = \mathcal{B}(S)$ of a Polish space S.

1) The homomorphism π maps $\mathrm{Aut}(B,I)$ onto $\mathrm{Aut}(B/I)$.

2) The kernel $N(B,I)$ comprises all automorphisms α supported in some $a \in I$.

Proof: 1) Given any automorphism α of B/I, we know, by a theorem of von Neumann (see [8; p.413]), that there is a Borel isomorphism $f : S \to S$ of S onto itself such that $\alpha[a] = [f^{-1}(a)]$. But then

$f^{-1} : B \to B$ is an automorphism in $\mathrm{Aut}(B,I)$ such that $\pi(f^{-1}) = \alpha$.

2) Clearly, any automorphism α supported in some $a \in I$ is an element of $N(B,I)$. Conversely, suppose that $\alpha \in N(B,I)$. Then $\pi(\alpha) = \mathrm{id}$. Again apply [8; p. 413] to find $f : S \to S$ with $\alpha = f^{-1}$. Then $\pi(f^{-1}) = \pi(\alpha) = \mathrm{id}$. Now if $g : S \to S$ is the identity map, we also have $\pi(f^{-1}) = \pi(g^{-1})$. For each $a \in B$, $f^{-1}(a)$ and $g^{-1}(a)$ are in the same equivalence class modulo I. By [8; p. 412], we have that $f = g$ except on a set $a \in I$. It follows that $\alpha = f^{-1}$ is supported in a.

Q.E.D.

3.2 Proposition: Let I_1 and I_2 be σ-ideals of sets in the Borel σ-fields $B_1 = \mathcal{B}(S_1)$ and $B_2 = \mathcal{B}(S_2)$ of Polish spaces S_1 and S_2, respectively. Suppose that (B_1,I_1) and (B_2,I_2) are homogeneous pairs with B_1/I_1 and B_2/I_2 non-trivial. Then the following conditions are equivalent:

1) $\mathrm{Aut}(B_1,I_1) \cong \mathrm{Aut}(B_2,I_2)$;

2) $B_1/I_1 \cong B_2/I_2$;

3) there is a Borel-isomorphism $f : S_1 \to S_2$ such that $A \in I_1$ if and only if $f(A) \in I_2$.

Demonstration: $1 \Rightarrow 2$: Suppose that $\varphi : \mathrm{Aut}(B_1,I_1) \longrightarrow \mathrm{Aut}(B_2,I_2)$ is an isomorphism. It follows from proposition 1.13 that φ maps $N(B_1,I_1)$ onto $N(B_2,I_2)$. Thus, φ induces an isomorphism of quotients

$$\mathrm{Aut}(B_1,I_1)/N(B_1,I_1) \cong \mathrm{Aut}(B_2,I_2)/N(B_2,I_2).$$

By lemma 3.1, this yields $\mathrm{Aut}(B_1/I_1) \cong \mathrm{Aut}(B_2/I_2)$.

Claim: Every automorphism of B_1/I_1 (and B_2/I_2) has a support.

Proof of claim: Applying von Neumann's theorem yet again, we find that each automorphism α of B_1/I_1 is of the form $\alpha[a] = [f^{-1}(a)]$ for some Borel-isomorphism $f : S_1 \to S_1$. Then

$$\text{supp}(\alpha) = [\{x : f(x) \neq x\}].$$

Identical reasoning holds for B_2/I_2.

By proposition 2.2, we have $B_1/I_1 \cong B_2/I_2$ as desired.

$2 \Longrightarrow 3$: Application of the von Neumann theorem [8; p.412] yields the existence of $N_1 \in I_1$ and $N_2 \in I_2$ and a Borel-isomorphism $g : (S_1-N_1) \to (S_2-N_2)$ such that $[B] \to [f(B)]$ is an isomorphism of B_1/I_1 onto B_2/I_2. Since (B_1,I_1) is homogeneous, it must be that since B_1/I_1 is non-atomic (lemma 1.4) it contains all countable subsets of S_1; also, if I_1 contains an uncountable set, then every uncountable set in B_1 contains an uncountable set in I_1. The same applies to (B_2,I_2). From this it follows that N_1 and N_2 above may be chosen of the same cardinality, and g may be extended to a Borel-isomorphism $f : S_1 \to S_2$ as in condition 3.

$3 \Longrightarrow 1$: Given such an $f : S_1 \to S_2$, define $\varphi : \text{Aut}(B_1,I_1) \to \text{Aut}(B_2,I_2)$ by $\varphi(\alpha) = g \circ \alpha \circ g^{-1}$. Then φ is an isomorphism of groups.

Q.E.D.

§4. Ideals with equal automorphism groups

Let X and Y be Hausdorff spaces with Borel structures $\mathcal{B}(X)$ and $\mathcal{B}(Y)$. Given a Boolean isomorphism $\varphi : \mathcal{B}(X) \to \mathcal{B}(Y)$, we know that for

$x \in X$, $\varphi(\{x\})$ is an atom of $\mathcal{B}(Y)$ and hence a singleton set. Define $f : X \to Y$ so that $\varphi(\{x\}) = \{f(x)\}$. Then f is a Borel-isomorphism such that $\varphi(A) = f(A)$ for each $A \in \mathcal{B}(X)$. In particular, if I is a σ-ideal in $B = \mathcal{B}(X)$, then each element of $\mathrm{Aut}(B,I)$ corresponds to a unique Borel-isomorphism $f : X \to X$. In this section, we shall identify such automorphisms with their corresponding point-maps.

4.1 Lemma: Let S be a Polish space and let I_1 be a σ-ideal in the Borel structure $B = \mathcal{B}(S)$. Suppose B/I_1 is non-trivial, and (B,I_1) is a homogeneous pair. If I_2 is another σ-ideal in B, then $\mathrm{Aut}(B,I_1) \subseteq \mathrm{Aut}(B,I_2)$ implies $I_1 \supseteq I_2$.

Proof: Suppose that $A \in I_2 - I_1$. As in the proof of lemma 1.7, write

$$S = S_0 \cup S_1 \cup \cdots \qquad A = A_0 \cup A_1 \cup \cdots$$

as disjoint sets in $B - I_1$ and let f_n be elements of $\mathrm{Aut}(B,I_1)$ such that $f_n(A_n) = S_n$, $n \geq 0$. But now each f_n is a member of $\mathrm{Aut}(B,I_2)$. Since $A \in I_2$, each $A_n \in I_2$, each $S_n \in I_2$, and $S \in I_2$, a contradiction.

Q.E.D.

4.2 Corollary: Let S be a Polish space and let I_1 and I_2 be σ-ideals in $B = \mathcal{B}(S)$ with B/I_1 and B/I_2 non-trivial, and (B,I_1) and (B,I_2) homogeneous pairs. Then

$$\mathrm{Aut}(B,I_1) = \mathrm{Aut}(B,I_2) \text{ if and only if } I_1 = I_2.$$

Note: These hypotheses are satisfied when I comprises all measure zero, first category, or countable subsets of $\mathbf{R}$.

4.3 Example: Let I_1 and I_2 be the σ-ideals in $B = \mathcal{B}[0,1]$ comprising all Borel sets of Lebesgue measure zero and Baire first category, respectively. Put $I = I_1 \cap I_2$. Then

1) $\mathrm{Aut}(B,I) = \mathrm{Aut}(B,I_1) \cap \mathrm{Aut}(B,I_2)$;

2) $\mathrm{Aut}(B,I) \subseteq \mathrm{Aut}(B,I_1)$, yet also $I \not\supseteq I_1$;

3) (B,I) is not a homogeneous pair.

Argument: 1) For any two σ-ideals I_1 and I_2, it is automatic that

$\mathrm{Aut}(B,I_1 \cap I_2) \supseteq \mathrm{Aut}(B,I_1) \cap \mathrm{Aut}(B,I_2)$.

Now suppose that $f \in \mathrm{Aut}(B,I) - \mathrm{Aut}(B,I_1)$. Then choose a set $N \in B$ with $mN = 0$ and $mf(N) > 0$: m is Lebesgue measure. (Necessarily, N is second category.) Define a measure m_0 on $[0,1]$ by $m_0(A) = mf(A \cap N)$. Then m_0 is non-zero and concentrated on N. There is [7; p. 4] a first category set $F \subseteq N$ such that $m_0(F) = m_0(N) > 0$. Since $F \in I_1 \cap I_2$, it must be that $f(F) \in I_1 \cap I_2$. So $0 = mf(F) = m_0(F)$ is a contradiction.

In a similar way, one shows that $\mathrm{Aut}(B,I) \subseteq \mathrm{Aut}(B,I_2)$.

2) Obvious.

3) This will follow from part 2) and lemma 4.1. It can, however, be proved directly by observing that if $F \subseteq [0,1]$ is a first category set of full measure, then $(B(F),I(F))$ and $(B(F^c),I(F^c))$ cannot be isomorphic pairs.

Q.E.D.

It should be noted that property 1) of the example does not always hold. We have

4.4 Example: Let m be Lebesgue measure on $\mathbf{R}$ and define measures m_1 and m_2 on $\mathbf{R}$ by

$$m_1(A) = m(A\cap(0,\infty)) \qquad m_2(A) = m(A\cap(-\infty,0)).$$

Let I_1 and I_2 be the σ-ideals of m_1 and m_2 null sets, respectively. Then $f : \mathbf{R} \to \mathbf{R}$ defined by $f(x) = -x$ is a member of $\mathrm{Aut}(B,I_1\cap I_2) - \mathrm{Aut}(B,I_1)$.

4.5 Lemma: Let S be an uncountable Polish space and suppose that I_1 and I_2 are σ-ideals in $B = \mathcal{B}(S)$. Suppose that I_2 contains all singletons and that $\mathrm{Aut}(B,I_1) \subseteq \mathrm{Aut}(B,I_2)$. Then either

1) $I_1 \subseteq I_2$

or

2) $I_1 \cap I_2$ contains no uncountable sets.

Proof: Suppose that both conditions fail. Then choose $N \in I_1 - I_2$ (necessarily uncountable) and $N_0 \in I_1 \cap I_2$ uncountable. Replacing N with $N - N_0$ if necessary, we may assume that $N \cap N_0 = \phi$. Let f be an automorphism of S with

$$f(N) = N_0 \qquad f(N_0) = N$$

$$f(x) = x \quad \text{for} \quad x \notin N \cap N_0.$$

Then $f \in \mathrm{Aut}(B,I_1) \subseteq \mathrm{Aut}(B,I_2)$, a contradiction.

Q.E.D.

4.6 Proposition: Let S be an uncountable Polish space and suppose that I_1 and I_2 are σ-ideals in $B = \mathcal{B}(S)$ each containing all singleton sets. If $\mathrm{Aut}(B,I_1) = \mathrm{Aut}(B,I_2)$, then either

1) $I_1 = I_2$

or

2) $I_1 \cap I_2$ is the σ-ideal of countable sets, and there are sets

$N_1 \in I_1 - I_2$ and $N_2 \in I_2 - I_1$ such that

i) $A_1 \subseteq N_1$ belongs to I_2 if and only if A_1 is countable;

ii) $A_2 \subseteq N_2$ belongs to I_1 if and only if A_2 is countable.

Demonstration: Applying lemma 4.5, we see that both

$I_1 \subseteq I_2$	$I_1 \supseteq I_2$
or	or
$I_1 \cap I_2 = \{\text{countables}\}$	$I_1 \cap I_2 = \{\text{countables}\}$

obtain. Thus, there are four cases to consider. E.g., if $I_1 \cap I_2$ contains only countable sets, and $I_1 \supseteq I_2$, then $I_2 = I_1 \cap I_2$. If possible, choose $N \in I_1 - I_2$ and let $f : S \to S$ be a Borel-isomorphism with $f(N) = S - N$. Then $f \in \mathrm{Aut}(B,I_2) - \mathrm{Aut}(B,I_1)$, a contradiction.

These four cases reduce to two: either $I_1 = I_2$ or $I_1 \cap I_2$ is the σ-ideal of all countable sets, and $I_1 - I_2$ and $I_2 - I_1$ are both non-void. Choose $N_1 \in I_1 - I_2$ and $N_2 \in I_2 - I_1$.

Q.E.D.

4.7 Corollary: Suppose that I_1 and I_2 contain singletons and that (B,I_1) is homogeneous. Then $\mathrm{Aut}(B,I_1) = \mathrm{Aut}(B,I_2)$ if and only if $I_1 = I_2$.

Proof: Apply the proposition along with lemma 4.1.

In proposition 4.6, it is indeed possible that $I_1 \neq I_2$, as the following example shows. Indeed, as the proposition indicates, however, this is in some sense the only example of such behaviour.

4.8 Example: Take $S = \mathbf{R}$ and $B = \mathcal{B}(\mathbf{R})$, and define the σ-ideals

$$I_1 = \{N \in B : N \cap (0,\infty) \text{ is countable}\}$$

$I_2 = \{N \in B : N \cap (-\infty, 0)$ is countable$\}$.

Clearly, I_1 and I_2 are distinct σ-ideals containing all singletons. We assert that $\mathrm{Aut}(B, I_1) = \mathrm{Aut}(B, I_2)$.

Suppose that $f \in \mathrm{Aut}(B, I_1)$. Then $f(0,\infty) \cap (-\infty, 0)$ is countable: if not, then both it and $(0,\infty) \cap f^{-1}(-\infty, 0)$ would be elements of $B - I_1$, a contradiction. Given $N \in I_2$, write

$$f(N) \cap (-\infty,0) = [f(N\cap(0,\infty)) \cap (-\infty,0)] \cup [f(N\cap(-\infty,0)) \cap (-\infty,0)]$$

$$\subseteq [f(0,\infty) \cap (-\infty,0)] \cup f(N\cap(-\infty,0)),$$

which is countable. Thus $f \in \mathrm{Aut}(B, I_2)$. Symmetry applies to show that $\mathrm{Aut}(B, I_1) = \mathrm{Aut}(B, I_2)$.

Q.E.D.

§5. Bibliography

[1] Anderson, R.D., On homeomorphisms as products of conjugates of a given homeomorphism and its inverse, in Topology of 3-Manifolds, M.K. Fort, ed., Prentice Hall (1961) 231-234

[2] Eigen, S., The group of measure-preserving transformations of [0,1] has no outer automorphisms, Math. Annalen 259 (1982) 259-270

[3] Fathi, A., Le groupe des transformations de [0,1] qui préservent la mesure de Lebesgue est un groupe simple, Israel Journal of Math. 29 (1978) 302-308

[4] Filipkiewicz, R.P., Isomorphisms between diffeomorphism groups, Ergodic Theory and Dynamical Systems 2 (1982) 159-171

[5] de Groot, J., Groups represented by homeomorphism groups I, Math. Annalen 138 (1959) 80-102

[6] Kuratowski, K., Topology, Vol. I, Academic Press-PWN, New York-Warsaw 1966

[7] Oxtoby, J.C., Measure and Category, Springer-Verlag, New York 1971

[8] Royden, H.L., Real Analysis, 3rd ed., Macmillan, New York 1988

[9] Whittaker, J.V., On non-isomorphic groups and homeomorphic spaces, Ann of Math. 78 (1963) 74-91

[10] Anderson, R.D., The algebraic simplicity of certain groups of homeomorphisms, Amer. J. Math. 80 (1958) 955-963

The Topology of Close Approximation

A.H. Stone

Department of Mathematics
Northeastern University
Boston, MA 02115, U.S.A.

This report describes joint work with the late Eric van Douwen. A detailed account, with proofs, is expected to be published elsewhere.

Although generalizations are easy, we restrict attention for simplicity to the set $\mathcal{L}$ of all Lebesgue measurable functions f: $R \to R$ where R denotes the set of real numbers. One familiar topology on $\mathcal{L}$ is the uniform topology $\mathcal{T}_U$, in which a neighborhood base at $f \in \mathscr{L}$ consists of all sets

$$U(f,\varepsilon) = \{g \in \mathscr{L} : |f(x) - g(x)| < \varepsilon \quad \text{for all } x \in R\},$$

where $\varepsilon > 0$.

A less familiar one is that of "close approximation" (introduced in Maharam and Stone 1980; for the name I am indebted to Fremlin 1985), $\mathscr{T}_C$, in which a neighborhood base at $f \in \mathscr{L}$ consists of all sets

$$C(f,\varepsilon) = \{g \in \mathscr{L} : |f(x) - g(x)| < \varepsilon(x) \quad \text{for all } x \in R\},$$

where $\varepsilon \in \mathscr{L}^+$, the set of everywhere positive measurable functions. The idea of letting ε depend on x has proved useful in other situations, for instance in integration theory (e.g. Henstock 1963, Pfeffer 1986 and 1987); and for continuous ε (and f) it gives the "m-topology" (cf. Gillman, Henriksen and Jerison 1954, Hewitt 1948), which goes back to E.H. Moore 1911-12. But the present use is rather different. The motivation (Maharam and Stone 1982) is partly that, in $(\mathscr{L},\mathscr{T}_C)$, the subspace $\mathscr{L}^1$ of summable functions is open-closed, and on $\mathscr{L}^1$ the operation of integration $(f \to \int_{-\infty}^{\infty} f(x)dx)$ is continuous. Moreover, arbitrary members of $\mathscr{L}$ can be approximated arbitrarily closely (in $\mathscr{T}_C$) by injective ones (Maharam and Stone 1982).

On the other hand, $\mathscr{T}_C$ has the drawback of resembling the box-product topology of R^R (which is just what results if the measurability requirements are omitted), and consequently it is rather pathological. Closed sets need not be G_δ (in fact $\mathscr{L}^+$ is open but not F_σ in $(\mathscr{L}, \mathscr{T}_C)$); and $(\mathscr{L}, \mathscr{T}_C)$ is not completely normal. Whether or not it is normal is an open question. It is completely regular, being (like $(\mathscr{L}, \mathscr{T}_U)$) a topological group under addition. Also like $(\mathscr{L}, \mathscr{T}_U)$, it fails to be a topological vector space (scalar multiplication is not continuous). But in both topologies the bounded functions form an open-closed subspace that is indeed a topological vector space.

An important property of both $(\mathscr{L}, \mathscr{T}_U)$ and $(\mathscr{L}, \mathscr{T}_C)$ is that they are Baire spaces: the intersection of countably many dense open sets is dense. Hence it makes sense to investigate which families of functions are "unusual", meaning "meager" or "of first category." We say that "most functions" (always understood to belong to $\mathscr{L}$) have a property P if the complementary set, of measurable functions not having P, is meager.

As one would expect, "most functions" are not "nice." The set of all continuous functions from R to R is a closed, hence G_δ, subset of $\mathscr{L}$ in the metrizable topology $\mathscr{T}_U$, and hence also in the finer topology $\mathscr{T}_C$. It is nowhere dense in $\mathscr{T}_C$, and hence also in $\mathscr{T}_U$. More surprisingly, it is discrete in $\mathcal{T}_C$. (This, like some other peculiarities of $\mathcal{T}_C$, depends on the fact that one can easily arrange that $\varepsilon(x)$ takes arbitrarily small values in every interval.)

The set $\mathcal{I}$ of all functions increasing in the wide sense (i.e. non-decreasing) is a closed nowhere dense G_δ set in both topologies. More surprisingly, the set $\mathcal{S}$ of strictly increasing functions is a closed nowhere G_δ set in $\mathcal{T}_C$, and so also is the set $\mathcal{I}\setminus\mathcal{S}$ of functions that increase but not strictly. The following families of (measurable) functions are also closed nowhere dense subsets of $(\mathcal{L}, \mathcal{T}_C)$, though possibly not G_δ: namely, those which are.

(a) continuous at at least one point, or
(b) of bounded variation on at least one interval, or
(c) having at least one fixed point, or
(d) surjective.

(The family in (c) is definitely not G_δ .)

A more interesting question concerns the injective functions. Are most functions in $(\mathcal{L}, \mathcal{T}_C)$ injective?

A preliminary answer is: yes, if the functions are suitably redefined on suitable null sets. More precisely, the set of $f \in \mathcal{L}$ such that, for some $E(f) \subset R$ of measure 0 the restriction $f|R\setminus E(f)$ is one-to-one, is a dense G_δ set in both $\mathcal{T}_U$ and $\mathcal{T}_C$. For $\mathcal{T}_C$, a better answer has been supplied, independently, by D. Preiss (privately communicated) and by D.

Fremlin (1985), the latter in a much more general situation: most functions are not injective. Both used the same powerful general method for the proof, showing that, in a suitable "Banach-Mazur" game (Oxtoby 1957) the second player has a winning strategy. By elaborating Preiss's method it can be shown that, in $(\mathcal{L}, \mathcal{T}_C)$, most functions take c different values c times each -- a result in striking contrast to the preliminary answer. A similar but simpler construction shows that in $(\mathcal{L}, \mathcal{T}_U)$ most functions are injective nevertheless.

The same general method can be applied to show that, both in $\mathcal{T}_U$ and $\mathcal{T}_C$, most functions have ranges that are of measure 0, and in fact of Hausdorff dimension 0. It can also be shown that, in $\mathcal{T}_U$, most functions f are such that f(q) is isolated in f(R) for every rational q (the rationals here could be replaced by any other countable subset of R). The analogous statement in $(\mathcal{L}, \mathcal{T}_C)$ is false, but it turns out that, in $\mathcal{T}_C$, most functions f have the following curious property.

Either f(R) has no isolated points,

or f(R) contains an isolated point with unique inverse (that is, for some $x_o \in R$ and $\delta > 0$,

$$|f(x) - f(x_o)| < \delta \quad \text{implies } x = x_o).$$

Here neither alternative by itself applies to most functions. This raises a question of some interest: Is it true that, in $(\mathcal{L}, \mathcal{T}_C)$, most functions take some value (depending on the function f, in general, and not necessarily isolated in f(R)) exactly once ?

It would also be interesting to know whether the injective functions form a Borel set in $(\mathcal{L}, \mathcal{T}_C)$, or even in $(\mathcal{L}, \mathcal{T}_C)$; and, if so, of what Borel class?

REFERENCES

Fremlin, D.H. (1985), The Topology of Close Approximation, Séminaire d'Analyse Fonctionelle 1983-1984 (B. Beauzamy, J.L. Krivine, B. Maurey, G. Pisier), Univ. Paris VII-VI pp. 33-44.

Gillman, L., Henriksen, M., and Jerison, M. (1954), On a theorem of Gelfand and Kolmogoroff concerning maximal ideals in rings of continuous functions, Proc. Amer. Math Soc., 5, pp. 447-455.

Henstock, R. (1968), A Riemann-type integral of Lebesgue power, Canad. J.M., 20, pp. 79-87.

Hewitt, E. (1948), Rings of real-valued continuous functions I, Trans. Amer. Math. Soc., 64, pp. 45-99.

Maharam, D., and Stone, A. H. (1980), One-to-one functions and a problem on subfields, Measure Theory Oberwolfach 1979, Lecture Notes in Math. 794 (Springer-Verlag, Berlin) pp. 49-52.

Maharam, D., and Stone, A. H. (1982), Expressing measurable functions by one-one ones, Advances in Math., 46, pp. 151-161.

Moore, E. H. (1911-1912), On the foundations of the theory of linear integral equations, Bull. Amer. Math. Soc., 18, pp. 334-362.

Oxtoby, J. (1957), The Banach-Mazur game and Banach category theorem, Contributions to the Theory of Games III, Annals of Math. Studies 39 (Princeton Univ.), pp. 159-163

Pfeffer, W. (1986), The divergence theorem, Trans. Amer. Math. Soc., 295, pp. 665-685.

Pfeffer, W. (1987), The multidimensional fundamental theorem of calculus, J. Australian Math. Soc., 43, pp. 143-170.

A Generalized Process of Stone-Čech Compactification and Realcompactification of an Arbitrary Topological Space

Hueytzen J. Wu

Department of Mathematics
Texas A & I University
Kingsville, Texas 78363

I. INTRODUCTION

For the terminologies in General Topology which are not explicitly defined in this paper we will refer the readers to [5] and [6].

Let A be a family of continuous functions on a topological space X. A net $\{x_i\}$ is an *A-net* in X, if $\{f(x_i)\}$ converges for each f in A. In the rest of this paper, C(X) and $C^*(X)$ will denote the spaces of all real continuous functions and all bounded real continuous functions on X, respectively. A topological space X is *C-complete* if every C(X)-net in X has a cluster point in X, and an

* This research was funded in part by the Organized Faculty Research Grant of Texas A & I University.

ordered pair (Z,h) is called a *C-completion* of a topological space X if h is an embedding of X as a dense subspace of Z and Z is C-complete. For convenience, we will use the Theorem 1 in [6,p.108] as the Theorem 1 in the following :

THEOREM 1 Let A be any family of continuous functions on a topological space X. Then X is compact if, and only if (1) f(X) is contained in a compact subset C_f for each f in A, and (2) every A-net has a cluster point in X.

From this theorem, we can easily see that a compact space is a C-complete space and a compactification (Z,h) of a topological space X is a C-completion of X.

II. A SPECIAL TYPE OF C-COMPLETE SPACES

From Theorem 4 in [6,p.111], we know that a topological space Z is C-complete such that C(Z) separates points of Z if, and only if Z is a Tychonoff space such that every C(Z)-net converges in Z. We will see from the following theorem that a realcompact space is a C-complete space.

THEOREM 2 A topological space Z is realcompact if, and only if Z is C-complete such that C(Z) separates points of Z.

PROOF If Z is realcompact (See [5,p.126]), w.l.o.g., assume that Z is a closed subspace of a product space $\prod_{\alpha\in\Lambda}\mathbb{R}_\alpha$ with $\mathbb{R}_\alpha = \mathbb{R}$ for each α in Λ. Since each projection map $\Pi_\beta : \prod_{\alpha\in\Lambda}\mathbb{R}_\alpha \to \mathbb{R}_\beta$ is a real continuous function and $\{\Pi_\alpha : \alpha \in \Lambda\}$ separates points of $\prod_{\alpha\in\Lambda}\mathbb{R}_\alpha$, thus C(Z) separates points of Z. Let $\{x_i\}$ be a C(Z)-net in Z. Then $\{\Pi_\alpha(x_i)\}$ converges in $\mathbb{R}_\alpha$ for each α in Λ. This implies that $\{x_i\}$ converges to a point t in $\prod_{\alpha\in\Lambda}\mathbb{R}_\alpha$. Since Z is closed, t is Z. Thus, Z is C-complete. For the converse, let $e : Z \to \prod_{f\in C(Z)}\mathbb{R}_f$ be a mapping such that $\Pi_g(e(x)) = g(x)$ for each g in C(Z), where $\mathbb{R}_f = \mathbb{R}$ for each f in C(Z) and Π_g the projection map of $\prod_{f\in C(Z)}\mathbb{R}_f$ onto $\mathbb{R}_g$. From Theorem 4 in [6,p.111], we know that Z is a Tychonoff space, thus e is an embedding of Z as a subspace of $\prod_{f\in C(Z)}\mathbb{R}_f$. Let t be a point in cl(e(Z)), the closure of e(Z). Then there is a net $\{x_i\}$ in Z such that $\{e(x_i)\}$ converges to t in $\prod_{f \in C(Z)} \mathbb{R}_f$. For each f in C(Z), $f(x_i) = \Pi_f(e(x_i))$ and $\lim\{\Pi_f(e(x_i))\} = \Pi_f(t)$, thus $\{f(x_i)\}$ converges for each f in C(Z);i.e., $\{x_i\}$ is a C(Z)-net in Z. Since Z is C-complete and C(Z) separates points of Z, Theorem 4 in [6,p.111] implies that $\{x_i\}$ converges to a point x in Z. Since $\Pi_f(e(x)) = f(x) = \lim\{f(x_i)\} = \lim\{\Pi_f(e(x_i))\} = \Pi_f(t)$ for all f in C(Z), e(x)= t. Hence, e(Z) is a closed subspace of $\prod_{f \in C(Z)} \mathbb{R}_f$.

III. A PROCESS OF C-COMPLETION OF AN ARBITRARY SPACE

For an arbitrary topological space X, let $A = \{f_\alpha : \alpha \in \Lambda\}$ be a subset of C(X). A collection N_x of subsets of X containing a point x is called *the neighborhood base at x induced by A* if $N_x = \{\bigcap_{\alpha \in H} f_\alpha^{-1}(f_\alpha(x) - \varepsilon_\alpha, f_\alpha(x) + \varepsilon_\alpha) : H \text{ is a finite subset of } \Lambda \ \& \ \varepsilon_\alpha > 0\}$. A point x in X is called an *A-Tychonoff point* (or *X is A-Tychonoff at x.*) if x can be separated from closed sets by functions in A; i.e., the neighborhood base at x induced by A is a neighborhood base at x in X. Let X_A be the set of all A-Tychonoff points in X and let S be the collection of all A-nets in X having no cluster point in the set X_A. Define an equivalence relation $\simeq$ on S by setting $\{x_i\} \simeq \{y_j\}$ if $\lim\{f(x_i)\} = \lim\{f(y_j)\}$ for all f in A. Let Y = $S/\simeq$, the equivalence classes in S, and set $X^* = X \cup Y$. For each f in A, let $f^* : X^* \to \mathbb{R}$ be defined as follows : $f^*(x) = f(x)$, if x is in X; $f^*(x) = \lim\{f(x_i)\}$, if $x = \{x_i\}^*$, an equivalence class in Y containing an A-net $\{x_i\}$. We will call f^* *the extension of f to* X^*. Equip X^* with the topology induced by the neighborhood bases β_x at points x in X^* defined as follows: β_x is an arbitrary neighborhood base at x in X, if x is in $X - X_A$; β_x is the neighborhood base at x induced by $A^* = \{f^* : f \in A\}$, if x is in $X_A \cup Y$. Clearly, this is a valid assignment of neighborhood bases in X^*, and every f^* in A^* is a continuous function on X^*.

PROPOSITION 3 Let $h^* : X \to X^*$ be defined by setting that $h^*(x) = x$ for all x in X. Then h^* is an embedding of X as a dense subspace of X^*.

PROOF By the construction of X^*, it is clear that h^* is an embedding of X as a subspace of X^*. Let y be a point in $X^* - X$. Pick an A-net $\{x_i\}$ in X such that $\{x_i\}^* = y$. Since $\lim\{f^*(x_i)\} = \lim\{f(x_i)\} = f^*(\{x_i\}^*) = f^*(y)$ for all f^* in A^*, thus $\{x_i\}$ converges to y. Hence $h^*(X)$ is dense in X^*.

COROLLARY 4 Let $\{x_i\}^*$ be the equivalence class in Y containing an A-net $\{x_i\}$ in X, then $\{x_i\}$ converges to $\{x_i\}^*$ in X^*.

PROPOSITION 5 Every A^*-net $\{y_j\}$ in X^* has a cluster point in X^*.

PROOF We prove this by cases. Case 1 $\{y_i\}$ contains a subnet $\{y_k\}$ in X : If $\{y_k\}$ has a cluster point y in X_A, then $\{y_i\}$ clusters at y; If $\{y_k\}$ has no cluster point in X_A, Corollary 4 implies that $\{y_k\}$ converges to $\{y_k\}^*$; i.e., $\{y_i\}$ clusters at $\{y_k\}^*$. Case 2 $\{y_i\}$ contains a subnet $\{y_k\}$ in Y : For each y_k in $\{y_k\}$, let $\{x_j^k\}$ be an A-net in X such that $\{x_j^k\}^* = y_k$. Since for each f_α in A, $\{f_\alpha^*(y_k)\}$ converges to a point r_α and $\lim\{f_\alpha(x_j^k)\} = f_\alpha^*(y_k)$, so that for every $\varepsilon > 0$, there is an x_α^ε in X such that $|f_\alpha(x_\alpha^\varepsilon) - r_\alpha| < \varepsilon$. Thus, for any finite subset H of Λ and $\varepsilon > 0$, there is an x_H^ε in X such that $|f_\alpha(x_H^\varepsilon) - r_\alpha| < \varepsilon$ for each α in H. Direct $\{(H,\varepsilon): H$ is a finite subset of Λ & $\varepsilon > 0\}$ by setting that $(H_1,\varepsilon_1) \geq (H_2,\varepsilon_2)$ if $H_1 \supset H_2$ and $\varepsilon_1 \leq \varepsilon_2$. Then $\{x_H^\varepsilon\}$ is a net such that $\lim\{f_\alpha^*(x_H^\varepsilon)\} = r_\alpha$ for each f_α in A; i.e., $\{x_H^\varepsilon\}$ is an A-net in X. If $\{x_H^\varepsilon\}$ has a cluster point x in X_A, by the facts that $\mathbb{R}$ is Hausdorff and $\{f_\alpha(x_H^\varepsilon)\}$ converges for each f_α in A, we can easily show that $\lim\{f_\alpha(x_H^\varepsilon)\} = f_\alpha(x)$ for each f_α in A; i.e., $\lim\{f_\alpha^*(y_k)\} = r_\alpha = f_\alpha^*(x)$ for each f_α^* in A^*. Thus $\{y_k\}$ converges to x, and therefore $\{y_i\}$ clusters at x. If $\{x_H^\varepsilon\}$ has no cluster point in X_A, then the equivalence class $\{x_H^\varepsilon\}^*$ containing $\{x_H^\varepsilon\}$ is in Y. Since $f_\alpha^*(\{x_H^\varepsilon\}^*) = \lim\{f_\alpha^*(x_H^\varepsilon)\} = r_\alpha = \lim\{f_\alpha^*(y_k)\}$ for each f_α^* in A^*, thus $\{y_k\}$ converges to $\{x_H^\varepsilon\}^*$; i.e., $\{y_i\}$ clusters at $\{x_H^\varepsilon\}^*$.

THEOREM 6 (X^*,h^*) is a C-completion of X. Furthermore, if A is a subset of $C^*(X)$, then (X^*,h^*) is a compactification of X.

PROOF Since every $C(X^*)$-net $\{x_i\}$ in X^* is also an A^*-net in X^*,

Proposition 5 implies that $\{x_i\}$ has a cluster point in X^*; i.e., X^* is C-complete. Thus, Proposition 3 implies that (X^*, h^*) is a C-completion of X. If $A \subset C^*(X)$, by the construction of f^*, we have that A^* is a family of bounded real continuous functions on X^*. Theorem 1 and Proposition 5 imply that X^* is compact; i.e., (X^*, h^*) is a compactification of X.

COROLLARY 7 Every topological space has a C-completion and a compactification.

IV. THE C-COMPLETIONS OF X INDUCED BY A AND $\hat{A}$

For convenience, the C-completion (X^*, h^*) constructed in III will be called *the C-completion of X induced by A*. In the following lemmas and propositions, $\hat{A} = \{f \circ h^* : f \in C(X^*)\}$, clearly, $A \subset \hat{A}$.

LEMMA 8 Let X_A and $X_{\hat{A}}$ be the sets of all A-Tychonoff points and all $\hat{A}$-Tychonoff points in X, respectively. Then $X_A = X_{\hat{A}}$.

PROOF It is clear that $X_A \subset X_{\hat{A}}$. If $X_{\hat{A}} \not\subset X_A$, let $x \in X_{\hat{A}} - X_A$. Pick a net $\{x_i\}$ in X converging to x. Case 1 Assume that $\{x_i\}$ has a cluster point y in X_A. Since $\mathbb{R}$ is Hausdorff and $\{f(x_i)\}$ converges for each f in A, $\lim\{f(x_i)\} = f(y)$ for all f in A. This implies that $\{x_i\}$ converges to y in X. Hence, $g(h^*(y)) = \lim\{g(h^*(x_i))\} = g(h^*(x))$ for all g in $C(X^*)$. By the construction of the neighborhood base at $h^*(y)$ in X^*, we have that for any g in $C(X^*)$ and $\varepsilon > 0$, there exist $f_1, f_2, \ldots, f_n$ in A and a $\delta > 0$ such that $\bigcap_{i=1}^{n} f_i^{*-1}(f_i^*(h^*(y)) - \delta, f_i^*(h^*(y)) + \delta) \subset g^{-1}(g(h^*(y)) - \varepsilon, g(h^*(y)) + \varepsilon)$; i.e., $\bigcap_{i=1}^{n} h^{*-1}(f_i^{*-1}(f_i^*(h^*(x)) - \delta, f_i^*(h^*(x)) + \delta)) \subset h^{*-1}(g^{-1}(g(h^*(x)) - \varepsilon, g(h^*(x)) + \varepsilon))$. And thus $\bigcap_{i=1}^{n} f_i^{-1}(f_i(x) - \delta, f_i(x) + \delta) \subset h^{*-1}(g^{-1}(g(h^*(x)) - \varepsilon, g(h^*(x)) + \varepsilon))$. This implies that x is in X_A, contradicting the assumption. Case 2 If $\{x_i\}$ has no cluster point in X_A. Then $\{x_i\}^*$ is in Y and $\{x_i\}$ converges to $\{x_i\}^*$. Thus, $g(\{x_i\}^*) = \lim\{g(x_i)\} = g(x) = g(h^*(x))$ for all g in $C(X^*)$. By the construction of the neighborhood base at $\{x_i\}^*$ in X^*, we have that for any g in $C(X^*)$ and $\varepsilon > 0$, there exist $f_1, f_2, \ldots, f_n$ in A and a $\delta > 0$ such that $\bigcap_{i=1}^{n} f_i^{*-1}(f_i^*(\{x_i\}^*)-\delta, f_i^*(\{x_i\}^*)+\delta) \subset g^{-1}(g(\{x_i\}^*)-\varepsilon, g(\{x_i\}^*)+\varepsilon)$; i.e., $\bigcap_{i=1}^{n} f_i^{*-1}(f_i^*(h^*(x))-\delta, f_i^*(h^*(x))+\delta) \subset g^{-1}(g(h^*(x))-\varepsilon, g(h^*(x))+\varepsilon)$. And thus, $\bigcap_{i=1}^{n} f_i^{-1}(f_i(x)-\delta, f_i(x)+\delta) \subset h^{*-1}(g^{-1}(g(h^*(x))-\varepsilon, g(h^*(x))+\varepsilon))$. This implies that x is in X_A, contradicting the assumption.

LEMMA 9 *For any two nets $\{x_i\}$ and $\{y_j\}$, $\lim\{f(x_i)\} = \lim\{f(y_j)\}$ for all f in A if, and only if $\lim\{g(x_i)\} = \lim\{g(f(y_j)\}$ for all g in $\hat{A}$.*

PROOF Let $\{x_i\}$ and $\{y_j\}$ be two nets such that $\lim\{f(x_i)\} = \lim\{f(y_j)\}$ for all f in A. Suppose first that one of the two nets $\{x_i\}$ and $\{y_j\}$ has a cluster point x in X_A. Since $\mathbb{R}$ is Hausdorff and both $\{f(x_i)\}$ and $\{f(y_j)\}$ converge for each f in A, we have that $\lim\{f(x_i)\} = \lim\{f(y_j)\} = f(x)$ for all f in A; i.e., both $\{x_i\}$ and $\{y_j\}$ converge to x in X. Thus, for all t in $C(X^*)$, $\lim\{t(h^*(x_i))\} = t(h^*(x)) = \lim\{t(h^*(y_j))\}$; i.e., $\lim\{g(x_i)\} = g(x) = \lim\{g(y_j)\}$ for all g in $\hat{A}$. Suppose next that none of the two nets $\{x_i\}$ and $\{y_j\}$ has a cluster point in X_A. Then $\{x_i\}^*$ and $\{y_j\}^*$ are in Y such that $\{x_i\}^* = \{y_j\}^*$. Since $f^*(\{x_i\}^*) = \lim\{f(x_i)\} = \lim\{f^*(h^*(x_i))\}$ and $f^*(\{y_j\}^*) = \lim\{f(y_j)\} = \lim\{f^*(h^*(y_j))\}$ for all f in A, so that $\{h^*(x_i)\}$ and $\{h^*(y_j)\}$ converge to $\{x_i\}^*$ and $\{y_j\}^*$ in X^*, respectively. Thus, for any t in $C(X^*)$, $\lim\{t(h^*(x_i))\} = t(\{x_i\}^*) = t(\{y_j\}^*) = \lim\{t(h^*(y_j))\}$; i.e., $\lim\{g(x_i)\} = \lim\{g(y_j)\}$ for all g in $\hat{A}$. For the converse, it is clear from that $A \subset \hat{A}$.

COROLLARY 10 *A net $\{x_i\}$ is an A-net if, and only if $\{x_i\}$ is an $\hat{A}$-net.*

COROLLARY 11 *Two nets $\{x_i\}$ and $\{y_j\}$ are A-nets such that $\lim\{f(x_i)\} \neq \lim\{f(y_j)\}$ for some f in A, if and only if $\{x_i\}$ and $\{y_j\}$ are $\hat{A}$-nets such that $\lim\{g(x_i)\} \neq \lim\{g(y_j)\}$ for some g in $\hat{A}$.*

PROPOSITION 12 *A and $\hat{A}$ induce the same C-completion of X.*

PROOF By Lemmas 8, 9 and the Corollaries 10 and 11, it is clear that $Y = Y_{\hat{A}}$, where $Y_{\hat{A}}$ is the collection of all equivalence classes of $\hat{A}$-nets having no cluster point in $X_{\hat{A}}$. Since the neighborhood bases at points x in $X_A \cup Y$ (= $X_{\hat{A}} \cup Y_{\hat{A}}$) induced respectively by A^* and $C(X^*)$ are equivalent. Thus, A and $\hat{A}$ induce the same C-completion of X.

V. THE TYPICAL C-COMPLETIONS OF AN ARBITRARY TOPOLOGICAL SPACE

For a C-completion (Z,h) of a topological space X, let $C(Z) \circ h = \{f \circ h : f \in C(Z)\}$ and let $X_{C(Z)oh}$ be the set of all C(Z)oh-Tychonoff points in X. A C-completion (Z,h) of a topological space X will be called *a typical C-completion of X* if every C(Z)-net in Z has a cluster point in $Z-h(X-X_{C(Z)oh})(=(Z-h(X)) \cup h(X_{C(Z)oh}))$ and C(Z) separates points of $Z-h(X-X_{C(Z)oh})$.

REMARK 13 From the construction of a C-completion in III and the proof of Proposition 5, we can easily see that the C-completion (X^*,h^*) of a topological space X induced by a subset A of C(X) is a typical C-completion of X.

THEOREM 14 Let X be a Tychonoff space. Then (Z,h) is a realcompactification of X if, and only if (Z,h) is a typical C-completion of X.

PROOF If X is Tychonoff, then $X_{C(Z)oh} = X$ and $Z-h(X-X_{C(Z)oh}) = Z$. The conditions for a typical C-completion of X are equivalent to that Z is C-complete and C(Z) separates points of Z. Thus, by Theorem 2, we have the conclusion.

COROLLARY 15 A Hausdorff compactification (Z,h) of a Tychonoff space X is a typical C-completion of X.

PROPOSITION 16 Let (Z,h) be a typical C-completion of a topological space X. Then the neighborhood base at any point x in $Z-h(X-X_{C(Z)oh})$ induced by C(Z) is a neighborhood base at x in Z.

PROOF For any x in $Z-h(X-X_{C(Z)oh})$, let β^* be the neighborhood base at x induced by C(Z), β a neighborhood base at x in Z. If β is strictly finer than β^*, then there is a U in β such that for all V in β^*, $V \not\subset U$. For each V^i in β^*, pick an x_i in $V^i - U$, then

$\{x_i\}$ is a C(Z)-net such that $\{f(x_i)\}$ converges to f(x) for all f in C(Z). The conditions for a typical C-completion imply that $\{x_i\}$ has a cluster point y in $Z-h(X-X_{C(Z)oh})$. Pick a subnet $\{x_{i_k}\}$ of $\{x_i\}$ such that $\{x_{i_k}\}$ converges to y. Then for all f in C(Z), $\{f(x_{i_k})\}$ converges to f(x) and f(y) simultaneously. The Hausdorffness of ℝ implies that f(x)= f(y) for all f in C(Z). Since C(Z) separates points of $Z-h(Z-X_{C(Z)oh})$, we have that x = y. This implies that $\{x_i\}$ clusters at x, contradicting the assumption.

COROLLARY 17 *Let $Z_{C(Z)}$ be the set of all C(Z)-Tychonoff points. Then $Z_{C(Z)}= Z-h(X-X_{C(Z)oh})=(Z-h(X)) \cup h(X_{C(Z)oh})$.*

COROLLARY 18 *If a C(Z)-net $\{x_i\}$ clusters at a point y in $Z_{C(Z)}$, then $\{x_i\}$ converges to y in Z and y is unique in $Z_{C(Z)}$.*

Let (Z_1,h_1) and (Z_2,h_2) be two typical C-completions of a topological space X. We will call (Z_1,h_1) and (Z_2,h_2) *two equivalent typical C-completions of X* if (1) any net $\{x_i\}$ in X is a $C(Z_1)oh_1$-net if, and only if it is a $C(Z_2)oh_2$-net, and (2) two nets $\{x_j\}$ and $\{x_k\}$ in X are two different $C(Z_1)oh_1$-nets (i.e., there is an f in $C(Z_1)$ such that $\lim\{f(h_1(x_j))\}\neq \lim\{f(h_1(x_k))\}$) if, and only if they are two different $C(Z_2)oh_2$-nets.

PROPOSITION 19 *Let (Z_1,h_1) and (Z_2,h_2) be two equivalent typical C-completions of X. Then $X_{C(Z_1)oh_1} = X_{C(Z_2)oh_2}$.*

PROOF For any x in $X_{C(Z_1)oh_1}$, let β_1 and β_2 be two neighborhood bases at x induced by $C(Z_1)oh_1$ and $C(Z_2)oh_2$, respectively. It is clear that β_1 is equivalent to the neighborhood base at x induced by C(X). Since $C(Z_2)oh_2 \subset C(X)$, so that β_1 is finer than or equivalent to β_2. If β_1 is strictly finer than β_2, then there is a U in β_1 such that $V \not\subset U$ for all V in β_2. For each V^i in β_2, pick an x_i in V^i-U. Then $\{x_i\}$ is a $C(Z_2)oh_2$-net such that $\lim\{f(h_2(x_i))\}= f(h_2(x))$ for all f in $C(Z_2)$. The condition (1) for the equivalence of two typical C-completions implies that $\{x_i\}$ is also a $C(Z_1)oh_1$-net. By the Corollary 18, $\{h_1(x_i)\}$ converges to a unique point y in $Z_1-h_1(X-X_{C(Z_1)oh_1})$. If $h_1(x)\neq y$, pick a net $\{x_k\}$ in X converging to x. Since $C(Z_1)$ separates points of $Z_1-h_1(X-X_{C(Z_1)oh_1})$, $\{x_i\}$ and $\{x_k\}$ are two different $C(Z_1)oh_1$-nets, and therefore $\{x_i\}$ and $\{x_k\}$ are two different $C(Z_2)oh_2$-nets. This contradicts that $\lim\{f(h_2(x_i))\}= f(h_2(x))= \lim\{f(h_2(x_k))\}$ for all f in $C(Z_2)$. Then $h_1(x)= y$, and this implies that $\{x_i\}$ converges to x, contradicting the picking of $\{x_i\}$. Thus, β_1 is equivalent to β_2, and therefore x is in $X_{C(Z_2)oh_2}$. Hence $X_{C(Z_1)oh_1} \subset X_{C(Z_2)oh_2}$. Similarly, we can readily show that $X_{C(Z_2)oh_2} \subset X_{C(Z_1)oh_1}$.

Let (Z_1,h_1) and (Z_2,h_2) be two equivalent typical C-completions of X. We define F, *the typical mapping from* (Z_1,h_1) *to* (Z_2,h_2) as follows : For any y in $Z_1-h_1(X)$, pick a net $\{x_i\}$ in X such that $\{h_1(x_i)\}$ converges to y. Then $\{x_i\}$ is a $C(Z_1)oh_1$-net, and thus is a $C(Z_2)oh_2$-net. The Corollary 18 implies that $\{h_2(x_i)\}$ converges to a unique point t_y in $Z_2-h_2(X-X_{C(Z_2)oh_2})$. If t_y is in $h_2(X_{C(Z_2)oh_2})$, then $\{h_1(x_i)\}$ will converge to $h_1(h_2^{-1}(t_y))$ in $h_1(X_{C(Z_1)oh_1})$, contradicting that $C(Z_1)$ separates points of $Z_1-h_1(X-X_{C(Z_1)oh_1})$. Thus, t_y is in $Z_2-h_2(X)$. If $\{x_k\}$ is another net in X such that $\{h_1(x_k)\}$ converges to y, then $\{x_i\}$ and $\{x_k\}$ are two equivalent $C(Z_1)oh_1$-nets, and thus are two equivalent $C(Z_2)oh_2$-nets. The Corollary 18 and that $C(Z_2)$ separates points of $Z_2-h_2(X-X_{C(Z_2)oh_2})$ imply that $\{h_2(x_k)\}$ converges to t_y also. Thus, we define $F(y)= t_y$, if y is in $Z_1-h_1(X)$; $F(h_1(x))= h_2(x)$, if x is in X. Clearly, F is well-defined.

PROPOSITION 20 For two equivalent typical C-completions (Z_1,h_1) *and* (Z_2,h_2) *of a topological space X, let F be the typical mapping from* (Z_1,h_1) *to* (Z_2,h_2)*. Then F is a one-to-one and onto function.*

PROOF It is enough to show that the restriction $F|_{Z_1-h_1(X)}$ of F is one-to-one and onto from $Z_1-h_1(X)$ to $Z_2-h_2(X)$. If y_1 and y_2 are two different points in $Z_1-h_1(X)$, then there is a g in $C(Z_1)$ such that $g(y_1)\neq g(y_2)$. Pick two nets $\{x_i\}$ and $\{x_j\}$ in X such that

$\{h_1(x_i)\}$ and $\{h_1(x_j)\}$ converge to y_1 and y_2, respectively. Then $\lim\{g(h_1(x_i))\}= g(y_1)\neq g(y_2)= \lim\{g(h_1(x_j))\}$; i.e., $\{x_i\}$ and $\{x_j\}$ are two different $C(Z_1)oh_1$-nets, and thus $\{x_i\}$ and $\{x_j\}$ are two different $C(Z_2)oh_2$-nets; i.e., there is an f in $C(Z_2)$ such that $\lim\{f(h_2(x_i))\}\neq \lim\{f(h_2(x_j))\}$. This implies that $F(y_1)= t_{y_1} = \lim\{h_2(x_i)\}\neq \lim\{h_2(x_j)\}= t_{y_2} = F(y_2)$. To see that F is onto, let z be an element in $Z_2-h_2(X)$, pick a net $\{x_k\}$ in X such that $\{h_2(x_k)\}$ converges to z. Then $\{x_k\}$ is a $C(Z_2)oh_2$-net, and thus is a $C(Z_1)oh_1$-net. Corollary 18 implies that $\{h_1(x_k)\}$ converges to a unique point y in $Z_1-h_1(X-X_{C(Z_1)oh_1})$. By arguments similar to those preceding Proposition 20, we have that y is in $Z_1-h_1(X)$ and for any net $\{x_j\}$ in X, $\{h_2(x_j)\}$ converges to z if, and only if $\{h_1(x_j)\}$ converges to y. Hence, $F(y)= z$.

REMARK 21 It is clear from the definition of typical mapping that F^{-1} is the typical mapping from (Z_2,h_2) to (Z_1,h_1).

PROPOSITION 22 Let (Z_1,h_1) and (Z_2,h_2) be two equivalent typical C-completions of X. The typical mapping F from (Z_1,h_1) to (Z_2,h_2) is continuous at points in $Z_1-h_1(X-X_{C(Z_1)oh_1})$.

PROOF For any y in $Z_1-h_1(X-X_{C(Z_1)oh_1})$, F(y) is in $Z_2-h_2(X-X_{C(Z_2)oh_2})$. Let $\varepsilon > 0$ and let f be in $C(Z_2)$. It is enough to show that there is an open neighborhood U of y such that $F(U)\subset f^{-1}(f(F(y))-\varepsilon,f(F(y))+\varepsilon)$. Let $D= h_1(h_2^{-1}(f^{-1}(-\infty,f(F(y))- \varepsilon/2)))$. We claim that $F^{-1}(f^{-1}(-\infty,f(F(y))-\varepsilon])\subset cl(D)$, where cl(D) is the closure of D. For any t in $F^{-1}(f^{-1}(-\infty,f(F(y))-\varepsilon])$, if $t= h_1(x)$ for some x in X, then $F(t)=F(h_1(x))= h_2(x)$ and is in $f^{-1}(-\infty,f(F(y))-\varepsilon]$. Thus, x is in $h_2^{-1}(f^{-1}(-\infty,f(F(y))- \varepsilon/2))$ and therefore $h_1(x)$ is in D; i.e., t is in cl(D). If t is in $Z_1-h_1(X)$, then F(t) is in $f^{-1}(-\infty,f(F(y))- \varepsilon/2)$. Since $h_2(h_2^{-1}(f^{-1}(-\infty,f(F(y))- \varepsilon/2)))$ is dense in $f^{-1}(-\infty,f(F(y))- \varepsilon/2)$, there is a net $\{x_i\}$ in $h_2^{-1}(f^{-1}(-\infty,f(F(y))- \varepsilon/2))$ such that $\{h_2(x_i)\}$ converges to F(t). This implies that $\{x_i\}$ is a $C(Z_2)oh_2$-net, and therfore is a $C(Z_1)oh_1$-net. The arguments preceding Proposition 20 imply that $\{h_1(x_i)\}$ converges to t. Since $\{x_i\}$ is a net in

$h_2^{-1}(f^{-1}(-\infty,f(F(y))-\varepsilon/2))$, thus t is in cl(D). Similarly, $F^{-1}(f^{-1}[f(F(y))+\varepsilon,\infty)) \subset cl(E)$, where $E=h_1(h_2^{-1}(f^{-1}(f(F(y))+\varepsilon/2,\infty)))$. Let $U= Z_1-(cl(D) \cup cl(E))$. Then U is an open set in Z_1 such that $U \subset F^{-1}(f^{-1}(f(F(y))-\varepsilon,f(F(y))+\varepsilon))$. Finally, we show that y is in U. Since $F(y) \notin f^{-1}(-\infty,f(F(y))-\varepsilon/2]$, so that for any net $\{x_i\}$ in $h_2^{-1}(f^{-1}(-\infty,f(F(y))-\varepsilon/2))$, $\{h_2(x_i)\}$ does not converge to F(y). The equivalence of (Z_1,h_1) and (Z_2,h_2), and the arguments preceding Proposition 20 imply that for any net $\{x_i\}$ in $h_2^{-1}(f^{-1}(-\infty,f(F(y))-\varepsilon/2))$, $\{h_1(x_i)\}$ does not converge to y. Thus, y is not in cl(D). Similarly, y is not in cl(E). Hence, y is in U.

COROLLARY 23 *F^{-1} is continuous at points in $Z_2 - h_2(X-X_{C(Z_2)oh_2})$.*

REMARK 24 If (Z_1,h_1) and (Z_2,h_2) are two equivalent typical C-completions of a Tychonoff space X, then $X_{C(Z_1)oh_1} = X = X_{C(Z_2)oh_2}$, $Z_1-h_1(X-X_{C(Z_1)oh_1})= Z_1$, and $Z_2-h_2(X-X_{C(Z_2)oh_2})= Z_2$. From Proposition 22 and Corollary 23, it follows that the typical mapping F from (Z_1,h_1) to (Z_2,h_2) is a homeomorphism.

From the Remark 13 and the Corollaries 10 and 11, it is clear that if (Z,h) is a typical C-completion of a topological space X, then (Z,h) is equivalent to the C-completion (X^*,h^*) of X induced by C(Z)oh.

PROPOSITION 25 *Let (Z,h) be a typical C-completion of a topological space X and let (X^*,h^*) be the C-completion of X induced by C(Z)oh. Then the typical mapping G from (X^*,h^*) to (Z,h) is continuous.*

PROOF From Proposition 22, we can see that it is enough to show that G is continuous at points $h^*(x)$ in $h^*(X-X_{C(Z)oh})$. For any x in $X-X_{C(Z)oh}$, let U be an open neighborhood of h(x) in Z. Then $h^{-1}(U)$ is an open neighborhood of x in X. By the construction of (X^*,h^*) and the definition of typical mapping, we have that $h^*(h^{-1}(U))$ is an open neighborhood of $h^*(x)$ and $G(h^*(h^{-1}(U)))= h(h^{-1}(U)) \subset U$. Thus, G is continuous at $h^*(x)$.

Remark 26 If (Z,h) is a typical C-completion of a Tychonoff space X and if (X^*,h^*) is the C-completion of X induced by $C(Z)oh$, then from the paragraph preceding Proposition 25 and Remark 24, it follows that the typical mapping G from (X^*,h^*) to (Z,h) is a homeomorphism.

Two C-completions (Z_1,h_1) and (Z_2,h_2) of a topological space X are said to be *homeomorphic* if there is a homeomorphism $H : Z_1 \to Z_2$ such that $H(h_1(X)) = h_2(X)$.

THEOREM 27 *Let (Z_1,h_1) and (Z_2,h_2) be two equivalent typical C-completions of a topological space X, (X_1^*,h_1^*) and (X_2^*,h_2^*) the C-completions of X induced by $C(Z_1)oh_1$ and $C(Z_2)oh_2$, respectively. Then (X_1^*,h_1^*) is homeomorphic to (X_2^*,h_2^*).*

PROOF From the paragraph preceding Proposition 25 and the definition for the equivalence of two typical C-completions, we can see easily that (X_1^*,h_1^*) and (X_2^*,h_2^*) are two equivalent typical C-completions of X. Let F be the typical mapping from (X_1^*,h_1^*) to (X_2^*,h_2^*). Then F is one-to-one and onto. From Proposition 22, we have that F is continuous at points in $X_1^* - h_1^*(X-X_{C(Z_1)oh_1})$. Similar to the proof of Proposition 25, we can easily prove that F is continuous at points in $h_1^*(X-X_{C(Z_1)oh_1})$. Thus, F is continuous on X_1^*. Since F^{-1} is the typical mapping from (X_2^*,h_2^*) to (X_1^*,h_1^*), in the same way, we have that F^{-1} is continuous on X_2^*. Finally, it is clear that $F(h_1^*(X)) = h_2^*(X)$.

COROLLARY 28 *Let (Z_1,h_1) and (Z_2,h_2) be two equivalent typical C-completions of a Tychonoff space X, (X_1^*,h_1^*) and (X_2^*,h_2^*) the C-completions of X induced by $C(Z_1)oh_1$ and $C(Z_2)oh_2$, respectively. Then any two of (Z_1,h_1), (Z_2,h_2), (X_1^*,h_1^*) and (X_2^*,h_2^*) are homeomorphic.*

PROOF This is obvious from Remarks 24, 26 and Theorem 27.

COROLLARY 29 *A realcompactification (or Hausdorff compactification) of a Tychonoff space X is homeomorphic to the C-completion (X^*,h^*) of X induced by $C(Z)oh$.*

VI. THE NICE C-COMPLETIONS OF AN ARBITRARY TOPOLOGICAL SPACE

A typical C-completion (Z,h) of a topological space X will be called a *nice C-completion of X* if for any typical C-completion (Y,k) of X equivalent to (Z,h), the typical mapping F from (Z,h) to (Y,k) is continuous.

REMARK 30 From Remark 24, it is clear that any typical C-completion of a Tychonoff space X is a nice C-completion of X.

THEOREM 31 (Z,h) is a nice C-completion of a topological space X if, and only if (Z,h) is homeomorphic to the C-completion (X^*,h^*) of X induced by C(Z)oh.

PROOF If (Z,h) is a nice C-completion of a topological space X. Let G be the typical mapping from (X^*,h^*) to (Z,h). From Proposition 25, G is continuous. Since G^{-1} is the typical mapping from (Z,h) to (X^*,h^*) and (Z,h) is a nice C-completion of X, thus G^{-1} is continuous. Hence G is a homeomorphism. If (Z,h) is homeomorphc to (X^*,h^*), let (Y,k) be any typical C-completion of X equivalent to (X^*,h^*), (Y^*,k^*) the C-completion of X induced by C(Y)ok. From Remark 13 and the paragraph preceding Proposition 25, it is clear that (Y^*,k^*) is a typical C-completion of X equivalent to (X^*,h^*). Let H and F denote the typical mappings from (Y^*,k^*) to (Y,k) and from (X^*,h^*) to (Y^*,k^*), respectively. From the definition of a typical mapping, it follows that H o F is the typical mapping from (X^*,h^*) to (Y,k). From Proposition 25 and Theorem 27, we have that H is continuous and F is a homeomorphism. Thus, H o F is continuous. Hence, (X^*,h^*) is a nice C-completion of X and so is (Z,h).

COROLLARY 32 Let X be a Tychonoff space. Then the following are equivalent : (1) (Z,h) is a realcompactification of X; (2) (Z,h) is a typical C-completion of X; (3) (Z,h) is a nice C-completion of X; (4) (Z,h) is homeomorphic to the C-completion (X^*,h^*) of X induced by C(Z)oh.

REMARK 33 :

(1) One-Point Compactification : Let X be a locally compact, noncompact Hausdorff space, $C_o(X)$ the space of all real continuous

functions f on X such that for every $\varepsilon > 0$, $|f|^{-1}[\varepsilon,\infty)$ is compact. Let A be a subset of $C_o(X)$ such that the topology on X is the weak topology induced by A, $\{x_i\}$ an A-net in X having no cluster point in X. For any f in A, if there is an $\varepsilon > 0$ such that $\lim\{|f(x_i)|\} > \varepsilon$, then there is an i_o such that $|f(x_i)| > \varepsilon$ for all $i \geq i_o$; i.e., the subnet $\{x_i : i \geq i_o\}$ is in the compact subset $|f|^{-1}[\varepsilon,\infty)$. Theorem 1 implies that $\{x_i : i \geq i_o\}$ clusters at a point in $|f|^{-1}[\varepsilon,\infty)$, contradicting the assumption that $\{x_i\}$ has no cluster point in X, Hence, $\lim\{f(x_i)\} = 0$ for all f in A. Let (X^*,h^*) be the C-completion of X induced by A. Then, by the construction of X^*, we have that $X^*-h^*(X)$ is a one-point set $\{y\}$ such that $f^*(y)=0$ for all f^* in A^*. Since for all f^* in A^* and all $\varepsilon > 0$, $|f^*|^{-1}[\varepsilon,\infty)=|f|^{-1}[\varepsilon,\infty)$ is a compact subset of X. Thus, (X^*,h^*) is the one-point compactification of X.

(2) Stone-Čech Compactification : Let $(\beta X,e)$ be the Stone-Čech compactification of a Tychonoff space X. Then $(\beta X,e)$ is a nice C-completion of X such that $C(\beta X) \circ e = C^*(X)$. By the Corollary 32, we have that $(\beta X,e)$ is homeomorphic to the C-completion (X^*,h^*) of X induced by $C^*(X)$.

(3) Hewitt Realcompactification : Let $(\nu X,e)$ be the Hewitt realcompactification of a Tychonoff space X. Then $(\nu X,e)$ is a nice C-completion of X such that $C(\nu X) \circ e = C(X)$. The Corollary 32 implies that $(\nu X,e)$ is homeomorphic to the C-completion (X^*,h^*) of X induced by C(X).

REFERENCES

[1] Dugundji, J., Topology, Allyn and Bacon, Boston (1966).

[2] Kelly, J. L., General Topology, D. Van Nostrand Co., Inc., Princeton, N. J.

[3] Semadeni, Z., Banach Spaces, Vol. 1 (Warszawa, 1971).

[4] Stone, M. H., Applications of the Theory of Boolean Ring to General Topology, AMS Trans. 41, 375-481 (1937).

[5] Willard, S., General Topology (Addison-Wesley, MA 1970).

[6] Wu, H. J., Extensions and New Observations of Tychonoff, Stone-

Weierstrass Theorems, Compactifications and Realcompactification, Topology Appli. 16 (1983), 107-116.

[7] Wu, H. J., A Vector Lattice Representation Theorem and a Characterization of Locally Compact Hausdorff Spaces, Journal of Functional Analysis, Vol. 65, No. 1, January 1986, 1-14.

Index